PROF. NICHOLAS MARCELLUS HENTZ.
The Father of American Araneology.

AMERICAN SPIDERS

AND

THEIR SPINNINGWORK.

———◦———

A NATURAL HISTORY

OF THE

ORBWEAVING SPIDERS OF THE UNITED STATES

WITH SPECIAL REGARD TO THEIR INDUSTRY AND HABITS.

———◦———

BY

HENRY C. McCOOK, D. D.,

VICE-PRESIDENT OF THE ACADEMY OF NATURAL SCIENCES OF PHILADELPHIA;
PROFESSOR OF ENTOMOLOGY IN THE PENNSYLVANIA
HORTICULTURAL SOCIETY.

VOL. III.

WITH DESCRIPTIONS OF ORBWEAVING SPECIES AND PLATES.

PUBLISHED BY THE AUTHOR,
ACADEMY OF NATURAL SCIENCES OF PHILADELPHIA,
A. D. 1893.

AUTHOR'S EDITION.

This Edition is limited to Two HUNDRED AND FIFTY copies, of
which this set is

SUBSCRIPTION No. *74*

AUTHOR'S SIGNATURE,

THE PRESS OF
ALLEN, LANE & SCOTT,
PHILADELPHIA.

THESE STUDIES IN NATURAL HISTORY
ARE DEDICATED TO
THE VENERATED MEMORY
OF MY FATHER,

JOHN McCOOK, M. D.,

A LOVER OF NATURE, A FRIEND OF SCIENCE,
A GOOD PHYSICIAN, A SERVANT OF
HIS FELLOW MEN,
WHOSE FAITH IN THE UNSEEN
NEVER FALTERED.

PREFACE.

WITH profound satisfaction the author gives to the scientific public the third and last volume of a work which has engaged his thoughts for more than twenty years. That he has been permitted to finish a labor prolonged throughout so great a period, and wrought upon amidst the many duties and burdens of a busy professional career, excites earnest gratitude. The fear that he might not finish his self imposed task, and thus leave an incomplete work, has caused sore anxiety, especially when, at sundry times, more or less serious illness has commanded pause. Happily this apprehension is now dismissed, and the duty at last ended is herewith submitted to the judgment of fellow workers in and lovers of Natural History.

Finished Work.

In the first part of the volume six chapters are taken to consider various natural habits and physiological problems for which there was no space in the two previous volumes. These topics are in the line of those studies in Œcology to which the author has heretofore especially given his attention. In addition thereto, and forming indeed the bulk of this volume, the second part thereof contains descriptions of many indigenous species of Orbweavers, illustrated by thirty lithographic plates, colored by hand from Nature. Most of these plates are of Orbweavers, the group to which the author has given special systematic study. But two plates are added, without descriptions attached thereto, of representative species of the other aranead groups, especially of those species whose habits have been presented in the foregoing volumes. This descriptive work has been thought necessary to complete studies which avowedly chiefly concerned habits and industry. The general forms, colors, and proportions of spiders as they present themselves to an observer's eye in Nature are important to the accurate understanding of their habits. One cannot appreciate in full the role which these creatures have to play in Nature until he have a just conception of how they look in the midst of the scenes wherein their life energies are spent. For this reason it formed part of the author's original purpose to present the subjects of his study as they appear in natural site, that his readers may have acquaintance not only with their life history but with themselves.

Scope of the Volume.

Moreover, in studying the habits of spiders it has been necessary to identify the species, and in many cases to describe them. It has seemed proper, therefore, that the work thus done should be preserved to science in connection with the descriptions of the animals' life history. But the author has to admit that this part of his work grew in his hands far beyond the bounds of his first intent, and finally shaped itself into the resolve to publish descriptions and plates not only of the Orbweavers whose habits he had described, but, of all accessible American species of that group. In this matter he has been led along step by step, adding species to species, page to page, and plate to plate, by a desire to make his work yet more and more complete. Working naturalists, at least, will sympathize with and appreciate this fact.

Growth of the Work.

This descriptive work has made the closing volume in many respects the most difficult one of the series. To one who has to deal with small animals, scientific description is always a laborious service. When it is impossible to mount these animals in any satisfactory way, as is the case with spiders, and one is compelled to labor with alcoholic specimens, many of which are minute and mutilated, and often with unique examples in hand which may not be broken up for convenient study, the ordinary difficulties are much increased. Nevertheless, the work has not been an unpleasant one; for there is a fascination about studies in classification which every true naturalist has felt. Dry and uninteresting as the details usually are to the general public, to the specialist they have peculiar interest. The comparison of species with species and genus with genus; the task of separating on this side and on that; of solving the numerous problems that are constantly arising, and other duties of a like kind, bring into play some of the most pleasing faculties of the intellect, and contribute largely to the enjoyment of the systematic naturalist. Nevertheless, to one who can only labor at odd hours, and who is thus apt to lose the connection established by long and careful comparisons, the pleasure is much marred. This has been the author's estate, and will add to the satisfaction which he will feel should it be judged that he has wrought with reasonable accuracy.

Descriptive Work.

In this connection it is proper to say that the increased cost of printing text and plates made it necessary two years ago to notify the public that the original price of ten dollars per volume, or thirty dollars for the entire set, including plates, must be increased to fifty dollars the set. All subscribers at the original price will be served with Volume III. without additional charge, but others must pay the advanced price. The author feels compelled to make this statement here in order to relieve himself from the painful duty of refusing requests, of which some have already come, to sell the work at the first named price. Even at the price now named, subscribers will receive the work at less than its actual cost; a statement which is made not in the way of complaint, for which there is no reason at all; nor to excite sympathy, which is neither required nor desired, but to give a plain and honest reason for a

Cost of the Books.

change which ought to be explained. For further business notice those interested therein are referred to the advertisement at the close of the book.

The most agreeable part of a preface to an author is his acknowledgment for kindly aid rendered by colaborers and friends. First of all, I express my gratitude to Dr. George Marx, of Washington, for the friendly and valuable service which he has given me throughout many years.

Personal Thanks. With a rare generosity and singleness of eye to the advancement of science, he placed at my disposal the Orbweavers in his notable collection. Not only so, but on all occasions he has cheerfully and freely given me the benefit of his advice and judgment. He has thus laid under lasting obligation, not only the author, but all who are interested in his work. I have also to thank others, in different parts of the country, who have contributed specimens and information. Among these are Professor and Mrs. George W. Peckham, of Milwaukee, Wisconsin, whose joint studies of the Attidæ have given to Araneology some of its most attractive and valuable chapters. Messrs. Orcutt, Davidson, and Blaisdell, and the late Mr. John Curtis, of California; Miss Rosa Smith, now Mrs. Eigenmann, and her mother, Mrs. Louisa Smith, of San Diego, California; Professor Orson Howard, of Utah, Mr. Thomas Gentry, of Philadelphia, and Messrs. Charles H. Townsend and Nathan Banks, of-Washington, have contributed material that has entered into this work. Among European naturalists I am indebted to Mr. F. M. Campbell, of Herts, England, for many courtesies; Mr. Thomas Workman, of Belfast, Ireland, and Mr. Frederick Enock, of London, have sent me specimens. To Professor Waldemar Wagner,.of Moscow, Russia, and Mr. Eugene Simon, of Paris, I am especially indebted for copies of their valuable papers and books, and for permission to engrave and use some of the figures with which they are illustrated. To the veteran araneologist, Professor Tamerlane Thorell, whom I gladly acknowledge as "magister," I am indebted for advice from time to time rendered.

I add an expression of my obligations to one who, unhappily for the interests of Science, no longer lives to prosecute his faithful and distinguished labors, the late Count Keyserling, of Germany. His descriptions·of American Spiders have been of great service in determining indigenous species, and many specimens personally examined and identified by him have passed through my hands in the course of these studies. The posthumous volume of his noble work, "Die Spinnen Amerikas," Part IV., edited by Dr. Marx, and which relates to the Epeiridæ, was not issued until a large part of my descriptions were already in print. For this reason some species here appear as new which are described by him in his last work, and have priority, inasmuch as their publication antedates my own. The names, however, are the same, inasmuch as the specific titles given *in litteris* by Count Keyserling to the examples in Dr. Marx' collection have been preserved by me. These discrepancies I have corrected as far as possible in the plate titles.

I count it a duty as well as a pleasure to place among the number of those entitled to my public thanks the name of Miss Elizabeth F. Bonsall, who has made the original drawings for nearly all the plates contained in the atlas. Her faithful and successful work has not always been correctly reproduced by lithographers and colorists, but for the most part it speaks for itself in the admirable rendering from life of the species which she has figured.

As the frontispiece of this volume I have printed a portrait of Professor Nicholas Marcellus Hentz, M. D., who may justly be regarded as the father of American Araneology. John Abbot was indeed before him in the field, and during the early part of this century made personal studies in South Carolina and Georgia of our American spider fauna. The results of these studies remain in the descriptions of Walckenaer and in the beautiful manuscript drawings now preserved in the Library of the British Museum of Natural History in Kensington, London, and to which fuller reference is made in the pages which follow. Some interesting notes upon the life of Professor Hentz, written by the late Mr. Edward Burgess, may be found in the preface to "The Spiders of the United States," published by the Boston Society of Natural History. I am indebted to Professor Henshaw, the Secretary of that Society, for a photograph of the likeness from which the phototype plate of Professor Hentz has been made. It has been reproduced as faithfully as the age and condition of the original photograph would allow.

Professor Hentz.

In reviewing this book it falls out as a matter of course that I note imperfections therein. Most of these, it may be said in all fairness, are due to the peculiar circumstances under which the work has been wrought. Some of the plates were finished, printed, and even colored, awaiting their place in the volume, as many as ten years ago. In the progress of study my views of certain species were modified, thus compelling some modification of the printed results. But this, as expressed in the plates, could not be done without rejecting and remaking the plates, a loss I did not feel it necessary to bear. Corrections and modifications have therefore been made in the text and in the plate descriptions, and no practical disadvantage need be felt by the student. Moreover, the detached manner in which all my work has been done, taking an hour here and there, or a week or so from a summer vacation, and the inability, because of professional obligations, to give close and connected oversight to the work of artists, lithographers, copyists, and colorists has resulted in some blunders which have indeed been easily corrected in the text, and would attract but little attention from the ordinary observer, but which none the less to an author are a blemish upon his work.

Errors and Blemishes.

Nevertheless, the author has at least the satisfaction of believing that he has honestly, faithfully, and impartially endeavored to meet every question, whether in the life habits or classification of spiders, to which he has directed his attention. He indulges the hope that he may at least have cleared the way for others to follow, in a field where the difficulties are undeniably great, but where the rewards to an earnest seeker after Nature's secret ways are abundant. They are had not only in the gratification of such pleasant toil, and in the consciousness of having added to human knowledge and enjoyment, but in the higher satisfaction of having contributed somewhat to man's knowledge of the works of his Creator.

The Creator's Works.

The author would count himself faithless to truth as well as to duty were he not to add that the last named consideration has been to him a continuous stimulus and support. He believes thoroughly in that view of Divine Providence taught him by beloved parents in his childhood which makes it to be God's "most wise and powerful preserving and governing all His creatures and all their actions." The smallest creatures and the lowliest adventures of their humble lives are within the care of the Good Father of all, the Lord of spiders as well as the God of men. To bring all knowledge uncovered from the secret places of the natural world, and lay it devoutly before the world's Creator as a tribute of worship and a token of spiritual fellowship, has been the chief motive which has urged the author to, has guided him through, and sustained him in, this work of twenty years, now happily ended.

Author's Chief Motive.

H. C. McC.

The Manse,
Philadelphia, July 3d, A. D. 1894.

TABLE OF CONTENTS OF VOLUME III.

PART I.—GENERAL HABITS, BIOLOGICAL MISCELLANY, AND ANATOMICAL NOMENCLATURE.

CHAPTER I.

TOILET, DRINKING, BURROWING, AND SOCIAL HABITS.

CHAPTER II.

MEMORY, MIMICRY, AND PARASITISM.

CHAPTER III.

BIOLOGICAL MISCELLANY.

CHAPTER IV.

WEATHER PROGNOSTICATIONS, SUNDRY SUPERSTITIONS, COMMERCIAL VALUE OF SPIDER SILK.

CHAPTER V.

MOULTING HABITS OF SPIDERS.

CHAPTER VI.

REGENERATION OF LOST ORGANS AND ANATOMICAL NOMENCLATURE.

PART II.—DESCRIPTION OF GENERA AND SPECIES.

PART III.—PLATES AND INDEX.

CHAPTER I.

TOILET, DRINKING, BURROWING, AND SOCIAL HABITS.

I.

CONTRARY to general opinion, spiders are tidy in their personal habits. They are indeed sometimes found in positions suggestive of anything but neatness, and occasionally their webs are much soiled with accu-

Toilet Habits. mulated dust, particularly those of tribes which spin sheeted webs in cellars, stables, barns, and like places. Even in such cases the creature rises above her environment and keeps her body clean. Orbweavers' webs are rarely seen much soiled by dust or floating refuse of any sort, a fact which of course is chiefly due to the transient life of the snare, which for the most part is limited to a single day. These webs, as we have already seen, are generally made in attractive surroundings among grasses, leaves, and flowers, which would prove a veritable aranead Eden were it not for obtruding evil spirits in the shape of raiding wasps, hungry birds, and other foes.

When spiders become covered wholly or in part with objectionable matter, whether dew, rain or dust, or soil as in the case of ground workers, **Toilet Im-** they soon proceed to cleanse themselves. Their brushes and **plements.** combs are the hairy armature of the legs and palps, together with the hairs and teeth that arm the mandibles, and these toilet implements are well adapted to the work. Did the habit of cleanliness arise from the possession of these implements? Or, were the implements developed out of the vital necessity for a cleanly person?

A large female Domicile spider suspended downward upon a series of cross lines, by her hind legs (Fig. 2), accomplished her toilet something in this wise: The fore leg was drawn up and placed at the tibia between the fangs. It was then slowly drawn outward, the mandibles meanwhile gently squeezed upon it (Fig. 4), until the whole leg had passed through the combing process, when it was stretched out and another leg substituted; and thus on until all had been cleaned. The palps were combed in the same way, and then were used for cleaning the face and fore part of the mandibles. In this act the palp, after having been drawn through the mouth, perhaps to moisten it, would be thrown to the top of the caput, which it overclasped in the position of Fig. 3, and then was gradually drawn down over the eye space and front of the mandibles, smoothing down and cleansing the surface thereof as it was moved along. The motion resembled that of a cat in the act of cleaning her face and the back part of her head and ears, after having licked her paw.

Spiders may often be seen making their toilets in the early morning. The heavy dews discomfort them and they brush away the drops which cling to them. The same act may be observed after showers of rain, after feeding, and often after making a snare. The viscid beads and bits of flocculent matter from her own web some-times entangle with the hairs and spines of the legs, after a more than usually vigor-ous effort in capturing and swathing a vic-tim. This is so disagreeable that the cap-tive will be trussed up in the open space of the broken orb until the tidy aranead removes the offending matter. Sometimes after a hearty meal Arachne will make her toilet, thus reversing the human mode of dressing before dinner.

Fig. 1.   Fig. 2.

Fig. 1. The Agricultural ant cleaning the tip of her abdomen. Fig. 2. Domicile spi-der cleansing her leg while suspended on a web.

One spider (Epeira vertebrata), captured in a large glass tube while eating a fly, kept hold of her food, deliberately adjusted herself to her new position, spun out a few lines which were rapidly attached to the sides of the glass, then turned over and with great sang froid concluded her meal. When she had finished she began cleaning her palps and feet, and gave me a fine opportunity to see the whole operation. I here observed that the mouth secreted freely a liquid which appeared to be a little mucilaginous, and that the paws were drawn through this. The stiff hairs upon the upper part and inner sides of the mandibles must materially aid the process of cleansing.

Hair-dressing the Feet.

The fangs are used as claspers in the process of cleansing. The leg is passed underneath one fang which clasps it around in the bent part at the articulation, thus holding it up to and within the mouth. The tendency of the legs to spring back from their unnatural position is probably thus overcome until they can be cleansed. The fangs may also serve to move the leg back and forth through the jaws. During this process the mandibles work back and forward like the jaws of vertebrate animals, only that they move horizontally instead of vertically. The fangs are used in the same manner to clasp and adjust the prey during the act of feeding. They thus serve, together with the palps, the purpose of fingers or hands.

Use of Fangs.

Fig. 3.   Fig. 4.

Fig. 3. Combing and washing the head with the palp.
Fig. 4. Combing a fore leg with the fangs.

When a hind leg is cleansed it is bent forward and downward beneath the abdomen and so into the mouth, where it is treated as above described.

The drawing Fig. 5 shows Argiope cophinaria as seen in this phase of making the toilet. The sides of the abdomen are cleansed by brushing them with the sides of the third pair of legs, which are pressed against the body and pushed downward, as one would stroke a cat's hair with his hand. The cleansing of the dorsal part of the abdomen is effected by throwing a hind leg over the top thereof and moving it downward towards the spinners, keeping it meanwhile pressed against the skin. The spines and bristles on the legs thus act as a comb or brush.

I have often had opportunity to note like habits of personal cleanliness in our American Mygalidæ. My longlived tarantula "Leidy" was remarkably tidy. Always after digging
Tarantu-
la's Toilet in its burrow it was quite sure to cleanse its person, and, by reason of its size, the use of its palps in wiping off the fore part of its body presented an amusing likeness to the familiar action of pussy when washing her face with her paws. The fore legs were cleansed by placing them against the palps and rubbing the two together. The toilet was also accomplished by overlapping one leg with the other, the second leg over the third, for example, and then rubbing the two as if a man were to scratch his legs by drawing the inner surface of one along the front surface of the other. The first leg was thus rubbed against the second, of course being pressed down upon it meanwhile. The palp was thrown back to the first leg, which it brushed off in the same manner.

Fig. 5. Argiope cleansing a hindermost foot by drawing it through the fangs.

It is interesting, and suggestive of the substantial unity in the primary functions of life which
Compar-
ed with
Ants. prevades living things, to note this community of habit and method between a vertebrate and an arachnid. The same may be remarked of the ants, whose toilet habits I have carefully observed and described in my "Agricultural Ant of Texas."[1] The methods of cleaning their persons practiced by ants and spiders are quite similar; more so, indeed, than one would suppose, considering the remarkable difference in the general life economy of the two creatures. It is not a particularly striking fact, but rather what one would expect, that a spider should hang herself up by a hind foot to comb, brush, and wash herself. But it strikes one as somewhat out of the ordinary that an ant should resort to the same turnverein process, yet it does so, as I have shown in the case of the Agricultural

[1] Chapter VIII. on Toilet, Sleeping, and Funeral Habits.

ant.[1] I reproduce a figure from the above work to show the likeness noted. (Fig. 1.) In sooth, one may go further up the grade of zoological life, even to the apex of the pyramid, and note that man himself in the act of combing his hair unconsciously adopts artificial implements which resemble the natural combs and brushes supplied in the tibial combing spur of ants, and the hairs, bristles, and tarsal scopulæ of spiders. The economic harmony, here at least, certainly threads vast intervals of being.

II.

The tidiness of spiders is further shown by the fact that they are extremely loth to sully with excrement the boxes in which they are imprisoned. I continually observe that, when emptying my collecting boxes in order to colonize spiders on my vines, the first act is to void excreta, which they often do with great freeness, in large white drops, showing that they have really done violence to nature by retaining the same rather than mar the little box in which they were confined. So, also, they are careful in this natural act to avoid fouling their webs. The abdomen is thrown so far outward that the voided matter never comes in contact with the web lines.

Tidy house-keeping.

It is interesting to observe an Orbweaver in the process of cleansing its web from material which has fallen upon it. I made a complete observation of a female specimen of the Shamrock spider engaged at this work. Several leaves of an ampelopsis vine on which her snare was spun, and two bits of the stem thereof, one at least four inches long, had become entangled in the lower part of the orb. The spider had just commenced the work of clearing away this extraneous material when my observation began.

House-cleaning.

She was hanging by a line which she had attached to the hub of her orb, and which dropped down upon the inside of the web, so that she faced the leaf that she was then about to remove. One hind foot reached upward beyond the abdomen and held to this line, which, of course, was also attached to the spinnerets. (Fig. 6.) During part of the operation the other hind foot was stretched backward, and clasped the line near the spinners, as though to give additional poise and security to her position. But throughout a large part of the entire operation of clearing away the debris she hung by one hind foot alone, and used the other one for the work of dragging out, revolving, and expelling the material. In this position, hanging thus opposite her point of endeavor, she reminded me of painters swinging upon their little seats by ropes fastened far above and engaged in painting the sides of a house; or of workmen let down from heights for the various purposes of their handicrafts. This position was never abandoned for the whole period of time, the spider being able to swing

---

[1] Pogonomyrmex barbatus. Ibid., page 129 and pl. xvii., Fig. 80.

herself back and forward across an arc of four or five inches, moving herself by the free legs, but always holding to the dragline by at least one foot.

Having thus secured a position for convenient labor she seized with her fore feet the intruding leaf, and began removing the spiral and radial lines upon which it was entangled. These were pulled away by the claws and bitten off by the mouth. To promote this purpose the leaf was turned over by the fore legs, assisted by the short third pair. When one end was released it was carried towards the spider's mouth, gradually passed underneath the face by the movement of the fore legs, and the clinging parts of the viscid lines in the meantime were gnawed away from the undetached portions of the leaf. Finally the leaf was freed from its entanglement, and held off a little space from the body by the legs, which were now bunched close underneath the jaws. Then, swinging herself outward a little ways from the orb, the spider passed the leaf away from her downward, and when nearly freed from her grasp gave it a joint push and fling with the fore part of the legs which cast it to the ground.

This process was repeated in the case of the other rubbish in the web.

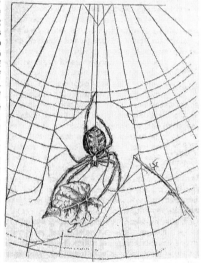

Fig. 6. A Shamrock spider cleansing her snare of an entangled leaf. The figure shows her in the last act.

The long twig, which hung crosswise of the orb, caused her much trouble, but she got rid of it very skillfully. Cutting away all the lines on either side, she seized the twig and gradually pulled it beneath her face as she hung head downward, and so passed it underneath her body and away from the orb little by little. Then she poised it for a moment in a convenient position, and with a quick fling cast it from her towards the ground, the fore legs being used for this act of expulsion. She experienced much difficulty at times with the sticky lines, and at various intervals was compelled to pause and clear her feet and legs of the viscid material. While cutting away the viscid spirals, the portions of the snare above or below had to be looked after lest the orb should collapse by the sundering of the supporting radii. This, however, was adroitly managed as in the case of cutting out entangled insects, the inevitable dragline being used to splice and stay from the spinnerets while the spider cut away with the fangs.

When the two leaves and two twigs had been cleared away, a vacant section was left in the web of about one-fifth the whole. At this point my observation ceased, and I cannot say whether the spider built a new orb immediately, clearing away all the rest, or patched the damaged section. On the following morning, however, she was resting within her nest, holding to a trapline attached to a perfect orb, on which were no traces of mending.

A female Epeira sclopetaria was observed clearing off a lot of straggling threads stretched across a window. These were gathered up with the

Scraps of Web Eaten.
second and third pairs of legs principally, which, aided by the palps, drew them towards the mouth, into which the spider put them. This is a common way of disposing of ragged bits and fragments of spinningwork, which no doubt yield some nourishment that may again be transformed into webs.

According to Mrs. Treat, the Turret spider is a neat housekeeper. She leaves no debris in her cellar under the tower. The remains of insects

Penalty of Untidiness.
are thrown from the top in the same manner that she throws excavated pellets. The Tiger spider, on the contrary, always leaves the skeletons of insects in the bottom of its tube, which in time makes a rich black mould. As the result of this, the occupant is often driven from its room by a great mushroom starting from the bottom of the burrow upward and completely demolishing it, forcing the tenant to seek new quarters. Such a catastrophe never happens to the neater tower builder.[1]

The advantages of cleanliness are certainly thus remarkably illustrated, and a sufficient reason given why, for the most part, spiders are careful to carry from their dens and snares the debris of insects eaten by them. This is not the universal rule, however, as other species besides Lycosa tigrina will sometimes overspin the remains of their feasts, entirely covering over with spinningwork the hard chitinous portions which are rejected. Nor does this act always result in such a calamity as that above recorded.

The Turret spider, after working upon her tower or in her burrow for an hour or more, is apt to stop and assume her favorite position, seated across the top of her tower, in order to make her toilet. First one leg and then another is passed between the palps several times, and all the while her mandibles are at work as if chewing, the moisture meanwhile working up between them.

---

[1] Home Studies in Nature.

The Purseweb spider, according to Mr. W. L. Poteat,[1] is scrupulously neat. The droppings of his captive spiders were deposited outside the nesting tube, and generally at such a distance as necessitated **Purse-** her leaving the nest. These deposits were observed only in the **web** morning, so that she quits her tube at night, at least for this **Spider.** purpose.[2] One usually finds a cluster of insect remains loosely adhering to the outer wall of the tube, a little below its upper extremity. These do not seem to be purposely attached to the tube, but to be accidentally entangled when being thrown out, as with excavated earth, for they are often seen on the ground at the foot of the tube. The leavings of a single feast are frequently seen bound together with silk. On one tube was recognized the remains of some Neuropterous insect and of two woolly-bear caterpillars, such as hair, bits of chitinous integument, mandibles, joints of legs, etc. The elytra of beetles are also common.

FIG. 7. A tarantula drinking water from a saucer.

III.

Spiders require water, as do most animals, for their health, comfort, and growth. They can, indeed, live long periods deprived of water, but unless **Drinking** supplied with an equivalent in the animal juices of their prey **Habits.** they perish from thirst. Even when insect food is abundant they enjoy fresh water, and habitually partake of it in nature. The dews which gather upon their webs during the hot months probably afford a common supply. In the morning after a heavy dew, or after a rain shower, spiders may be seen brushing away the moisture accumulated upon the hairs which clothe their bodies. This is done by passing the fore legs forward over the head and cephalothorax, and the hind legs over the abdomen backward. The legs, which gather the moisture upon their armature of hairs and spines, are then doubled under the body and drawn between the two mandibles, or between the mandibles and lip, thus brushing off the water, a part of which, however, remains and is taken into the mouth.

[1] A Tube Building Spider, page 16.   [2] Ibid., page 15.

Again, I have often seen the mouth parts applied directly to water, which appeared to be appropriated in the usual way of feeding by pressing the liquid into the gullet. Spiders of all tribes have been seen drinking in this way, and this is the method continually practiced by my tarantulas in confinement as shown in the sketch at Fig. 7. I frequently receive living spiders sent long distances in boxes or bottles, and my first act is to give them fresh water, which they usually rush upon and at once eagerly apply their mouth parts thereto as here shown.

A brood of young Zillas kept in my study were given water daily by throwing it in spray above the greatly extended fine web upon which they were domiciled. They were often observed to take the moisture by passing the legs to the mouth in the manner above described. On one occasion I observed one of the brood carrying a goodly sized globule of moisture in her jaws, which were spread out (Fig. 8) upon the drop over which, on either side, the palps were also extended. These organs seemed to be inserted into the globule, which, however, probably only adhered to them by means of the delicate hairs upon them. At all events the young aranead climbed over her web, carrying the particle with her.

At the same time a young Agalena nævia, which had wandered from her little tent spread on the table beneath, and was promenading the broad

sheeted commons of the Epeïroids, had seized one of the largest drops of spray and was making off with it. The water was attached to the mouth parts, as in the above instance, but in addition the animal had thrown one fore leg (Fig. 9) around the side of the globule,

FIG. 8.   FIG. 9.
FIG. 8, a young Zilla, and FIG. 9, a young Agalena carrying a drop of water.

and thus trudged along, literally carrying an armful of water. I watched her until she had gone eight inches in this way, when the drop, which had gradually diminished in size, had nearly disappeared. It was certainly a curious sight, this little spiderling trampling over the gossamer highway carrying in jaw and claw this strange drinking cup, which shone like a silver ball against the black body of its wee porter.

The same behavior was noticed in another individual of a brood of Epeïroids, similarly confined. One of the young had taken or become entangled with a drop of water, which it encompassed in part by one of its second pair of legs, and with the remaining legs strode, back downward, along the web. The moisture did not adhere to the lines, although frequently in contact with them, and the drop was carried along several inches to a tall box. As soon as the drop touched the wood it was absorbed, and the spiderling returned to the lines, whereon she suspended herself and began licking the dampness from her legs. Such facts strengthen the probability that the dew furnishes a supply for satisfying aranead thirst.

IV.

An interesting note upon the feeding habits of spiders has been communicated to me by the Philadelphia entomologist, Mr. P. P. Calvert. While
Feeding Habits. studying the habits of dragonflies he observed early in May a species of spider, which appears to be a young Dolomedes sexpunctatus, feeding upon newly transformed imagines of these insects. The spiders were lurking upon tall grasses and water plants, on the margin of a small pool near Bartram's Garden in Philadelphia. The dragonflies had come to these plants to transform, and before their wings were dried and ready for flight, while they were yet helpless, the young Dolomedes seized them and sucked their juices. The two species which were thus preyed upon are Ischnura verticalis Say and Nehalannia posita
Dolomedes. Hagen, both of them small species. Dolomedes, as heretofore described (see Index of Vol. II.), is a semiaquatic species, running rapidly upon the water to seize insects, and remaining for a considerable length of time underneath the surface. The mother deposits her cocoon in a large leafy nest among the bushes, within which the young are hatched. The specimen shown me by Mr. Calvert as taken while in feeding on dragonflies was not more than half grown. We thus have a glimpse of one of the methods in which this Citigrade species, and doubtless many others, obtain food. It shows also the disadvantages and perils of insects during transformation, when they are exhausted by the process and have not acquired the natural facilities for escape or defense.

V.

Various spiders run fearlessly on the surface of the water; some even descend into it spontaneously, the time during which they can respire,
Water Habits. when immersed, depending upon the quantity of air confined by the surrounding liquid among the hairs with which they are clothed. In this manner the European Argyroneta aquatica is able to pursue its prey, to construct its dome shaped dwelling, and to live habitually in water. There are, however, a few exceptions of extremely small spiders, Neriena longipalpis and Savignia frontata, for example, which, though they do not enter water voluntarily, can support life in it for many days, and that without the external supply of air so needful to the existence of Argyroneta under similar circumstances.[1] This is certainly a remarkable fact. I have known spiders that seemed to be drowned by long immersion in water to revive shortly after being taken out; even those plunged in alcohol, if not kept therein too long, will recover from seeming death. But that these small and delicate creatures should live several days in water surely strains one's belief in even so trustworthy an observer as Blackwall.

[1] Blackwall, Spiders Gt. B. & I., Introduction, page 9.

Argyroneta has not been found in America, and no spider with habits in anywise resembling it, but our spider fauna contains a number of species, principally limited to the Citigrades, that are much at home either on or within the water. Several species of Dolomedes habitually **Rafting Dolo- medes.** live in the neighborhood of water, and may be seen continually running about upon the surface in search of prey. They avail themselves sometimes of floating material in order to rest during their predatory excursions. This incidental occurrence, in the case of Dolomedes fimbriatus of England, seems to have been specialized into the habit of constructing a rude sort of raft by lashing together floating leaves. This raft is utilized as a point of departure for raids upon water insects, and as a "lunch room" in which the captured prey are fed upon. It floats upon the fens of England, apparently at the sport of the wind.

Dolomedes sexpunctatus, in the neighborhood of Philadelphia, is able to remain for a long time beneath the surface of water. I have on various occasions timed the period of submergence, and one specimen remained underneath the water forty-two minutes. While thus submerged the spiders are surrounded more or less completely with bubbles of air which have the appearance of a silvery coat of mail, as one looks down into the water. I have alluded elsewhere[1] to the habit of certain Lycosids, as reported by Dr. Alan Gentry, to live underneath frozen water during winter, and pass from point to point by means of threads strung upon water plants. This single observation opens up a new and strange chapter in the winter life and amphibious habit of these animals, which invites investigation. No doubt the ability to exist while surrounded with water is **Long Submer- gence.** of special value during periods of heavy rain, when their burrows must be inundated, and when they are themselves submerged for hours or perhaps days together. It is probably true that all spiders can endure a good deal of submerging; they seem at least to be able to survive under the heaviest and longest continued showers. How easily even Orbweavers can adapt themselves to the water habit may be found by reference to Vol. I., page 160, where it is seen that Tetragnatha habitually sails over the surface of Deal Lake, New Jersey, by means of outspun filaments of thread; and where also (page 161) it is shown that Epeira can avail herself of an accidental float in the shape of her own flossy ball of cocooning silk.

VI.

Lycosa tigrina digs a tube in the earth from six to twelve inches in depth, which is bent in a little elbow near the surface. The upper part beyond the bend forms a sort of vestibule,[2] which assumes the shape of a broad, silk lined funnel at the mouth of the burrow. The background

[1] Acad. Nat. Sci., Philadelphia, 1884, page 140.
[2] See Vol. I., Figs. 305, 306, page 323.

is composed of whatever material Tigrina can reach with her long hind legs, while her fore legs rest in the edge of her tube. This funnel is the foundation of a concealed room, which it sometimes takes the

Burrowing Methods. spider several nights to build. It seems to refrain from working during the day. Mrs. Treat says that the burrow of Tigrina is uniformly straight. My observation is entirely different; that of Arenicola is uniformly straight down, but Tigrina builds a bent burrow as above described.

A female of this species had a nest in a bed of green moss, and the cover looked like a moundlet of moss and leaves. The longest diameter measured five inches, and the shortest four and a half inches.

Lycosa Tigrina. The base cover was made of acorn cups and sticks firmly held together with threads of silk. Then a silken canopy was spun, and over this were laid green moss, dry leaves, and sticks held fast with spinningwork. This made a neat little upper room, the walls of which were smooth, silk lined within, but showed natural inequalities on the outside.

Maternal Ingenuity A window was left in the room, the use of which soon appeared. The builder had an egg cocoon attached to her spinnerets, and would put herself in position to let this rest against the window where it received the rays of the sun. For three weeks this was her daily occupation, patiently holding her egg sac in the sunlight. Was she not conscious of the fact that this aided the healthful development of her progeny?

On the 20th of May the observer removed the cover from the burrow, and toward evening Tigrina began to restore it. She reached out her hind legs, feeling for material, and first drew in an acorn cup and proceeded to fasten it. On the following morning (May 21st) a broad funnel shaped ring had been built around the tube, but not covering it. By May 24th the spider had made a room above her burrow lightly covered with moss.[1]

The male of Tigrina is a handsome fellow and nearly as large as the female. In color he is a light snuff brown, with dashes of dark purple, while the legs are striped like a tiger's. The female is nearly black. The male takes as much pains in building its domicile as the female; indeed, one confined in a jar entirely outdid the female in making a tasteful retreat. He utilized a little twining plant by winding his web around it, thus making a living green bower over his tube.

A New Hampshire Lycosa whose species is unknown was taken from a burrow sixteen inches deep by Mrs. Treat, and placed in a glass jar with

Lycosa Digging. five inches of moist earth well pressed down. It soon commenced to dig a burrow next the glass, giving a fine opportunity to see it work. It dug the earth loose with its mandibles and with the fore feet compressed it into a pellet. It again turned, seized the ball in its mandibles, necessitating a third turn, and then came to the edge of the

---

[1] American Naturalist, August, 1879, page 488.

tube, always with its back to the glass, and adjusted its fore feet so that the tips touched beneath and partly behind the ball of earth. Then with a sudden movement, like snapping the fingers, she shot the earth forth with sufficient force to make it hit the opposite side of the jar. These pellets were held together with a kind of mucilage, sometimes mixed with web lines, showing that in massing the pellets preparatory to deporting them, probably some secretion from the salivary glands is used, and that occasionally filaments of thread are also utilized for binding the particles of earth together.[1]

A Turret spider was kept in confinement by Mrs. Treat and furnished with sticks and moss in order to observe the manner of erecting her tower.

Turret Spider Building. When she had carried down her burrow about two inches in depth she commenced to build the tower above it. She would take the sticks from the lady's fingers and place them at the edge of her tube. She worked while inside her burrow, holding the stick with her forefingers until it was arranged to suit her. She then turned and fastened it with a strong web. She took another stick, proceeded in the same way, and continued thus until she had laid the foundation of a five sided wall.

Next she went to the bottom of the tube and brought up a pellet of earth which she placed at the top of the sticks, and proceeding thus erected a circle of pellets which she next overspun. They were so arranged as to cover the sticks on the inside, leaving the inner walls perfectly rounded and silk lined. Then the spider was ready for more sticks, which she continued to alternate with pellets until the tower reached a height two-and-one-half inches above the burrow. Sometimes bits of moss an inch or two in length were given her by the observer, which were used by fastening them to a stick with threads of silk. This made a wall fringed on the outside with moss.

If the spider were not in a mood for building, and a stick were offered her she would take it in her mandibles, and with her fore feet give it a **Flinging Dirt Pellets.** quick blow, often sending it away with force enough to hit the enclosing jar. When she was digging and bringing up pellets of earth which she did not wish to use upon her tower, she would throw these from the top of the walls with sufficient force to send them a foot or more from the burrow, had it not been for the intervening glass. This habit accounts for the fact that the observer could never find fresh earth near the burrows of Turret spiders.

What motive could the aranead have for thus casting the fresh earth away from the immediate vicinity of her nest? No doubt, had the soil been permitted to accumulate by dumping it directly from the tower, the protective uses of the tower would thus have been destroyed. Can it be, moreover, that the secretive instinct which is observed in the building

---

[1] Harper's Magazine, 1880, page 862.

habits of other Arthropods is responsible for this habit? Our American Carpenter ants, Camponotus pennsylvanicus, when burrowing their halls

Secret-iveness. and galleries within a tree, cast out the fresh chippage from the openings through the bark. But these are invariably gathered up by a squad of workers at the foot of the trunk, and carried away to a distance, as though it were thus intended to conceal all traces of the neighborhood of the formicary. At least, I can find no other satisfactory motive for such behavior. May it not be that the Turret spider is moved by a similar sentiment?

VII.

Atypus digs obliquely a deep tunnel of fifteen to twenty centimetres, the size of its own body. It drapes the tunnel with a straight silken tube,

Atypus. of close tissue, of which the upper part, longer than the subterranean gallery, is laid horizontally upon the soil and terminates in a point. Near its lower extremity the tube presents a considerable enlargement, forming a quite spacious chamber in which the spider dwells. At the entrance or throat of this enlargement the cocoon is suspended by a number of threads. M. Simon says that he has many times taken Atypus holding worms in its mandibles, and he believes that these worms form the substance of their nourishment. In effect, if one examines the silken chamber he remarks a place where the tissue is much thinner and transparent, and Simon thinks it probable that Atypus removes the silk lining and thus readily procures prey without the necessity of mounting to the surface of the earth. However this may be, the actual observations heretofore made upon the feeding habits of Atypus show that she subsists on insects which she captures by seizing them through her tube as they rest or crawl upon the outer surface thereof. She then drags her prey inside to feed thereon and repairs the rent.

Mr. Bates describes Territelarian spiders (Mygale Blondii and M. avicularia) as inhabiting broad tubular galleries smoothly lined with silken webs.

Mygali-dæ. The galleries are two inches in diameter and run in a slanting direction two feet.[1] Again he speaks of them as spreading a thick web beneath a deep crevice in trees, and having their cells under stones.[2] Once more, in alluding to their diversified habits, he says that some species construct among the tiles or thatch of houses dens of closely woven web which resembles fine muslin in texture. From these domiciles they invade the house apartments. Others, according to Mr. Bates, build similar nests in trees.[3] I believe it will be found that the creatures that burrow in the earth are identical with those which spread sheeted webs among the trees. Numbers of tarantulas come to our port (Philadelphia) in fruiting vessels, and are often found in the great pendant

---

[1] Bates' "The Naturalist on the Amazon," Vol. II., page 58.
[2] Ibid., Vol. I., page 61.   [3] Ibid., Vol. I., page 106.

bunches of bananas, to which they had no doubt resorted as a convenient field for capturing prey, and were themselves captured and shipped, hidden away among the clusters of fruit.

In the case of the spider "Leidy," described in Vol. II., page 428, the only effort made at nest building was a rude burrow which was excavated against one side of the box, and which in the course of time was extended

downward to the bottom of the box, and laterally along the bottom either way, thus forming an irregular cavity. Into this it frequently descended, dividing its time between the cave and the outside surface. This burrow was entirely destitute of a silken lining, although occasionally the opening at the surface would be overspun with a thin sheet of spinningwork. I have seen the same habit in other individuals of the species kept in confinement. The only attempt at a nest ever observed by me has been this burrow, with an occasional sheeted closure, and more rarely a slight silken lining of the interior of the burrow. I believe, therefore, that the popular theory that the tarantula makes a trapdoor like the California Cteniza is without foundation in fact, and that its ordinary habitat is a plain burrow like that made by most Lycosids.

Fig. 10. A tarantula (Mygale) digging out her burrow.

The mode of making the burrow was well observed by me at various times.[1] In the act of digging the spider first used the two leg like palps, the digital brushes of which are well adapted for that

Fig. 11. Tarantula (Mygale) carrying dirt from her burrow.

service. Then the two front feet were brought into play to gather up the loose pellets of soil and scrape them into a ball. The first and second pairs of legs then closed up around and under the balled mass, compressing it inside the mandibles. (Fig. 10.) When the pellets had thus been gathered and squeezed into a mass, they were held within the extended

---

[1] Acad. Nat. Sci., Philadelphia, 1887, page 381.

mandibles, the palps in the meantime girdling them at the side and beneath, and so were carried away from the burrow to the dumping ground. (Fig. 11.)

I never observed any scratching and scraping the dirt backward, in the fashion of a dog digging in a rabbit burrow, which is also the action of bees and wasps when excavating the earth. Always the pellets were deliberately loosened as I have indicated, squeezed together into a ball and carried off. During the act of digging, and indeed quite habitually during all actions such as eating, etc., tarantula kept her spinnerets curved above the posterior end of the abdomen, while a diverging ray of threads issued therefrom to the surface beneath.

VIII.

Miss Estelle Thomson, a correspondent of a weekly journal,[1] gives an interesting account of the nesting and burrowing habits of the California Trapdoor spider (Cteniza californica), which contains some observations worthy of a more permanent place and wider circulation among araneologists. The spider's location of her nest is carefully planned. It is never made in a hollow, but invariably upon high, dry, sloping knolls so placed that moisture from the winter rains drains off in every direction. This accords with other observations of nesting site communicated to me.

California Trapdoor Spiders.

A young man in the neighborhood of San Diego made a number of experiments to determine if the occupants of the trapdoor nests would replace the doors of their burrows. He removed as many as sixty in the course of a week, unhinging them at night, marking the site, and going to the nests in the morning to note results. Without exception the spider completed and hung a new door in the interval. There was, however, a limit to this industry, and a remarkable series of progressive deterioriation in the quality of the successive doors. The second door was always of coarser fibre than the first, its proportion of silk being smaller. The third was in about equal proportion of silk and earth; the fourth largely of mud; the fifth of mud, with barely sufficient webbing to coat and hinge it. The sixth was a poor attempt at forming a mud closure without any webbing, and no instance was observed when a single spider completed more than five new doors, with perhaps half of the sixth. One may attribute this behavior either to the natural exhaustion of the spinning material required for replacing such continuous losses, or to the physical exhaustion of the spider, with a strong element of intellectual disgust and discouragement over such an unusual series of accidents. Did the spider's mind at last reach the conclusion that it had come across an experience quite separated from the realm of accidents; and dimly apprehend

Repairing Doors.

---

[1] The Christian Union, New York, May 20th, 1893.

that she was in conflict with a power beyond that which controls ordinary misfortunes, and which therefore it was quite useless to further oppose?

Another experiment with interesting results was the fastening of the trapdoor with a pin or peg into the adjacent soil, so as to prevent exit therefrom. Invariably when this was done in the evening, a little side branch was excavated over night, with an opening at the nearest point to the original mouth of the tube, and a new door hung upon it. It is possible that some of the various nests, first described by Mr. Moggridge[1] as branched nests, may have been due to accidental stoppage of doors. It has been supposed that spiders created these branches as a refuge from enemies, or perhaps from aggressions of the elements. Miss Thomson's record would indicate that the branch tube is simply a natural effort of the spider to provide an exit from her burrow whenever the ordinary mode of departure has been prevented. The curious thing about it, perhaps, is that the inmate did not attempt to burrow out the obstructed door, instead of taking the roundabout and more laborious course of making for herself a side exit. What could have caused this peculiarity of behavior? Can we account for it by the general suspicious temper which characterizes spiders, and many other animals, when brought in contact with a new experience?

Branching Nests.

Miss Thomson attributes to Cteniza californica the secretive tendency which naturalists have observed in other Trapdoor spiders. She conceals her abode from observation by causing it to mimic the adjacent site. The door corresponds so closely to the character of the surrounding surface that it is difficult to discover it. If the bank is bare the top of the door is also bare; if the bank is covered with lichens, the spider cuts a crop of minute lichens and glues them with nice judgment to the outside of her door, thus disguising the entrance.

Site Mimicry.

When one nest is discovered it is comparatively easy to discover two, as they are almost always in pairs, and many times so close that their lids touch when open. The observer does not state whether these contiguous burrows are occupied by the different sexes, and it would be interesting to know the facts in regard to this.

When leaving her burrow Cteniza simply allows the door to drop of its own weight. When returning she scampers off at a smart pace for her dwelling, and apparently lifts up the door with the fangs of her mandibles, and as she backs down the burrow allows the trap to fall behind her.

After the hatching of the eggs from seventy-five to a hundred black-and-green spiderlings will be found occupying the maternal nest. When these are a few weeks old they leave the native burrow, and begin to excavate in sunny places minute tubes of their own. Often a dozen such

---

[1] Harvesting Ants and Trapdoor Spiders.

small abodes will be clustered about the old trapdoor. These vary greatly in size, but all are quite perfect in form. The smallest nest measured by Miss Thomson was barely three inches in depth, yet this was fitted with a diminutive circular door no larger than the nail of a lady's little finger. The largest adult nest measured was twelve inches in depth.

IX.

Heretofore I have considered the nesting habits of spiders[1] and the influence of enemies upon their architecture (Vol. II., Chapter XIII.).

Taren-
tula
Opifex.
Elsewhere I have tried to trace the relations between the nesting habits of the two great tribes, Citigrades and Tunnelweavers.[2] A discovery lately made by Mr. W. A. Wagner, of Moscow, gives new interest to these statements and enables me to complete the chain of resemblances pointed out. The connecting link between the industry of the two tribes is found in Mr. Wagner's Tarentula opiphex,[3] a Russian spider of the family Lycosidæ.[4] The nesting habits of this spider are thus described by Wagner. It was observed in numbers in the Russian province of Orel, and dwells among the tufted vegetation of fallow lands, its principal habitation being fields of wheat and potatoes. The species is agile in movement, active in habit, and comparatively small in size, having a body length of less than one-half inch, ten millimetres. (Figs. 12, 13.) The burrow is not deep, that of the adult usually not exceeding two and a half inches; is enlarged at the bottom, giving it a bottle shape (Fig. 15); is silk lined throughout, but the lining is extremely thin except toward the entrance; the walls are smooth and more carefully finished than usual with known Lycosids, as, for example, Trochosa singoriensis.

But the most remarkable and distinct feature is the covering of the burrow, which is constructed after the well known type of the Trapdoor spiders, Figs. 14, 17, 18. This door consists of a single layer

Trapdoor
making
Lycosid.
of silk covered externally with a coating of soil, whose pellets are bound together by a mesh of threads and spread unequally upon the surface, being much thicker in front than behind. It has the usual shape of the Trapdoor spider's door, something more than semicircular, or a circular plate cut squarely across the end by which it is hinged to the burrow. (See Figs. 17, 18.) Instead of being beveled along the edge like the door of our Cteniza californica, and thus fitting into the burrow like a cork into a bottle, it rests when closed upon the surface edge of the burrow like a basket lid upon a basket. The front, or entrance end, projects beyond the burrow (Figs. 15, 16), making a sort

---

[1] Vol. I., Chapter XVIII.
[2] Proceed. Acad. Nat. Sci., 1887, Philadelphia, page 377, sq.
[3] Opifex?
[4] Bulletin Soc. Imper. des Naturalistes de Moscow, No. 4, 1890.

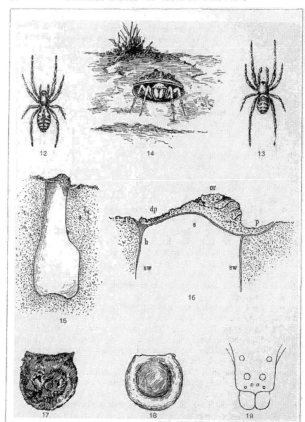

Nesting Architecture of Tarentula opifex. (After Wagner.)

Fig. 12. Female. Fig. 13. Male. Fig. 14. Watching for prey at the bottom of its burrow, the door partly raised. Fig. 15. Section view of the burrow, showing its bottle shape. Fig. 16. Side view in outline of door and throat of burrow; s, silk lining of same; h, hinge of silk; cr, crest of the door in front; p, portico of same; sw, silken walls. Fig. 17. Outside of door from above. Fig. 18. Inside of door, silk lined. Fig. 19. Grouping of eyes.

of portico for the spider when it is on guard. (Fig. 14.) The silk lining of the walls of the burrow (Fig. 16, sw) is continued along one side of the under surface (s) of the door by a thickened ribbon of silk (h), which serves the purpose of a hinge upon which the lid turns when it opens and shuts; its motion backward, however, is limited, for if one tries to bend it beyond the vertical the hinge is fractured.

It will be seen from the section view of the upper part of the burrow (Figs. 15, 16) that the lid is much thickened toward the front, forming a crest (cr), while the hinder part next the hinge has only a thin **The Lid and Hinge.** coat of soil. This arrangement, Mr. Wagner points out, serves to bear down the free end of the lid, and closes it rapidly and tightly when the spider enters or goes forth; it has, in fact, the advantage of the strong and elastic hinge of the Trapdoor spider's nest, which unites with the gravity of the door to bring it down into the burrow's mouth.

Some time before sunset, and probably during the day, Tarentula opifex may be seen on guard at the mouth of its den (Fig. 14); its head and fore

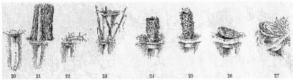

COMPARATIVE VIEW OF TERRITELARIAN ARCHITECTURE.

Fig. 20. Simple burrow (Mygale), unlined or lined only at top. Fig. 21. Purseweb spider's tubular nest supported on trees; burrow sparsely lined, covered with sand, wood, and mould. Fig. 22. Atypus piceus; low hung tubular nest without opening, covered with vegetable miscellany. Fig. 23. Silken aerial tower, Leptopelma elongata. Fig. 24. Conglomerate tower, doorless. Fig. 25. Conglomerate tower, with wafer lid. Fig. 26. Burrow, with lid at the surface; silken lining. Fig. 27. Thick door, many layers, beveled edge, burrow completely lined with heavy silk.

legs are then thrust over the margin of the door, which hangs ajar, and is supported upon the head and back. Here it will remain for a while as though on sentry duty until it ventures forth in search of prey. It is only when thus awaiting at its partly open door that it can be well seen; at the least movement of the observer or at sight of his approach the spider plunges into its burrow, the lid drops heavily, being borne down by the mass of soil accumulated at the crest (cr), and when closed it so closely resembles the surrounding surface that it is nearly impossible to discover it.

Here, now, is the one link which was wanting to entirely connect the architecture of the Lycosids with that of the Tunnelweavers, and complete that resemblance which I had pointed out. The series as thus completed may be arranged as represented at Figs. 20–32. Mr. Wagner has referred to the general likeness between the nest of his Opifex and that of the typical Trapdoor spiders, Nemesia or Cteniza, but has dwelt even more upon the differences. He calls attention to the fact that Cteniza's door consists of a series of superimposed layers of silk and mud, amounting sometimes to thirty,[1] is thick, of equal width, and beveled at the edge; while Opifex makes a thin door composed of a single layer of silk and soil, much thicker in front, and with unbeveled edge. The hinge of the former is also tough and elastic, while the latter is feeble and with little elasticity. All this is true, but Mr. Wagner appears to have lost sight of the fact that the Territelariæ embrace many species besides Nemesia and Cteniza whose industry is greatly varied in form, and furnishes examples much nearer that of Wagner's species than the one with which he compares it. Moggridge has called attention to these in that form of trapdoor which he calls the "wafer" type as distinguished from the "cork" type.[2] The latter is the form with which alone Mr. Wagner appears to have been familiar, while the former more closely resembles his own interesting discovery.

The eminent French araneologist, M. Eugene Simon, has added greatly to our knowledge of these aranead architects, and I have quoted freely[3]

COMPARATIVE VIEW OF LYCOSID ARCHITECTURE.

FIG. 28. Lycosa scutulata, simple burrow in the ground; a temporary closure for moulting and cocooning. FIG. 29. Funnel shaped tube of silk, more or less supported and disguised by vegetable foliage and debris; Lycosa tigrina. FIG. 30. Turret of silk protected by armor of twigs or grass bits; Lycosa arenicola. FIG. 31. Vestibule of silk, armored with moss, etc., with a rude door, Lycosa tigrina. FIG. 32. Silk lined burrow, with wafer trapdoor at surface; Tarentula opifex.

from his papers and given a number of illustrations exhibiting the apparent development of this peculiar industry, from the mere straight burrow to the beautiful silk lined tube of Cteniza, crowned with its admirable hinged trapdoor. In this series is one, the nest of Stothis astuta,[4] a South American species, which in its general characteristics resembles that of Wagner's Basketlid spider. The same wafer door is found upon the nest of another species, Dolichoscaptus Latastei Simon, which builds a columnar turret covered with a movable lid.[5] We are thus able to construct a double comparative series of nests, one from the Lycosids, the other from the Territelariæ, which will show the following facts: First, a progressive advancement from the simple tubular burrow in the ground to a silken lined burrow covered by a hinged lid. Second, the various stages of the two series

[1] See Vol. II. of this work, page 249 and Fig. 264.   [4] Vol. II., page 412, Figs. 347-349.
[2] "Harvesting Ants and Trapdoor Spiders."   [5] Vol. II., page 411, Fig. 346.
[3] See Vol. II. of his work, page 409, sq.

on the one hand correspond generally with the several stages on the other. Third, on the whole, the mechanical skill of the Tunnelweavers gives a more finished product. Fourth, progression in both series is from an equally primitive habitat upward to the most complex and complete, thus making the two series entirely independent, and not the one a continuation of and development from the other.

I am quite aware that this language is analogical and nothing more, as there is no information in my possession which permits us to think of the Tunnelweavers and their industrial habits as actually developed from the Lycosids (or the reverse) in any sense known to science. Nor is there evidence of improvement in the nesting skill of any single species, within which the character of the architecture is persistently unchanged. Nor is there proof of a gradation in the architecture of any species corresponding with the faunal position of the architect. My purpose is simply to point out the marked analogies which present themselves in the study of the architecture of the various species of the two suborders. For a grouping of facts which seem to extend this analogy, as to the essential factor of a tubular web, over the wider field of the entire order Araneæ, the reader is referred to my Vol. I., Chapter XVIII.

X.

In a preceding volume of this work[1] I have considered with some detail the tendency of spiders to assemble in communities. The observations

Social Spiders.

of Darwin, Azara, and others on what they supposed to be the gregarious or social habits of adult spiders are there noticed, and the opinion expressed that the examples cited were accidental assemblages of individuals held together in close neighborhood by various favorable circumstances, but with individual metes and bounds more or less distinctly marked. Nevertheless, in view of the possibilities of Nature, such a conclusion was held with reservation. I there further show that, in the babyhood of numerous species, spiderlings quite invariably maintain assem-

Baby Communities.

blages, and dwell together peacefully, or at least with few breaches of fraternity; a state of amity which is maintained until Nature prompts the individuals to a wider individual life, at which time the assemblages are broken up, and the natural solitary habit and ferocity of the order assert themselves.

I have often pondered whether this strong habit, fixed upon the early life of spiders, might not have formed a basis for the development, in adult life, of some such companionship, fraternity, and unity as mark the social Hymenoptera, so well illustrated in ants and wasps. It does not at first thought seem strange that a habit so marked in babyhood should be carried forward and become permanent in adult character and life. Yet, so

[1] Vol. II., pages 230-241.

far as my observations warranted, I could find nothing to justify such conjecture, and the records examined did not seem sufficiently clear to permit an opposite opinion.

Since the issue of my work, however, some remarkable and most interesting observations have been published by the eminent araneologist, M. Eugene Simon, of Paris,[1] which have induced me to review the subject.

Fig. 33. Common incubating nest of Epeira bandelieri.
Fig. 34. A single cocoon. (After Simon.)

M. Simon's Studies. In a paper presented to the Entomological Society of France, February, 1891, he relates and illustrates the habits of certain so called sociable spiders representing several families, observed by him during his voyage to Venezuela, South America, during the winter and spring of 1887–'88. This sociability presented several degrees. It was sometimes temporary and limited to the period of reproduction; sometimes permanent. In some cases the work executed was absolutely common and alike for all individuals of the community; in others, the common work did not exclude some portion of individual work. With these qualifications he proceeds to classify the sociable spiders of Venezuela in three categories.

Epeira bandelieri, ordinarily, does not appear to differ in habits from typical Epeiras. Its web is the normal solitary one, but at the time of laying their eggs several females unite and construct in common, upon a bush, a large shell or cocoon case, of a yellow and woolly tissue, in which they proceed to lay their eggs and fabricate their cocoons. (Fig. 33.) These

---

[1] Observations Biologiques sur les Arachnides, Soc. Entomol. de France, 1891. By M. Eugene Simon.

are of a thick fibre, analagous to the cocoons of Argiope,[1] are rounded upon one face, almost flat upon the other, and attached to the walls of the incubating chamber by a short pedicle. (Fig. 33.) M. Simon **Sociable Epeira.** had found many of these shells enclosing as high as ten cocoons, and "five or six females sharing together the cares of maternity." He did not know what might have transpired at the moment of hatching, but thought it probable that the shell would be found at that time filled with a large number of young Epeiras. He had received from Quito another species of the same genus, of which the societies ought to be more numerous, if one may judge by the series of cocoons disposed in wreaths, which had been sent to him. But the cocoons of this species are spherical, and tied together by a loose wadding without being enclosed within a case.

This last named feature is not uncommon, as may be seen by reference to my observations of cocooning habits.[2] There seems to be a decided, and in some cases entire, suspension of pugnacity and **Pugnacity Suspended.** ordinary appetite in the females at the time of ovipositing. So intense are they upon discharging the functions of Nature, and so pressing the necessity which is upon them, that they appear to have no place in their organism for any other passion or appetite, but push straight on, before whatever difficulties or dangers, in the discharge of their maternal duties. At such times they appear quite indifferent to the presence of other spiders engaged in like work; and as it falls out that the same retreats are sought by various mothers of the same or of different species and genera, they often do come together in such places, as, for example, under the canopy of a bit of bark (Fig. 55, Vol. II.), or in the angle of a convenient wall or cornice (Fig. 60, Vol. II.).

In such cases, one mother will lay her cocoon close by that of another. The first made cocoon will be overlapped in part by the spinningwork of the second, the second by the third, and so on till a series closely wrapped together may be produced. (Vol. II., Fig. 60.) All this, however, as is manifest, is done without any collusion; it is a fortuitous result, and is wrought by spiders whose solitary habits are undoubted, and therefore it is no proof of sociability.

It would be more difficult, however, to explain on such a principle the preparation of the silken shell of Epeira bandelieri as described by **A Common Incubating Chamber.** M. Simon. It certainly does present at first view the seeming of an intentional provision, made in common by a number of individuals, who must have been moved by some common impulse which contains some element of sociability. Had the distinguished French naturalist observed the construction of this common enclosure or incubating chamber there would be little room for doubt; but as he appears to reason rather from the specimens as they

---

[1] See Vol. II., page 76 of this work.   [2] Vol. II., page 85.

had been collected by him after construction, I can hardly forbear the feeling that even this structure may be accounted for on the same principle as above, without bringing to bear upon it the theory of a social community. It is a most interesting point which can only be elucidated by future observation.

Simon does indeed say that the incubating chamber, with its included ten cocoons, has five or six females who have assumed the duties of maternity. Did he observe these in the joint act of constructing the case? If not, did he reach this conclusion by finding the dead bodies of several females enclosed within the chamber along with their cocoons? If he did not observe the actual construction of the outer case, his inference that the enclosed cocoons must have been the work of several females could only have come from the fact that five or six adult females were found inside. In the absence of this definite information one is perhaps justified in suspecting that the cocoons, as described, may have been enclosed by one mother.

The number is indeed large as compared with that produced by the ordinary Orbweaver, but by no means peculiar; for, as I have shown,[1] Cyclosa bifurca produces as many as thirteen cocoons, which are

Mothers with many Cocoons. bound together by a flossy string. The Basilica spider[2] encloses five cocoons within an exterior case of like tissue, which she spins above her snare. Theridium serpentinum[3] will produce as many as eight cocoons, which are assembled close to one another at the top of her meshed net, and enclosed within thickened walls of spinningwork. Yet more striking, perhaps, is the cocoon string of Segestria canities,[4] which contains as many as twelve cocoons overlapping one another like the tiles upon a roof, and overlaid with a thick sheet of spinningwork, which is further protected by a rude thatch of leaves collected from the bush upon which it hangs. The spider's tubular home is woven at one side of her treasures, and the whole is surrounded by an external maze of network supported upon the branches of adjoining shrubbery. These cases at least demonstrate that the example of spinningwork described by M. Simon might have been the product of one mother's industry.

Whatever may be the truth as to the above point, these two facts are clear, viz., first, that the exhibition of sociability, if Simon's view be accepted, is limited to a few hours, or at most days. It is an inci-

Conclusions. dental characteristic, and does not entitle the species to be called social any more than the fraternal communism of young spiders during the first few days after issuing from the egg. Second, even if the incubating chamber of Epeira bandelieri be the product of joint labor, the fact is only faintly comparable to the highly organized communal indus-

---

[1] Vol. II., page 103, Fig. 96.
[2] Ibid., page 105, Figs. 98, 99.
[3] Ibid., page 112, Fig. 108.
[4] Ibid., page 136, Figs. 165, 166.

tries of the social Hymenoptera, and the use of the word "social" in this restricted sense is deceptive.

XI.

The element of sociability appears to be much more highly developed with a Lineweaver which Mr. Simon describes as Anelosimus socialis, a species belong-

Social Theridi-oids. ing to the family Theridiidæ. Many hundreds, perhaps thousands of this species spin a common web, soft and transparent, but of a compact tissue analogous to that woven by Agalena. This snare is of indeterminate form, and sometimes attains immense dimensions, even enveloping an entire coffee tree. At first sight it appears more like the spinningwork of sociable caterpillars than of spiders. When one opens the exterior envelope he sees that the interior is divided by silken partitions into irregular lodges; within these the spiders freely move about, and upon meeting touch one another, as do ants, with their antennæ, and sometimes a number of them will be seen feeding upon the same prey. The cocoons

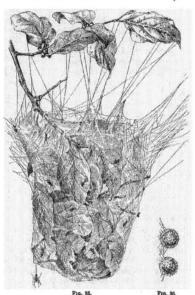

Fig. 35. Fig. 36.

Fig. 35. Common leafy tent of Anelosimus socialis.
Fig. 36. Cocoons of same. (After Simon.)

are rounded, formed of flocculent wadding of an iron gray color, are without pedicles, and are fixed to the common web by threads which form a soft net.

Upon this statement and the figure of M. Simon, which I reproduce in Fig. 35, I remark that the phenomenon is explicable by the ordinary habits of young spiders. This I have fully illustrated in Section V., Chapter VIII., Vol. II. Such an assemblage as there shown (Fig. 251) differs from that of Anelosimus socialis simply in extent. The habit of young spiders, immediately after their exit from the cocoon, is to surround themselves on all sides with a close tissue of just such spinningwork as M. Simon describes. One of the most remarkable of these I have described (Vol. II., page 227, Fig. 254), where the enclosing tent of the young brood covered a large working table and extended upwards to the ceiling of the room. I have seen some colonies covering a space eight or ten feet in length and four or five in width. I have also observed a large space of a vine or bush enclosed in a similar manner by broods of Epeira, one of which is described in this volume, in the chapter on Moulting Habits.

Tenting Commons of Spiderlings.
Now, it is only required that the broods of several cocoons, left by mothers in the same neighborhood, should issue at one time, to produce the results figured by the French savant. These colonies would certainly, as I can affirm from observation, unite their spinningwork, while retaining a degree of separation, and so enclose an immense space, overweaving leaves, and uniting the interspaces by a soft but compact fibre, precisely of the sort made by the Venezuela species.

I am at a loss to determine from Simon's language whether he is describing the work of several broods of spiderlings or the work of adult females. It is true that he speaks of the cocoon, and indeed figures it (Fig. 36); but he may have done this from an empty cocoon found in the midst of the colonial tent after abandonment by its inmates, as I have many times found cocoons. Until this point is settled, I feel constrained to say that there is nothing peculiar in this habit of this species, and nothing to justify one in regarding it as more sociable than other Theridioid spiders.

One fact, indeed, looks in the opposite direction; namely, that the spiders when meeting touch one another after the fashion of ants, who are well known to cross their antennæ for purposes of recognition. I have frequently observed a habit similar to this in the case of young spiders while in the period of assemblage immediately after issuing. They do touch each other with their fore paws, and even with their palps; though I should say that the manner is not strictly homologous with that of the ants, but it is only in a general way analogous thereto. If, however, the spiders described by M. Simon as engaged in constructing the tented domicile which he figures, were adult females, and if they have so far sunken their voracious and pugnacious habits as to recognize each other by palpal touch, and thereupon pass by without a hostile attempt, we have, indeed, a most remarkable fact, and one which relates the habits of spiders to those of the highest of the insects, in one of their most interesting features. The uses of a spider's palps are indeed various; but, as far as I know, the above observation stands alone in attributing to those organs such a function.

XII.

A third type of assemblage was observed by Mr. Simon in a species of Uloborus (U. republicanus), which is declared to be much more perfect, because it presents at the same time a common snare contributed

Uloborus Republicanus. by all the partners, and an individual snare proper to each one. Many hundreds of this Uloborus live together; they spin between the trees an immense web, formed of a central net quite compact, upon which many individuals of the two sexes hang side by side, but these assemblages are chiefly composed of males. This net is suspended by long threads diverging in all directions, and attached to surrounding objects. In the intervals of the open spaces formed by these large threads other Ulobori hang upon their orbicular snares, in rays and circles, each one of which is occupied by a single individual. One may see from time to time a spider detach itself from the central group, in order to seek among the upper cables a suitable place for the fabrication of its orbicular web.

It is the central net that appears to serve as the place of pairing, as far as the observer was able to judge by the quantity of males which were there gathered together. There, at least, it is certain that the hatching of the eggs takes place. This appears to occur almost simultaneously among all the females of the same colony. At that time the males have disappeared, the

Guarding Cocoons.

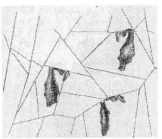

FIG. 37. Females of Uloborus republicanus with their hanging cocoons.

females have ceased to spin their regular snares, and hang motionless upon their central net, a few centimetres distant from each other, each one guarding her cocoon. (Fig. 37.) The cocoon itself is most singular in shape, and resembles more a bit of vegetable fibre accidentally fallen upon the snare than the spinningwork of a spider. In its general features it resembles the cocoons of our American Uloborus. (Vol. II., page 107, Fig. 103.)

The habits of this Uloborus differ little in their general characteristics from those of our American species. I have elsewhere described their tendency to hold rather closely to the neighborhood in which they were hatched, so that their snares may be seen in close contiguity. Indeed, other species of Orbweavers have the same tendency; and I have observed a number of small snares of young specimens spun upon the broad sheeted common which had been woven by the brood in their babyhood assemblage. Snares of the Labyrinth spider are often seen thus closely placed, and in point of fact the same may be said of almost all Orbweavers under favoring conditions. I have described an assemblage of Zillas whose snares were as closely placed as those of Uloborus republicanus, the foundation lines thereof being supported by the iron railing and columns upon the footway at the sluice where the waters of Loch Katrine pass down into Loch Achray, in the Trossachs Glen of Scotland. So also with the snares of our most common indigenous Epeiras, which one may see at times so closely placed along the surface of a stable wall or other favoring site that the foundation lines thereof are interblended, on the one side and the other, giving to the casual glance the appearance of a widely distributed colony. Yet, in point of fact, all these are simply examples of the contiguous placing of snares by individuals known to have no particle of social habit, and which are as absolutely distinct as though miles apart. I am constrained to believe that there is no more evidence of a really social community, analogous to that established among ants, wasps, and other social Hymenoptera, in •the assemblages of U. republicanus described by M. Simon, than in the examples which I have thus cited.

(margin notes: Orbweaving Neighborhoods. Not Social Communities.)

The assembling of the two sexes upon the outlying threads surrounding the orbicular snares has something more the appearance of friendliness. It would seem, indeed, that here we have an evidence that individuals are drawn together by some social tendency. Yet even in this fact one can see nothing absolutely conclusive of a really social habit; for it must be remembered that M. Simon notes that most of these individuals, thus found congregated upon the netted suburbs of the true snares, were males, a fact which is quite in accordance with the habits of that sex. I have elsewhere shown (Vol. II., page 21), that as many as three or four males have been observed by me hanging upon the outer precincts of the orb of an Argiope or of Epeira labyrinthea. I am inclined to think that the examples described by Simon may be thus explained; for although males of Orbweavers are disposed to quarrel with each other at times, they do also exhibit a remarkable degree of good temper, or at least absence of pugnacity, when thus waiting at the gates of their lady's bower. In view of all these observations, which appear to carry the habits of araneads nearer to those of social insects than any yet published, I am compelled to say that further facts are required before we can pronounce the author's conclusions to be well established. It is much to be hoped that M. Simon may have the opportunity to reëxamine the facts which he has communicated, and thus add the unquestionable solution of this most interesting problem to the brilliant service which he has rendered in that branch of natural science of which he has made himself a master.

Another observation looking in the same direction has been recorded since my work went to press. Rev. O. P. Cambridge[1] describes a spider of the family Eresidæ whose nest came under observation in the London Zoological Society's Gardens. It was sent thither by Colonel Bowker from Durban to Lord Walsingham, who at Mr. Cambridge's suggestion sent it to the Gardens. The nest contained from one hundred to one hundred and fifty living individuals of both sexes, some adult, some immature, and remained in their temporary home for some time in an active and thriving state.

A Gregarious Saltigrade in London.

The nest filled a box two feet long by nine inches wide and five inches deep. No detailed description of the home or habits of the colony is given, but it would appear to have been simply a mass of threads so thickly woven that they formed in places a close tissue something like that which Phidippus opifex (McCook) of California makes for herself on a much smaller scale. (Vol. II., page 150, Fig. 185.) That this spider is a social one may be inferred from the above scant description, and Mr. Cambridge's statement that the species is "unique in its gregarious habits." He notes that the individuals "appear to devour cockroaches and crickets, tearing them to pieces in concert." This statement, however, is somewhat neutralized by the additional remark that each "carries off his share of the prey, like a pack of hounds breaking up a fox." It is earnestly to be hoped that detailed notes and drawings of this "family" have been kept by some trustworthy observer in the Zoological Gardens. Such a rare opportunity ought to have yielded data for definitely determining this most interesting problem.

---

[1] Proceed. Zool. Soc., London, 1889, pages 34, 42, pl. ii., Figs. 4 and 5.

CHAPTER II.

MEMORY, MIMICRY, AND PARASITISM.

I.

An interesting example of the power of spiders to adapt themselves and their industry to circumstances occurred under my observation in the case of a Turret spider, Lycosa arenicola. Wishing to preserve a nest to study the life history of its occupant, I carefully took up the sod containing the tube, carrying away with me several inches in depth of the burrow. The upper and lower openings were plugged with cotton to retain the spider during transit. Upon arrival

Intelligence and Memory.

Fig. 38. Cotton lined nest of Turret spider.

of the nest in Philadelphia the cotton plug guarding the entrance was removed, but the other was forgotten, and thus allowed to remain. The nest with the enclosing sod was imbedded in the soil inside of a tub, and the spider left to work out naturally its industrial instincts. It immediately began removing the cotton at the bottom of its burrow, and cast some of it out upon the surface. But finally, guided, apparently by its sense of touch, to the knowledge that the softer fibres of the cotton would be an excellent material with which to line its tube, she put it to that use, and had soon spread a smooth layer over the inner surface and upon the opening. In this manner the interior was padded for about four inches from the summit of the tower downward. It may be taken for granted that this Turret spider for the first time had come in contact with such material as cotton, and had immediately utilized its new experience by substituting the soft fibre for the ordinary silken lining, or rather by adding it thereto. This nest, with the cotton wadding, is represented at Fig. 38. The cotton was distributed quite evenly over the walls of the burrow and tower, and had evidently been beaten down and pushed in after the manner of Lycosids and Agalenads when beating in the spinningwork of their cocoons and the silk lining of their burrows and tubes.

Mrs. Treat having learned how this spider, which had been taken from her grounds, had used the cotton, was led to make several experiments. She placed cotton by the side of seventeen burrows, both of the Turret and Tiger spiders, situated upon her lawn, and found eight of the number used the cotton as a lining, but none as artistically as the one above described. She then went to the edge of a wood, some distance away, and placed cotton by the side of eleven burrows there located. None of the occupants availed themselves of the artificial lining. This seems a curious fact, but the theory which the author uses to account for it, namely, that the individuals upon the lawn must have been descendants of species colonized from New England in the neighborhood of a cotton manufactory, can hardly be accepted. My recollection is that all these creatures were natives of New Jersey. I am sure at least that the one which wove the cotton lining for me was a native New Jerseyman.

One specimen of those situated upon the lawn not only used the cotton fibre for the lining, but also for a cover or door of its dwelling. This door she made smooth on the side, and fastened it firmly down on the outer edge of her wall. (Fig.

A Door of Cotton.

Fig. 39. Cotton utilized for a door.

39.) She did not make the same use of the cotton that she would of soft moss, which she sometimes uses in building. The fibre of the cotton was drawn out and interwoven among the sticks around the upper portion of the tower, and made to take the place of ordinary web work.[1] I have this nest in my collection and give a drawing thereof, Fig. 39. The use of the cotton is curious and interesting, and remotely suggestive of a door, perhaps. But such use is not clearly shown. These examples suggest no little elasticity of intellect on the part of these spiders, for they were at once able to perceive the usefulness of the new material brought within the range of their experience, and easily adapted it to their special needs in lining the interior of their towers. Were they conscious that such soft, pliable material permitted economy of silk secretion, and could that have been a motive for its use? These facts start a most interesting train of reflection and conjecture, and suggest a fruitful field of inquiry and experiment to one who may have both disposition and opportunity to engage therein.

---

[1] My Garden Pets, page 82.

II.

I took a female Epeira trifolium from her nest in order to observe the changes of color. She was kept within a glass vessel for forty-eight hours, and then returned to her old web and placed upon it near the centre. The web was about as when she left it. She paused about a half minute, seized the trapline, taking precisely the position and in the exact spot which she had occupied for several days before. Did she remember her nest after the forty-eight hours' interval? The same fact as to memory of local snare and nest was tested upon another Trifolium, with the same results. The example above quoted indicates that Epeira trifolium preserved during twenty-four to forty-eight hours a recollection, or at least a perception of some sort, of its old quarters within its home nest.

Memory of Trifolium.

Yet stronger examples may be cited of spiders remembering their homes. The Trapdoor spider, for example, that constructs its ingenious hinged door upon a bed of moss or lichen, and then covers the lid with plants precisely like those surrounding it, when it leaves its den and goes out upon excursions for food, and returns without difficulty to its home, certainly has preserved distinct recollection of the location of that home.

Again, the Tiger spider makes a burrow underneath beds of moss, erecting over it a vestibule or dome composed of the material everywhere surrounding the spot. From this she sallies forth into the vicinage, often making wide excursions after prey, and returns either by day or night to her nest, notwithstanding its general likeness to the environment. This is true of Lycosids generally. The mother Lycosa in the cocooning period oftens erects a cell or cave, underneath a stone or in like positions, which is partly lined with silk, and sometimes has a pretty approach to the surface between the sprays of grass, clover, or other vegetation, as may be seen in Vol. II., page 144, Fig. 175, with the nest of Lycosa scutulata.[1] From this retreat Lycosa will sally forth after food, dragging her egg sac behind her. It may seem a little strange that she should do so, and one might be inclined to think her rather stupid not to leave this treasure at home. Nevertheless, she attaches it to her spinnerets, and carries it with her in all her excursions, and thereby, no doubt, saves it from parasitic and other enemies. Having secured her food she returns to her cell, and notwithstanding the manner in which it is secreted, finds it without difficulty.

Sense of Location.

So also Saltigrade spiders, and others of like habit, who issue from their silken cells to stalk their prey on walls and trees, appear to find their way to their homes without difficulty. These facts indicate on the part of

---

[1] This species is there erroneously given as Lycosa saccata.

these spiders a good memory of the location of their domiciles, and a sense of direction sufficiently developed to bring them upon the return path with accuracy.

Of course there is nothing remarkable about this, for such facts are true of the insect world generally. It is known by all bee hunters, and by all keepers of bees, that either the wild or the hive bee will find its way home after a long excursion in search of honey. So also I have frequently observed the mud dauber wasp erecting its clay cell upon a wall or building, making excursions to all points to secure mud for her masonry, and invariably return with her mandible hod full to complete her nidus. The same accuracy of memory, again, is shown when, having finished her nest, she prowls through all the neighborhood in search of spiders, winging her course to all points of the compass, prying into nooks and crannies and out of the way places, mousing under leaves and diving into flowers, and yet always directing her return course without the slightest hesitation to her mud daub cell. It is needless to multiply such examples, and I only allude to them to show that in this respect the spider is not peculiar, but is gifted, like other Arthropods, with a memory sufficient for all the purposes of its life.

Insect Memory.

III.

Mr. F. M. Webster recently wrote me from the Ohio Agricultural Experiment Station (February 17th, 1892) a note which contributes an interesting item to the subject of mimicry as discussed Vol. II., Chapter XII., especially under Color Mimicry and Mimicry of Environment, page 367. A spider observed by him had mimicked the white excreta of birds so perfectly as to deceive this thoroughly trained and accurate observer. The color of the spider was whitish, with the dorsal abdominal portion clouded with blackish, exactly resembling a mass of bird droppings. The deception was further carried out by the spider having spun a thin irregular sheet of white web on an elm leaf, in the midst of which it was situated with legs drawn up. At the distance of a few feet the observer was completely deceived; he thought it the excrement of a bird until he had the leaf in his hand. The appearance of the semisolid mass within the white splash of semifluid matter was so closely counterfeited, that Mr. Webster says he was truly provoked that such an animal could so befool his eyes after all his years of training. Judging from the general description sent, I infer the species to be our old friend Misumena vatia, so famous both in America and elsewhere for its color mimicry.

Color Mimicry.

Mr. Webster's observation is all the more interesting because of its exact correspondence with one of which he was not informed until I called his attention to it, which has received considerable attention from naturalists.

Mr. Henry O. Forbes relates a similar experience of deception.[1] He had been allured into a vain chase after a large stately flitting butterfly (Hestia) through a thicket of Pandanus horridus, when on a bush that obstructed further pursuit he observed one of the Hesperidæ resting on a leaf upon a splash of bird dropping. He had often observed small Blues at rest on similar spots on the ground, and had wondered what the members of such a refined and beautifully painted family (Lycænidæ) could find to enjoy in food so seemingly incongruous for a butterfly. He approached **Mimicry of Bird Excreta.** with gentle steps and ready net to see, if possible, how the present individual was engaged. It permitted him to get quite close, and even to seize it between his fingers. To his surprise, however, part of the body remained behind; and in adhering, as he thought, to the excreta it recalled an observation of Mr. Wallace's on certain Coleoptera falling a prey to their inexperience by boring in the bark of trees, in whose exuding gum they became unwittingly entombed. He looked closely at the excreta to find if it were glutinous, and finally touched it with the tip of his finger. To his delighted astonishment he found that his eyes had been most perfectly deceived, and that the excreta was a most artfully colored spider lying on its back, with its feet crossed over and closely adpressed to the body.

The appearance of recent bird droppings on a leaf is well known. Its central and denser portion is a pure white chalk like color, streaked here and there with black, and surrounded by a thin border of the **Ornitho-scatoïdes decipiens.** dried up more fluid part, which, as the leaf is rarely horizontal, often runs a little way towards the margin. The spider observed by Mr. Forbes, like that seen by Mr. Webster, was pure chalk white in general color, with the lower portions of its first and second pairs of legs and a spot on the head and abdomen jet black.[2] It had woven on the surface of the leaf, after the fashion of its family, an irregularly shaped web of fine texture, which was drawn up towards the sloping margin of the leaf into a narrow streak, with a slightly thickened termination.

This form, as described by Mr. Forbes, was doubtless determined by the concavity of the leaf, and the facility which the slightly turned up edges gave for making good points of adhesion, at the same time leaving a little space between the netted spinningwork and the leaf's surface over which it was stretched. According to Mr. Forbes, the spider takes its place on its back upon this irregular spinningwork, holding itself in position by means of the spinal armature of the legs thrust underneath the web, and crosses its legs over its thorax. I do not remember to have observed any Thomisoïds in this position, but have always seen them crouching with back upward when not using a web. I would suppose that if they used

---

[1] A Naturalist's Wanderings in the Eastern Archipelago, 1885, pages 63, 64.
[2] Rev. O. P. Cambridge describes it as Ornithoscatoïdes decipiens.

a snare at all they would rest underneath the same with their backs downward toward the leaf, after the fashion of many other spiders. However, Mr. Forbes is so precise in his statement that one feels it necessary to accept it. As thus arranged, he speaks of the whole combination of spider and web as being "so artfully contrived" as to deceive a pair of human eyes intently examining it.

This similarity of habit in spiders of the same family, at such widely separated points as Java and the United States, is in itself interesting. It is also interesting to notice that two thoroughly trained observers should have independently named the same rather outre object as the one suggested by the spider's mimicry. Nevertheless, one is inclined to think that the suggestion of mimicry is a bit of anthropomorphism. Because suggested to the mind of the observer, it does not follow that any such deception had been devised by the spider. All the individuals of this family and tribe are in the habit of seeking their prey chiefly upon trees and plants of various sorts, stones, et cetera. They may often be found upon the white or whitish gray spots, or upon dried bits of lichen or moss found on plants. In such positions they certainly strongly suggest the idea of intentional mimicry. However, in point of fact, the combination might have been accidental so far as the spider is concerned. In the cases observed by Messrs. Forbes and Webster there appears no reason to infer that so subtle a process as that attributed to the spider could have found lodgment in its mind.

A Case of Anthropomorphism.

At least, if we accept Mr. Forbes' theory of an artful contrivance, we must conclude that this lowly organized animal could intelligently survey the field, put this and that together, select certain spots for settlement after determining its own color resemblance to such spots; then deliberately proceed to spin a web which, in its general contour, would resemble the particular form which semifluid masses are wont to assume in various positions, according to the inclination of the plane upon which they fall; and then, further, arrange its own body in such relation to the web as to present the appearance of mottled black and gray characteristic of bird droppings under like circumstances. The ability for such mental processes would undoubtedly establish an order of intellect and powers of observation and reasoning beyond those which we are at present warranted in attributing to a spider. All the facts can be accounted for, and are most naturally accounted for, without introducing a factor so strongly imaginary. I am disposed to think that the web as described was spun in an ordinary position, in the form habitual to the species, and in such locality as it usually frequents; moreover, that this was done without any intention to perpetrate a mimicry such as the observer fancied, and which in fact existed simply as an analogy in his mind.

Mimicry Improbable.

But excluding the idea of intentional deceit, the mimicry of Ornithosca-toïdes has been explained as a product of evolution through natural selection and survival of the fittest. The explanation implies that the existence of those individuals practicing the mimicry at first and accidentally had been preserved by the greater abundance of food or other advantage gained thereby, until it became a permanent habit. But we have only an inference that the habit is permanent in Ornithoscatoïdes. The facts recorded by Mr. Forbes stand entirely alone, and it seems more likely than otherwise that an extended observation of that spider would show that the same condition obtains as to its habits that we have observed in Misumena, and that it will be found to spin a web substantially as described by Mr. Forbes, in many positions which would preclude the supposition of usefulness through resemblance to the excreta of birds; as, for example, on the under side of leaves, underneath limbs of trees, between and under stones, hedges, etc.

Evolu-tion and Mimicry.

As to Misumena, we know that the case reported by Mr. Webster is abso-lutely unique as yet; knowing somewhat the general economy of this species we can confidently affirm that the incident is exceptional. Misu-mena spreads her cocoon nest in various positions, and is by no means limited to such locations as described by Professor Web-ster. One of these positions has been described and figured in Vol. II., page 152, Fig. 188. The general form of the cocoon nest is as there exhibited, and in choosing the site thereof the mother appears to give herself as wide range as do other species of the order. In other words, she puts her cocoon and the nesting tube surrounding it in such place as is most convenient to herself when the maternal function urges her to action.

The Mim-icry Ex-ceptional.

The spinningwork described by Mr. Forbes I take to be the cocoon nest of Ornithoscatoïdes, for it is not the habit of the Laterigrades generally to get their food by means of snares. They belong to the Wandering group of spiders, and stalk their prey along shrubbery, branches of trees, rocks, walls, etc. If this inference be correct, the peculiar web observed by Mr. Forbes is limited to the cocooning period, at which time Laterigrade spiders are usually found within or lurking around the little tent which overspins their egg sac. It must follow that the effect of industrial mimicry upon the preservation of that species must have been confined to a brief period at the end of the spider's life. To account for such mimicry as a product of the survival of the fittest through natural selection, one needs to con-ceive of the selection as operative in the early and most impressible stages, or at least during the active period of life, and not during the few days immediately preceding death.

IV.

In the second volume of this work (Vol. II., page 346) I ventured to express a suspicion, which I have sometimes entertained, that the color surroundings of the spider, in some manner not now explicable, may so rapidly influence the organism of the creature that a change of color is produced in harmony with its environment. I there raised this query: Can a spider have the power to influence at will the chromatophores or pigment bodies, so that she may change her color with the changing sites?

Mr. H. H. J. Bell has recently communicated an observation[1] which appears to be confirmatory of this suggestion. While traveling along the
Flower Mimicry. West Coast of Africa between two small towns on the Gold Coast (August, 1892), he was attracted by the appearance of what he supposed to be flowers upon the bushes bordering the path. On examining these he found that they were the webs of an orbweaving spider, whose spinningwork, according to the published description, resembles that of Argiope, as heretofore fully described by me. The spider's body was a light blue color; and the legs, which were symmetrically disposed in the shape of an X across a white ribboned hub, were yellow, ringed with brown. The body of the spider resembled the corol of a flower, and the crossed legs gave it the semblance of petals. Mr. Bell speaks of the illusion as remarkable, and supposes this mimicry of an orchidlike flower serves not merely to protect the spider, but rather as an attraction to butterflies and other flower frequenting insects on which the aranead preys.

The most interesting part of the observation, however, is the strange facility which the spider possessed of changing its color. Mr. Bell captured
Volition- al Color Changes. her in a white gauze collecting net, which was placed beneath her and into which she dropped when disturbed, as is the custom of many species. As soon as she touched the net the blue body color became white. On being shaken her body turned to a dark greenish brown. She was then placed in a glass tube, and gradually resumed her blue tint, but when shaken up always turned to a greenish brown. When placed in spirits the spider's color became a gray brown, and so remained. The observation was repeated with like result, except that the second individual did not turn white, but passed immediately from her normal blue into a dark greenish brown. I have no comments to make upon this interesting and, as it seems to me, important observation, but give it place here, as of undoubted value in its bearing upon the interesting and perplexing problem of mimicry as it is presented in the life history of the various aranead species. I have never observed and do not remember ever to have read of such instant and volitional changes of color. The slight changes which I have noted in spiders having metallic colors have been due to the play of light falling at and seen from different angles. The numerous and striking changes in the Shamrock spider (Epeira trifolium), so well illustrated in Plate I. of Vol. II., are produced gradually, and cannot be compared with the chameleonlike changes of the blue Orbweaver observed by Mr. Bell.

[1] Nature, April 13th, 1893, page 558.

V.

The trapdoor building habit is remarkable in its distribution over nearly every part of the globe in tropical and semitropical regions. From whatever part reported, Europe, Africa, Asia, Australia, or North and South America, the nest shows the same peculiarities of structure, and the architect appears to live in the same way. Even the habit of mimicking the surrounding surface by attaching sundry small plants is cosmopolitan. Moggridge in his charming book has already made us familiar with this form of mimicry in the European species; it seems also to characterize our southwestern trapdoors, and appears in Cambridge's Idiops Colletti, a Burmese (India) spider.[1] In the interesting description of General Collett, which Mr. Cambridge publishes, it is stated that the upper surface of the door is often covered with a dry black lichen growth. There are generally a few withered grass blades worked into the edge of the door, or into the edge of the mouth of the burrow, so as to form a kind of semicircular fringe, which often catches a practiced eye and leads to the detection of the hole. The grass blades are probably inserted to aid in assimilating the outside of the door to its surroundings, a purpose in which, General Collett opined, they certainly fail so far as the human animal is concerned. In a few cases he noticed also grass blades wrought into the general surface of the door, which in the dry season, when the grass is everywhere withered, certainly aid in its concealment. But during the season when the adjacent grass is green one would think that yellow withered grass blades on or near the burrow mouth would tend to make it conspicuous. I have considered at some length, Vol. II., pages 354, 355, the point thus raised independently by this intelligent observer, and its relation to so called mimicry of environment. I need only add here that one can hardly be asked to consider protection against human intelligence as a factor in the action of a spider's mind. The hostile elements which influence it are of quite another sort from the curiosity of a naturalist and the plundering of a collector.

Architectural Mimicry.

A Rude Door Garden.

VI.

Since publishing my observations on the parasitic enemies of spiders (Vol. II., page 391) a number of facts have come to my hands which I have thought well to embrace in these studies, and to add some general conclusions suggested by the subject. Mr. George Carter Bignell[2] has favored me with an account of the manner in which Drassus lapidocolens Walckenaer was attacked by an ichneumon (8th October (1890). While walking in the woods he noticed the spider suspended by a silken drop thread from the bough of a large oak. Looking for the

Parasitism.

---

[1] Proceed. Zool. Soc., Lond., 1889, page 37, pl. ii., Fig. 2.
[2] 7 Clarence Place, Stonehouse, Devon, England.

cause of such a situation, he found an ichneumon fly walking cautiously
down the thread towards its victim. When close to the spider she touched
it with her antennæ, whereat Drassus dropped a few inches lower.
Method of Ovipositing. This movement gave better opportunity to see the ovipositing of
the parasitic egg. The fly, having apparently ascertained that
she had found a suitable subject, turned round and walked backwards until close to the spider, where she paused a few moments, and
then deposited her egg on its abdomen close to the cephalothorax.

Mr. Bignell boxed both insect host and aranead guest, and took them
home. Two days afterwards the egg had hatched, and the larva was about
one line in length. A day or two thereafter both larva and spider were
found dead. The strangest fact in the above story is that the spider would
permit the fly to thus approach it without attack. The insect would seem
to have been within its power, yet it forebore to strike. Was it made
inactive by fear? By what spell did the mother parasite procure this rare
exemption?

The same gentleman, while beating for larvæ of Lepidoptera, found an
Orbweaver, Epeira cucurbitina Clerck, which had been attacked by an external parasite. This lay like a sack across its guest's
back, reminding one of a miller's man carrying a
bag of flour. It was taken May 22d (1882), and on
the 24th it was full fed. On the 23d it was figured
and described (Fig. 40), and found to have no legs,

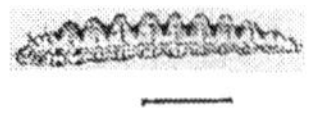

Fig. 40. Parasitic larva of Polysphincta tuberosa.

but in place thereof had sucking discs, two on the second segment and
four on the third and fourth. Six of these occupied the usual
A Parasite Larva. place of the legs of larvæ; the other four were half covered with
the skinfold usually seen on lepidopterous larvæ. On its back
were tubercles, the first on the fourth segment, the others on
the seven following ones; each tubercle was surmounted with two rings of
hooklets, with three or four in the centre. These served the larva to suspend itself from the round snare while feeding on its victim, and to hold
on to the web after it was consumed. When all the juices of the spider's
body had been extracted the legs and empty skin were allowed to fall down.

The larva then commenced to make itself a cocoon which was finished
by the third day, during which time the tubercles performed a prominent
role, having to do the work of the claspers of an ordinary caterpillar.
When a tentacle, attached to the silken cord, had to be removed, the
hooklets were withdrawn into the tentacle, which at once became disengaged and ready to make another attachment.

The anal segment often played an important part by being brought
round to the assistance of the mouth; this act was first seen while the
larva was feeding, and its purpose was to disengage some adhering portion
of the spider from its jaws. Afterwards it was frequently used while
spinning to unite the silk to some narrow part of the cocoon, where the

blunt head of the larva was too large to make an attachment. When full
fed the larva was about three-eighths of an inch long, and had fourteen
segments, counting the head as one. The cocoon was shuttle shaped,
whitish and thin, the spider's original web forming its suspending cords;
the movements of the larva and pupa were perceptible through the cocoon.
The perfect fly appeared on the 12th of June, and proved to be a female
of Polysphincta tuberosa Gravenhorst.

VII.

Mr. John L. Curtis,[1] of Oakland, California, has written me an interesting account of a new body parasite taken upon a species of Theridioid
spider, Labulla inconstans.[2]

Fig. 41. Parasite cocoon in site, natural size.
Fig. 42. The same, enlarged greatly. Fig.
43. The pupa.

**Body Par-
asite.** The spider is quite
common in the re-
gion surrounding
San Francisco, and domiciles
in large leaves, the edges and
ends of which it bends downward, and
fastens with a sheet of web composed of
many white threads spun across from side
to side. The cocoon, which is round and
white, is woven within this maze, and is
jealously watched by the mother, a small
spider of a light gray color, with pinkish
tints on the legs and a tinge of yellow on
the abdomen.

On July 13th (1890) a specimen was
found upon whose abdomen was fixed a
large yellow larva. By the 15th the larva had entirely consumed the spider
and spun itself into its cocoon. (Figs. 41, 42, 43.) On the 17th it changed
to chrysalis, and on the 27th of the month hatched. The cocoon is a cylindrical case of loose fibres, and hung suspended, head downward, upon lines
stretched within the bottle wherein it was bred. Mr. E. T. Cresson, of Philadelphia, has identified the parasite insect thus hatched as a probably new
species of Polysphincta, a genus of Pimplinæ, a subfamily of the great
family Ichneumonidæ. It is obvious that in order to deposit her egg upon
the body of the spider, this mother Ichneumonid must have successfully
threaded the labyrinth of intercrossed lines and obtained a favorable position underneath the abdomen. A delicate piece of aranead scouting this!

[1] I regret to record the death of this promising young naturalist since writing this note.
Though an incurable invalid he was an enthusiastic lover of araneology, and had already
done some good work therein.
[2] Dr. George Marx thus determines the genus, and the specific name is that suggested
by Mr. Curtis.

I here refer to another figured body parasite, which was overlooked when preparing the material on parasitic enemies for Vol. II. Mr. L. O.

The Dictyna Parasite. Howard, of the Entomological Department of the Bureau of Agriculture, Washington, has described and figured[1] Polysphincta dictynæ and its parasitic larva feeding upon Linyphia communis.

(Figs. 44 and 45.) The fly was raised from a larva found on a young Dictyna volupis Keyserling. When taken, May 15th (1887), the larva was about half as long as the spider's abdomen and about one-fourth as thick. It was attached by the mouth to the front of the abdomen. May 18th the host was dead and the larva full grown, larger than the spider had been, and had begun to spin a cocoon. May 25th it changed to a pupa, and the fly came out June 1st following. The adult parasite is a beautiful little male two and five-tenths millimetres long. The sketch,

as copied from " Insect Life," shows the position which the parasitic larva assumed on the spider. In the collection in which the above species was taken Mr. Howard found five other small spiders, four of which supported parasitic larvæ upon the dorsum of the abdomen, and one delicate cocoon from which a parasitic larva had been taken.

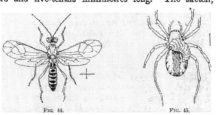

FIG. 44. Polysphincta dictynæ much enlarged. FIG. 45. Larval parasite on young spider, Dictyna volupis. (After Howard.)

In the same journal[2] Mr. Howard reported another species of Polysphincta found by Dr. W. H. Fox, of Washington, D. C., upon a young specimen of Steatoda borealis. The larva was slender, cylindrical, white, one millimetre long, apparently less than half grown, and was attached to its host substantially as above described. It was taken in February, which would indicate a larval hibernation of the parasite.

Mr. Howard has described still another species of Polysphincta, P. strigis, whose habits were quite fully observed. Mr. Nathan Banks found

Parasite on Epeira Strix. the larva of this species feeding externally upon Epeira strix, at Sea Cliff, Long Island, May 11th, 1891. At the time of capture the parasitic larva was considerably larger than the spider; it spun up May 14th. When brought to Mr. Howard (May 18th) the cocoon was completed in the vial in which Mr. Banks had placed the specimen; it was spun of dense yellow brown silk, was six millimetres long, cylindrical, two millimetres in diameter and rounded at both ends. It

---

[1] Insect Life, Vol. I., page 106.   [2] Insect Life, Vol. I., page 42.

was suspended by a loose band of darker colored coarser silk, seven millimetres long, and from the end of this band a few threads reached to the
bottom, twenty-seven millimetres, and sides of the vial. The abdomen of
the spider was reduced to the merest fragment, but the cephalothorax and
legs remained. May 25th the adult issued from one end of the cocoon.
The cocoon of P. dictynæ mentioned above was about the same size and
had a smaller supporting band, but was composed of white silk, and was
much more delicate, nearly transparent.[1]

The beautiful Argiope argentata may be added to the list of spiders
whose cocoons are parasitized. Dr. A. Davidson sent me from Los Angeles,
California, a fine large cocoon, which upon opening the box I
Parasite in Argiope Cocoon. found to be occupied by a numerous brood of small black Ichneumonids, which proved to be a species of Eupelmus.[2] It is
jet black, glossy, with a metallic lustre. The female is much
larger and stouter than the male. These insects were domiciled
within the yellow floss which pads the interior of the cocoon, and had
probably been reared from naked pupæ, as no pupal cases were found.
Several dead pupæ in at least two stages of development, a number of
infertile spider eggs, and a quantity of castings and disjecta membra of the
pupæ were strung throughout the padding surrounding the central cavity
in which the Ichneumons were congregated. Among these interlopers, or
crawling upon the surrounding fibres, were a few (six or eight) living spiderlings, the feeble remainder of the original colony. These as well as the
parasites were probably hatched en route from California in the mail bag.

Dr. Davidson took the spider cocoon on Catalina Island, California, where
the species is abundant. It spins its webs in cacti, and the cocoons are
always placed some distance from and behind the snare ("around
Argen- tata Co- coon. the corner, as it were"), upon one of the plants to which the
foundation lines are attached. The cocoons are never hung upon
the orb as with the specimen spun for me in captivity, and described Vol. II., page 84. The same gentleman remarks that the cocoons
so closely resemble in coloring the cactus as to make them almost difficult
to find; and he thinks it an example of concealment by mimicry of colors.
However, cocoons of the species received from various parts of California
have the same general hues, yellow, with more or less green,[3] quite regardless of their situation. In the above case the "protective" resemblance did
not protect, as appeared from the vigorous brood of invading Ichneumonids.[4]

---

[1] Howard: "The Hymenopterous Parasites of Spiders," Proceed. Entom. Soc., Washington, Vol. II., No. 3.

[2] Eupelmus piceus. Described by Mr. L. O. Howard.

[3] See Examples, Vol. II., pl. iv.

[4] Mr. Howard expresses the opinion, through a letter to Mr. E. T. Cresson, that although
many species of Eupelmus are egg parasites, still others are hyperparasitic; and judging
from the size of the above species he thinks that it is probably a parasite upon some Pimpla
which was the primary parasite in the spider cocoon.

From the above examples of life history, which are fairly characteristic, one may get the following summary of habits, which, while open to correction at some points, is substantially accurate : 1. The parasite mother passes from point to point in search of a suitable host with a rapid, jerky, intense action ; 2. discovering her victim she deposits an egg upon the abdomen, 3. and in so doing she will creep along the lines of spinningwork to the point where the spider is suspended. 4. When near her victim she backs down thereto and deposits but one egg upon each host, which is placed upon the abdomen close to the cephalothorax. 5. In about two days the egg hatches into a footless, white larva, which in some cases creeps within the body, probably· by some of the natural openings, and becomes an internal parasite. 6. The external parasite fastens upon the dorsal base or dorsum of the anterior part of the spider's abdomen, and in two or three days wholly consumes that soft organ, meanwhile growing rapidly, and then begins to spin its cocoon. 7. The cocoon is cylindrical, about six millimetres long and two thick, woven of open and loose white or yellowish silk, and is suspended to the spider's web or other object by a slight band, braced beneath by a similar support. 8. The spinning occupies from one to three days, several days are spent in the pupal change, and in about a week the imago appears. 9. It is probable that some larval parasites which are hatched late in the autumn hibernate with their hosts.

The habits of those parasitic Ichneumonids which infest spider eggs cannot be summarized with even as much satisfaction as the body parasites, but the following may at least suggest something better. **Egg Para-sites: Summary** 1. Soon after the spider mother has laid and enclosed her eggs, and sometimes probably during oviposition, the Ichneumon mother inserts her eggs, penetrating the cocoon case when necessary with her ovipositor, and leaving a· number of eggs or the entire brood within a single cocoon. 2. Each egg harbors one parasite, which in certain species[1] enters it as soon as the larval appetite awakes and entirely destroys it. 3. In some species it would appear that the larval parasites feed upon the eggs indiscriminately, and then spin stiff, close, white cocoons, through which the imago gnaws a hole and escapes from the spider cocoon in the same way. 4. These parasites are in turn exposed to parasitism from other members of the Ichneumonid family.[2]

VIII.

Mr. Howard has kindly furnished me a list of known American and European hymenopterous parasites of spiders.[3] Some of these, together with those heretofore referred to, I have arranged as below, with a view

---

[1] Acoloïdes saitides Howard.   [2] See Vol. II., page 395.

[3] Since this manuscript was prepared Mr. Howard has published his revised list and my table has been corrected thereby.

to a readier comparison from which to get clearer light, if possible, upon some questions on the general relations between hosts and guests.[1] The first column gives the name of the spider, when known, the second column the name of the parasite, the third column the character of the parasitism :—

ORBWEAVERS.

Argiope cophinaria.	Pezomachus (?) n. sp.	Egg parasite.
" "	Pimpla rufopectus.	" "
" "	Pimpla scriptifrons.	" "
" "	Mestocharis wilderi.	" "
" "	Chrysocharis banksii.	" "
" "	Chrysocharis pikei.	" "
Argiope argentata.	Eupelmus piceus.	" "
Epeira cavatica.	Large unknown larva.	Body parasite.[2]
Epeira (unknown).	Pimpla rufopectus.	Egg parasite.
" "	Holcopelte nitens.[3]	" "
" "	Bæus americanus.	" "
" "	Pezomachus dimidiatus (?).	" "
Epeira cavatica.	Pezomachus gracile (?).	" "
Epeira strix (?).	Polysphincta kœkelei.	Body parasite.
" "	Polysphincta strigis.	" "
Epeira diademata.	Hemiteles similis.*	" "
" "	Hemiteles tristator.*	Egg parasite.
" "	Pimpla oculatoria.*	" "
" "	Polysphincta boöps.*	Body parasite.
" "	Polysphincta carbonator.*	" "
" "	Polysphincta rufipes.*	" "
Epeira cucurbitina.	Polysphincta carbonator.	" "
Epeira antriada (?).	" " " *	" "
Epeira cucurbitina.	Polysphincta tuberosa.*	" "
Epeira angulata.	Pimpla aquilonia (?).	Egg parasite.
" "	Mestocharis wilderi.	" "
Epeirid cocoon.	Tetrastichus banksii.[4]	" "

LINEWEAVERS.

Theridium spirale (?).	Polysphincta n. sp.	Body parasite.
Theridium (unknown).	Pezomachus fasciatus.*	" "
Steatoda borealis.	Polysphincta n. sp.	" "
Linyphia communis.	" " "	" "
Labulla inconstans.	" " "	" "
Theridium	Polysphincta theridii.	" "
Theridium	Polysphincta boöps.*	" "

TUBEWEAVERS.

Dictyna volupis.	Polysphincta dictynæ.	Body parasite.
Micaria (unknown).	Pezomachus micariæ.	Egg parasite.

---

[1] Those species marked with stars (*) are European species.

[2] Within the body.

[3] Probably hyperparasitic, primarily infesting some Ichneumonid in the spider's cocoon. Howard.

[4] Hyperparasitic. Howard.

Micaria (unknown).	Pezomachus obscurus.	Egg parasite.
" "	Hemiteles micarivora..	" "
Prosthesima furcata.	Hemiteles prosthesimæ.	" "
Drassid (unknown).	Hemiteles drassi.	" "
Drassus lapidicolens.	Polysphincta tuberosa.*	Body parasite.
Agrœca brunnea.	Hemiteles tenerrimus.*	Egg parasite.
" "	Hemiteles aranearum.*	
" "	Hemiteles formosus.*	Body parasite.
" "	Pezomachus fasciatus.*	Egg parasite.
" "	Pezomachus corruptor.*	" "
" "	Pezomachus proximus.	" "
" "	Pezomachus zonatus.*	" "
Drassid (unknown).	Eupelmus drassi.	" "

LATERIGRADE, SALTIGRADE, CITIGRADE.

Laterigrade cocoon.	Pezomachus gracile.	Egg parasite.
Icius (unknown).	Polysphincta n. sp.	Body parasite.
Saltis pulex.	Acoloïdes saltides.	Egg parasite.
Phidippus morsitans.	" "	" "
Pardosa luteola.	Polysphincta .	Body parasite.

IX.

It is difficult to make any correct generalizations from the data in hand on this most interesting chapter in the biology of spiders, since the species or even genus of the host is in so many cases unknown, even when the parasite has been determined. But a few hints appear from the study of the above lists and preceding facts which may serve to at least open the way for others who in the future may have more perfect information.

Generalizations.

First, *it is evident that the exclusive occupation of a specific host by a specific guest is not the fixed rule.* For example, the eggs of Argiope cophinaria are parasitized by one species of Pezomachus, two of Pimpla, and two (though perhaps as hyperparasites) of Chrysocharis. Again, Epeira diademata of Europe serves as body host for three species of Polysphincta, and its eggs as host of one each of Hemiteles and Pimpla. Once more, Agrœca brunnea of Europe has one species of Hemiteles for a body parasite (?), and for egg parasites two species of Hemiteles and three of Pezomachus. Thus it would appear that the occupation of any specified host is not limited to any specific guest, but has a wide possible range, both as to species and genera.

Second, *the preference of any specific guest is not always confined to one specific host.* Thus, Polysphincta carbonator, a European body parasite, is reported as guest upon three species of Epeira; and Polysphincta boöps upon both Theridion and Epeira.

Third, *the guests are not absolutely separated by their habits into distinct groups of body parasites, on the one hand, and nest parasites on the other.* Thus, Polysphincta carbonator is a body parasite upon three Epeiroid species, but is reported as a guest upon unknown spider eggs; P. rufipes is parasitic on Epeira diademata and also upon supposed spider cocoons; Pezomachus fasciatus is guest within the egg nests of Agrœca brunnea, and on a species of Theridion; Hemiteles similis parasitizes Epeira diademata and an unknown spider cocoon. Hemiteles rufocinctus, H. fulvipes, and H. formosus are reported as body parasites, while other species of the genus are parasitic upon spider eggs. This generalization appears to me remarkable, and contrary to what one would naturally suppose from the ordinary specialization of such instincts. If correct, it shows much elasticity of habit among these parasitic Hymenoptera, for there would seem to be an immense distance on the maternal side between the instincts which prompt to the several acts; and on the side of the offspring, between development within a spider's egg and within a spider's body.

Mr. Howard, who was kind enough to look over this part of my manuscript, expresses the doubt here suggested in a more positive way. He thus writes me: "None of the spider Pimplas are typical, but belong to a restricted group, possibly a subgenus, all of which, I believe, are parasitic in spider egg bags. All of the Polysphincta, in my opinion, are external parasites of spiders; and this genus, with the genus Acrodactylus, I believe contains all of the external hymenopterous spider parasites." This opinion from such an authority would justify the omission of the above inference; but as the facts as reported are not changed in Mr. Howard's published lists, I permit the item to remain as qualified. The observations of early date which here compose the list are probably erroneous; and, indeed, they are confessedly defective.

Fourth, *the general spinning and other characteristics of spider species appear to make no marked difference in liability to parasitic attack.* Such sedentary spiders as Epeira, Theridion, and Agrœca are not exempt from body parasites by their occupation of snares. Yet, in such cases, one wonders how the ingenuity of the insects'· maternal instinct could overcome, as we know it does, the local difficulties, and drop an egg upon the spider's body while swinging upon or lurking within its web. The difficulty is not lessened, but rather enhanced, if we suppose that some of the parasitic larvæ are traveling parasites, and seek their hosts after independent hatching. The condition of the wingless female parasite would appear to make her task more formidable, but does not limit her ovipositing to periods when the spiders are off their snares and reposing on adjacent objects, or within their dens and tents.

Relations of Spinning Habits.

Fifth, *the special cocooning habits of spiders appear to have no relation to their exposure to parasitic attack.* Here, also, one meets apparently contradictory facts. For example, the egg cocoon of Argiope cophinaria is one of the most skillfully constructed to protect the enclosed young. (See Vol. II., pages 76–80.) The eggs are encased in a silken sheet, overlapped by a thick, compact oval blanketing, which in turn is encased in a tough glazed sac. The whole is

Of Cocooning Habit.

then encompassed with a maze of closely intercrossed lines, which to the human judgment seems impenetrable by the mother hymenopter. Yet this cocoon is among those most frequently parasitized.

Of course, one is always at liberty to offer as explanation the suggestion that this very exposure to attack has, in the struggle for survival, produced the evolution of the more perfectly armored cocoons. In short, **Effect of Struggle for Survival.** there has been a conflict between the mother aranead and the mother hymenopter, something like that between defensive plate armor on battle ships and offensive projectiles. The difficulty with such an explanation is to establish a point of contact between mother, offspring, and the facts of life from which to postulate conceivable interaction and reaction. The theory craves an available pou sto. The danger, if apprehended at all, would only be discerned by the eggs, which are the objects attacked; moreover, the impression must be conveyed through those eggs within the parasitized cocoon which were fortunate enough to escape destruction. To suppose that the reactionary influence of environment could thus operate through a sensitive egg to a conscious spiderling, and so to the maternal instinct of the adult female, and thus on by heredity, through infinitesimal increments of the protective wards and armor of a silk enswathed or mud daub cocoon, lays a rather heavy burden upon the scientific imagination, even in this heyday age of evolutionism. The possibility of cocoon evolution by transmittal through the adult mother need not be considered; for in the case of the above named species, and most others upon the list, the mother is necessarily eliminated from the problem. In point of fact, she usually dies immediately, or soon after ovipositing and cocooning, and knows nothing of her offspring or their dangers save by anticipative instinct, that foreordination in Nature which is everywhere so manifest in life habits.

Perhaps one might be permitted to approach the problem from another direction, and suggest that instead of the instinct of the parasitic hymenopter reacting upon the spider mother to cause increased cocoon safeguards, that instinct simply perceives the well armored cocoon and selects it as the one most secure for her own progeny.

The cocoon of Agrœca brunnea (see Vol. II., page 124) is also well devised to preserve the enclosed eggs, and withal is overcoated by the little mother with a plaster of mud, which adds to the protection of armoring that of mimicry. Yet this spider's eggs give hospitality to no less than six parasitic species, showing it to be especially assailable, or at least especially available for such purposes. In this case, however, we can suggest a reason; for although the spider's cocoon is so well armored, the mode of suspending it (see Figs. 134, 135, Vol. II.) permits the mother parasite to approach it with comparative ease, by crawling along the foot stalk from the plant on which it hangs. Indeed, speaking generally, one would infer that the cocoons of Tubeweavers and such other species as are placed

directly upon various surfaces without tented enclosures, or like special protection, would be most exposed to hymenopterous assault. In point of
fact, these genera are numerously represented in the lists of parasitized
spider eggs; but until more facts are in hand it is impossible to say
whether the proportion is greater or less than, for example, with the Orbweavers and Lineweavers, whose methods of protection are in this respect
so different.

Sixth, *the personal carriage of the cocoon by Lycosids would seem to be an
important factor in preserving the eggs.* At least I have not found a single
reported case of parasitization in cocoons of species having this
Cocoon   habit, which might be owing to poverty of observation rather
Carriage. than of existing facts. Some of the reported parasitized cocoons
we know to be personally guarded by the mother, as in the case of Salti-
grades, who usually stay within their silken cells with their eggs, and, for
awhile after hatching, with the young also. Occasional excursions for
food, however, might afford the required opportunity to the mother parasite.

CHAPTER III.

BIOLOGICAL MISCELLANY.

I.

RECENT observations extend our knowledge of the Hymenoptera that store spiders in the maternal nest. Mr. William J. Fox has published a list[1] of five species belonging to the one family of Pompillidæ. These **Spider Enemies.** are Pompillus Æthiops Cresson, P. biguttatus Fabr, P. marginatus Say, Priocnemis pompilius Cresson, and P. germanus Cresson. The first named species was taken in the act of capturing a large Lycosid; the second while carrying a small silvery spider, apparently a young Argiope argyraspis. It is probable that all of this family prey upon spiders, a fact which vastly enlarges the number of hymenopterous enemies which wage warfare upon the order Araneæ. It is so much the fashion to look upon spiders with disfavor, as a cunning, fierce, and relentless enemy of insects, to whom our sympathy rather goes forth, that we are apt to forget that poor Arachne is herself the victim of a remorseless fierceness and cunning on the part of insect families which far exceeds her own.

Mr. Francis R. Welsh has favored me with a vivid description of the manner in which the arachnophagous wasps pursue their prey. The pursuer in this case was a black wasp whose name was not known, and the spider was Agalena nævia. Agalena's web was spun about four inches **Wasps: Pursuit of Spider.** above the ground among plants and was of the usual form, that is, a horizontal sheet pierced by a tube, and having numerous stay lines rising several inches above the sheeted floor. For fully fifteen minutes the wasp pursued the spider under and over the leaves and web. On a straight dash the former was quicker in movement, but the latter beat her whenever the conditions would not permit her to fly, though she also ran surprisingly fast. At the end of the first round the spider escaped from the web unseen, and hid among some plants and stones about three feet away. The wasp beat about the web for a few minutes, and thence passed to some neighboring webs, and finally traversed the ground near the original web until she flushed her quarry. The observer thought that the spider bolted before the wasp saw it, but the pursuer must have been very close at the time. Agalena got safely back to its web, and the second round began and lasted about ten minutes, when it

[1] Entomological News, Vol. I., page 145.

again escaped. unseen, ran in between some stones a foot distant, and disappeared. The wasp beat about as before for five minutes, and gave up the chase.

The wasp seemed to have more staying powers than the spider, but the latter made up for this by hiding and resting. The spider would hide until the wasp spied it or came near; it would also rest under its web, hanging thereto, but never stopped elsewhere longer than a second or two, except on the two occasions when it entirely left the web. The wasp nearly caught the spider a number of times; it certainly must have touched it five or

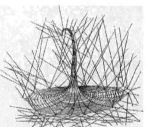

six times, but could never get a firm hold. When the spider hid, the wasp started to search over and under the leaves, and seemed very keen sighted. The spider did some excellent dodging over the edge of its web and over the leaves, but never attempted to double upon its track. The wasp was always in motion; the spider hid and rested when it could, and seemed to know pretty well where its pursuer was and when it was seen by her. The wasp tried several times to get over the web at the spider, but without success. Sev-

FIG. 46. Snare and nest of Epeira beccarii.

eral times the spider escaped by slipping through its web. The wasp evidently depended on sight alone; the spider on sight and, as the observer believes, largely on the vibration of the air and of the web.

II.

The tendency of orbweaving spiders to develop the hub of their snares into a tubular passageway between the orb and the nesting tent has been referred to as characteristic of Epeira labyrinthea. (See Vol. I., page 141.) The Messrs. Workman[1] have noted and figured another illustration of this, which is even more decided than that displayed by the Labyrinth spider. Epeira beccarii Thorell was found in considerable numbers growing at the side of the Deli Road, about two miles from Singapore. Their snares (Fig. 46) are dish shaped horizontal orbs, with the concavity upward. The hub is elevated into a cornucopia like shape, the top of which is slightly curved. About half way up this tube the egg cocoon is placed. Below the snare is a network of retitelarian lines larger than the orb, which is eight inches in diameter. The central tube reaches a length of three inches. It is braced to the surrounding twigs and leaves and thus kept upright.

A Tube-making Orb-weaver.

---

[1] Malaysian Spiders. By Thomas and M. E. Workman, Belfast, Ireland, 1892.

Mr. Workman says that the wonderful regularity of the circular snare and the beautiful curve·of the ascending tube, together with its perfect adaptation to the means of supplying food and protection to its constructor and her progeny, make,it a most interesting object. He justly compares the web to that of the Labyrinth spider of Hentz, and the resemblance certainly is striking. But the web of Epeira labyrinthea is always placed in a vertical position. The retitelarian labyrinth is therefore behind and at the side of the orb, instead of before it. The thickly lined gangway between the tent and the centre of the orb of Labyrinthea I have never seen developed into a complete tube, such as Mr. Workman describes, but .it is often much thickened next the hub, and is apt to assume a somewhat tubular form. At the tenting end, however, the spider frequently occupies a bell shaped silken

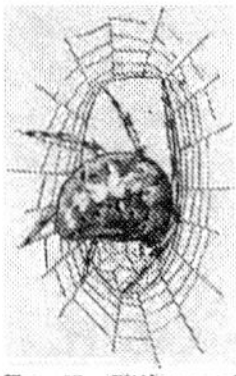

FIG. 47. Sitting position at the hub of a snare.

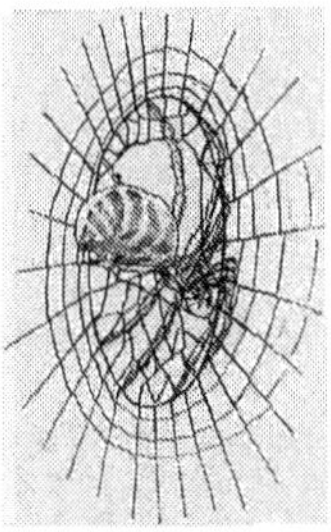

domicile. The reader in this connection is referred to Vol. I., Chapter XIX., in which the tubiform web is referred to as being the rudimentary one which appears more or less distinctly, and with greater or less development, in the spinningwork of all the principal groups of spiders. Let the reader compare Mr. Workman's drawing of Epeira beccarii with my figures of Agalena nævia (Vol. I., page 345, Fig. 336), or with examples of the spider's funnel shaped snare upon the hedges, lawns, and fields of America. He will then observe that by reversing Workman's figure he has before him substantially the outlines of Agalena's funnel shaped snare. We thus have represented in the spinningwork of this one species the typical webs of the three great sectional groups : Orbweavers, Lineweavers, and Tubeweavers.

FIG. 48. Epeira sitting upon a broken hub.

III.

Orbweavers are sometimes seen as represented at Figs. 47 and 48, in an attitude which might properly be called sitting within their hub. The upper part of the central meshes is usually removed, and the abdomen is thrust through the opening, and is supported by the remaining meshwork against which the venter rests. The hind legs are extended upward, and hang upon the margin of the opening, while the fore feet are more closely approximated and clasp the margin at its lower part. The radial lines are centred upon these fore feet, thus giving the spider full command of her nest. The moment an insect strikes the orb the spider draws up her abdomen, and with inconceivable rapidity spreads out her legs, straightens out her body, and faces towards the direction from which the agitation of the web had been signaled.

This sitting position is not unusual, but is not common; I have noticed several species which appear at times to indulge in it. The attitudes of the body, while substantially like those figured, vary somewhat according to circumstances. Epeira vertebrata, E. arabesca, and E. benjamina are some of the Orbweavers observed in this posture. No doubt the position gives a grateful relief from the ordinary attitude, by changing the direction of tension, if nothing more.

IV.

Under examples given in Vol. I., illustrating the physical and mechanical powers of spiders, was a case of a small fish captured by a species of Dolomede or Lycosid spider. The case has excited much interest, and it is gratifying to have it supported by a like well authenticated instance. Mr. Francis R. Welsh, of Philadelphia, writes me that a spider once killed two sun fish, each about two inches long, that he had in a basin in his room. After having attacked the first fish it ran over the water and fastened upon the second, which was also at the time apparently well and vigorous. Mr. Welsh drove the spider off, but the fishes died in a few hours. The basin was kept in a second story room of a country house at Chestnut Hill, and climbing vines covered the outer walls, nearly approaching a window close by the basin. It was inferred that the aranead entered the window from the vines. Mr. Welsh could not identify the spider, and could describe it only in a general way; but judging from the figures in my books he supposed it might have been either a Dolomedes or Agalena nævia. It would be entirely possible for the former animal to accomplish the deed, for in the cocooning season Dolomedes lurks in the bushes near her egg nest (Vol. II., page 145), spun among the leaves. One would hesitate, however, to think of Agalena as a creature of amphibious habits.

The same gentleman, referring to the engineering habits of spiders, as described Vol. I., page 211, and especially of the use of counterpoise, tells me that he saw a spider one day cut a dead leaf out of its web, which was dropped and hung by several threads. It then cut it down again twice, the leaf each time dropping and hanging lower. Mr. Welsh could not see if the spider attached any special threads to the leaf, but she ultimately used it as a counterpoise to her snare.

A Fish Killing Spider.

Counterpoise in Webs.

V.

Many Orbweavers have a curious habit of moving themselves rapidly upon their webs when excited by any cause which attracts their attention or awakens fear. It is particularly noticeable in the genera Argiope and Acrosoma. This action occurs when the spider is stationed at the centre

of her hub, and is in the following manner: The body is bowed or arched a little more than is usual when in that position, and then is set
Swaying the Body. in motion in such wise as to cause the web to move back and forth swinging upon its foundation lines. The motion is at first slow, and is rapidly increased until communicated to the whole web, which is oscillated at times so violently that the form of the spider becomes indistinct. This motion is continued for several minutes, the period varying in length. Mr. Muybridge, distinguished for his observation upon the motions of animals, informed me that he once watched a female Argiope cophinaria in continuous oscillation on her snare for a half hour.

The oscillation of the orbs by Argiope appears to be accomplished by lifting the abdomen up and down, or rather back and forth, from the web, and the alternate slackening and drawing taut of the line which connects the spinnerets with the central silken shield. At the same time the legs are alternately bent together and stretched out, thus drawing the orb in and repelling it. These movements of the spider of course put the orb in oscillation, and when rapidly repeated it moves back and forward with great rapidity.

Among Lineweavers this habit takes the form of a rapid whirling of the body within the snare. The species most addicted to this habit is our
Whirling Movement. common long legged cellar spider, Pholcus phalangioides. This creature hangs upon its snare of large loosely netted cross lines with its back downwards. Its feet, which are stretched upward from the body, by reason of the great length of the legs clasp these lines well together. Then begins a circular motion of the body. This at first is slow, but is rapidly increased until the body whirls around in circles so rapidly that it can scarcely be distinguished from the whole mass of agitated lines, legs, and body, which resembles a revolving cone.

The manner in which the motion is produced is probably the same in both tribes here mentioned, but the nature of the Orbweaver's web is such that the motion is limited by the upper and lower foundation lines within which it is set, and is so compounded as to compel the orb to a lateral or pendulum like movement; that is, a movement perpendicular to the plane of the orb. In the case of Pholcus, however, the web of right lines permits the body to revolve in circles which are parallel to the plane of the horizon.

This habit is doubtless protective, and probably serves to confuse attacking Hymenoptera, birds, and other enemies, and thus avert their aim or
Uses of the Habit. drive them from the position by the unwonted agitation of the web. But its origin and chief purpose are probably connected with the taking of prey. The swaying of the web must aid to entangle more completely a victim which strikes it, since it stimulates the captive to increase its struggles and thus more effectively to fasten itself to

the treacherous viscid or adherent lines. The action must also tend in some measure to locate the position of a victim. Can it possibly at times tend to produce a current of air into which insects are attracted? At all events, when Pholcus phalangioides is annoyed by being touched with pencil or finger, she is pretty sure to begin moving her body, and in a little while is swinging herself around, her body describing a circle which represents the base of a cone, of which the point whereat her clustered feet hold on to her line will be the apex.

VI.

Professor Wilder relates an interesting example of the tendency of young Argiope cophinaria to make excursions from the egg enclosure when opportunity presents. Numerous cocoons

Baby Spiders. were found at James Island, South Carolina, which had been torn open by birds to get nest building material. From the breach fragmentary wads and rolls of silk floss protruded, along which the spiderlings crawled, and finding themselves in the fresh air and sunshine concluded to enjoy the same. They would swing down from the projecting roll in long festoons, singly, or sometimes clinging to one another like bees when they swarm, but always retained their connection with the cocoon, to which they returned when satisfied with their taste of sunlight and liberty.[1] One is inclined to note here a judicious mingling of conservative caution with youthful sportiveness.

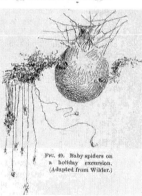

FIG. 49. Baby spiders on a holiday excursion. (Adapted from Wilder.)

VII.

In Vol. I. of this work were considered the poison apparatus of spiders and the effects of their poison upon animals and man. Since then numerous contributions upon the general subject have been made by

Spider Poison. many persons, from widely separated parts. Some of the most valuable of these have been published in "Insect Life," the official journal of the Entomological Bureau of the United States Department of Agriculture. These leave the facts substantially as generalized by me, and my own view thereof is therefore unchanged, as follows: The poison secreted by spiders is sparingly used, and it is not necessary for securing prey; its object is probably chiefly defensive, and its effect upon creatures of its own rank and size may often be serious and fatal; the effect upon

---

[1] Harper's Magazine, March, 1867, page 458.

human beings has no doubt been greatly exaggerated, and by far the greatest number of species are harmless to the vast majority of men; nevertheless, there is no reason to doubt that a few species secrete a poisonous fluid which is unusually virulent in its action upon men, producing painful, serious, and, rarely, even fatal wounds. These exceptions are so few that they give no warrant, or the very slightest, for the almost universal dread of spiders of all sorts and sizes. Man has no more and probably much less reason to fear injury from these animals than from ordinary irritating insects, and on the other hand is immensely their debtor by their services in holding in check the increase of the insect hordes.

It will suffice here to note one well authenticated instance of injurious results from spider poison, as reported by the distinguished arachnologist, Professor Bertkau, of Bonn. This naturalist has recently found *Professor Bertkau's Experience.* on grass and leaves, near Bingen on the Rhine, a number of species of Chiraianthium nutrix, which occurs also in Switzerland, France, and Italy, and has assured himself that the species is poisonous. He was bitten on his finger ends three times by these spiders. The pain was severe, burning, and extended almost instantaneously over the arm and to the breast, and was most intense in the wound itself and in the armpits. On the second morning after the second bite the pain disappeared, but returned upon pressure of the bitten spot, and changed to an itching sensation. When he was bitten again, four days later, both the pain and afterward the itching, especially, returned in the first two bites, and this time continued for almost a fortnight, when all unusual feeling had disappeared, while the later bites, which had festered, were still visible. The results immediately following the bite were a slight swelling and inflammation, which gradually disappeared. It is highly satisfactory to find a case of this kind described by a careful and distinguished specialist like Professor Bertkau, and indubitably traced to a definite species, instead of attributed to the inevitable and indefinite black spider.

Per contra, a well known fellow citizen, and intelligent associate of our Philadelphia Academy of Natural Sciences, Mr. Francis R. Welsh, thus *Per Contra.* writes me, apropos of the observations contained in my first volume: "When I was a small boy one of my friends and myself used to catch, carry about, and tease spiders with impunity. I was never bitten, but my friend was once by a 'black spider'—you know how vaguely the term is used. I can only say that it was almost certainly not an Orbweaver, but was probably either a Tubeweaver or Wanderer. The spider, as well as I can remember, was about two inches long when fully extended. The bite simply caused a little inflammation, and can be best compared to a severe mosquito bite."

It is probable that this expresses the ordinary effect of a wound inflicted by the indigenous species of this geographical province; while that of Professor Bertkau exemplifies the worst that may be apprehended. in cases of bites by our largest species. Nevertheless, it is not denied that under certain constitutional conditions of the bitten person, the stroke of a spider's fangs may produce far more serious results, or, in extremely rare cases, even death, a consequence, however, which might issue under like conditions from the sting or puncture of acculeate or irritating insects.

VIII.

The following facts in the courtship of Lycosa tigrina are recorded by Mrs. Treat.[1] It should have been inserted among mating habits in Vol. II., Chapter II., but was inadvertently omitted. The males **Tigrina's** of this species are different in appearance from the females, being **Court-** a yellowish color with dashes of dark brown; the body is some-**ship.** what smaller, though the legs are longer. In August the males are abundant. They were often seen bounding over grass and weeds, making long strides, fairly flying before the passer by. At such times it is extremely difficult to capture them.

One of these vagrant males was observed to approach the burrow of a female, who had stretched above the vestibule of her den a projecting cover, something like a hood or the top of a baby coach. When the female was within the burrow, the male stood at the door sometimes hours together. Nothing would induce him to venture within, and he was wonderfully oblivious of the observer's presence. An effort was made to push him into the den, but he would back out and find refuge in one's hand rather than be driven into the burrow. At last the female slowly advanced to meet him, and he slowly retreated from the mouth of the den, moving backward while she moved forward, just reaching him with the tips of her fore legs, as if caressing him. She followed him in this way a foot or more, then left him and returned quickly to her den, he in the meanwhile following her to the door where he kept his post until she came forth, when the same performance was repeated.

At the next visit the male was found on the back of the female with their heads within the burrow, and their long hind legs sticking out. This is the position assumed by the spider when he fertilizes the eggs, as may be seen by consulting Vol. II., Chapter II., Fig. 50. The two remained perfectly still until they were picked up by the observer and dropped into **Her** a wide mouthed glass bottle. This action displaced the male, who **Partner** crouched in a helpless sort of way as if paralyzed with fear, not **Eaten.** trying to make his escape at all. For a few moments the female paid no attention to him, but made vigorous efforts to escape. Soon, however, she observed her partner, pounced upon him, seized him on the underside of the head, literally by the throat. He made but feeble

[1] American Naturalist, August, 1879.

efforts to resist, in fact acted as if he rather enjoyed being eaten. The observer shook the bottle, but the female would not let go her hold. In a little while she had doubled up her partner into a ball, and with great relish proceeded to suck the juices of her slaughtered mate. The mouth of the bottle was now uncovered, whereupon the female disappeared into her burrow taking with her the remains of her lover. .In a day or two after this another male was at her door, behaving in a manner similar to that above described. His movements were not interfered with, and his fate was unknown.

IX.

It has been a question with araneologists to what extent spiders mend their nets. I would say that as a rule very little mending is done, except **Mending Snares.** to repair the damage wrought by the agitation of entrapped insects. The habit of Orbweavers is to produce a web in the early evening to serve for the ensnaring of prey during the night. If this happens to be well worn out by the abundance of victims entrapped, instead of repairing the snare the ragged remnants will be cut away, and upon the foundation lines a new web will be woven in the early morning for the capture of day flying insects. In brief, the Orbweaver adapts her spinningwork to the conditions of insect life, having in some rude way reached the generalization that there are nocturnal insects and diurnal insects.

It has been a question in which I have been interested, whether certain spiders did not adapt themselves to the one class, and others again to the **Noctur- nal and Diurnal Spiders.** other; so that we might speak of night preying araneads as we do of night flying insects; but my observations on this point have not been satisfactory. I have an impression that some spiders prefer the night to the day, as, for example, Epeira strix, who almost persistently abides within her den during the day and goes out at night to engage in the capture of insects. Indeed, this is the habit of many spiders. All those that live in leafy domiciles or silken tents hold quite persistently to these in the daytime, and forsake them in the evening, and take their station upon the centre of the orb. To this rule, however, such spiders as Epeira labyrinthea and E. globosa are exceptions. These are rarely found upon their web, but both day and night remain within their dens and capture their prey almost exclusively by means of the vibratory communications sent along the trapline, as I have heretofore fully explained.[1] On the other hand, large numbers of spiders, especially those of the genera Argiope, Acrosoma, Argyrœpeira, and Abbotia, keep their positions upon their orbicular web both day and night, and seem to have no regard, one way or another, for diurnal changes.

---

[1] See Vol. I., page 137.

As a rule, the last named genera will remain upon their webs until they are worn out, and repair them whenever the exigency requires, making the changes ordinarily either in the morning or in the evening, and not attempting to mend the broken lines, except as above stated, in a casual way, to prevent the collapsing of the snare after the disruptions caused by violent insect struggles. They will hold on to the snare and let it do what duty it may until toward the close of the day, when they proceed to cut away the fragments and build anew. It will thus be seen that the act of mending a web in the case of Orbweavers is not common, but confined to the damages done by struggling victims

Time for Repairs.

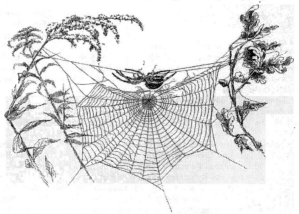

Fig. 50. A Shamrock spider mending a wind wrecked web.

and to the slight impairments due to incidental casualties, such as the dropping of leaves, twigs, or the effects of the wind.

An interesting illustration of the behavior of the spiders under the last named condition was seen during a gale of wind at Niantic, Connecticut. An adult female Shamrock spider, Epeira trifolium, had stretched her large vertical orb from a clump of young oak bushes on the one hand to a cluster of tall golden rods on the other, the intervening space over which her foundation line was stretched being six or seven feet. The lower parts of the snare were stayed to herbage on the ground. A strong gale from the sea was blowing, and whipped the tops of the golden rod back and forth so violently that the main foundation line was snapped and the upper part of the web collapsed,

A Wind Wrecked Web.

as shown in Fig. 50. It, however, hung partially by the radial supports attached to either end of the foundation line. I was fortunately on the spot when the accident occurred, and saw Trifolium issue in the most excited manner from her leafy den among the oak bushes and rush across her ruptured foundation line. From one of her hind feet she held out a stout thread, which evidently had been instantaneously extruded. In some way, which was accomplished too rapidly for my eyes to take in **Mending** the details thereof, she flung her body across the gap, and reached **a Founda-** with her fore feet the opposite end of the sundered foundation **tion Line.** line, to which she clung, while with the aforementioned thread and with other spinningwork flung rapidly from her spinnerets she spliced the break and thus restored the foundation line. During all this time she was swaying back and forth upon the thread, like a sailor upon a yard arm reefing sails in a heavy gale.

The whole process struck me as exceedingly ingenious, and it was accomplished with such rapidity, without the slightest hesitation as to what ought to be done and the method of doing it, that I concluded that this experience was not a new one, but that Trifolium was quite well used to and thoroughly prepared for such emergencies. Her next step was to gather up the radial supports attached to the foundation line, and thus make good her orb for its intended uses. I could not stay to see the details of this action, but have no doubt that the web was entirely repaired and, after the subsidence of the wind, was as good as ever. It may be well, however, to say that the foundation line[1] is regarded as the most important part of a spider's real estate. Given a foundation line there is no trouble, ordinarily, in swinging thereto a snare; but often spiders are sorely put to it to secure this, for which they are usually dependent upon the condition of the wind. It may, therefore, be that the rupture of the foundation line, in the case above described, was regarded as a capital accident, which called for special energy and prompt action. It is probable that an ordinary rupture in any portion of the snare itself would have been regarded with indifference, and that Trifolium would have remained snugly ensconced within her domicile until the gale had overblown, and perhaps would have taken no notice of it at all.

Of course, in considering this matter of repairing snares, the observer will distinguish between the comparatively ephemeral web of the Orb-weaver and the more permanent snare of the Tubeweaver. Such **Patching** spiders as Agalena nœvia and Tegenaria medicinalis[2] build almost **Tube-** permanent abodes for their occupants. The tubular portion of **weavers.** the snare is their home, and the snare is rarely rebuilt, never indeed, I believe, unless it is completely destroyed. That these spiders do mend their webs I know, having observed the same. As they grow they

---

[1] See Vol. I., page 66. [2] See Vol. I., page 230, Fig. 221.

will enlarge the margins of the sheeted parts thereof, and will also enlarge the tubular tent. The supporting lines above are also continually strengthened. These spiders may be seen in these acts from time to time by one who cares to watch their habits. We thus reach the conclusion that those web making spiders that dwell in permanent homes, to which snares are attached, are in the habit of frequently repairing the same; while those which make more ephemeral snares, as the Orbweavers, permit them to be quite worn out as a rule, whereupon they proceed to build new webs. There are, however, exceptions to the above, and occasionally Orbweavers will be found patching their webs. The reasons for this difference, of course, are found in the different characters of the webs themselves, as has heretofore been fully explained in Vol. I. of this work, and need not be here further entered upon.

X.

Mr. Walter Titus, a youth engaged upon a ranch in the neighborhood of Los Angeles, made a number of intelligent observations upon Trapdoor spider nests, which are furnished me through the kindness of Miss Estelle Thompson.[1] This lady is decidedly of the opinion that the nests of the California Trapdoor spider are quite commonly placed in such positions as to allow of good drainage, that is to say, so that the nests are protected from excessive rains.

Intelligent Location.

In attempting to lift the lid it was invariably found to be held down, as though by suction from underneath. In one case a spider which was holding down a trapdoor did not let go until the lid was lifted, when she slid into the tube as though going down a well. A full grown spider can force up the door of its house, even when there are three ounces of lead on the top thereof. The manner of entrance to and exit from the burrow is well described by Mr. Titus. When the spider wants to leave home it lifts the lid by pushing from beneath; and when, on the contrary, it wants to reënter its nest it lifts the lid with its mandible, fastening its hooked fangs thereinto, then places its two front legs down into the hole, as though to stay up the hinged trap, and thereupon darts within. It is evident that before retreating she reverses her position, for it is stated that she backs down the nest, and the lid closes by its own weight.

Entering the Nest.

The first door that Cteniza makes after she is old enough to set up housekeeping is composed almost wholly of silk. The next one will not contain so great a proportion of silk, and succeeding ones each a less

---

[1] I have given on page 29 an abstract of some observations on the California Trapdoor spider made by Miss Estelle Thompson. Having written for more accurate and fuller details, the following facts came too late to be used in connection with the above, but in time to add to this chapter of Miscellany.

proportion, until she has made her fifth lid. In this the inside alone is lined, and all the rest is made of clay and a sort of glue or mucilaginous material, which the spider secretes from its mouth parts, and which will dissolve in water.

Miss Thompson had frequently found burrows with doors that were bare and others covered with lichens, according to the condition of the bank on which they were located. She had never in a single **Mimicry of Site.** instance found a nest on a lichened hillock that did not contain lichens upon the lid. She had cut the upper portion of many from the ground in order to preserve them, and after the door was thoroughly dried the lichens usually fell off. She had not seen the spiders in the act of fastening on the plants, but had no doubt that in some way this was accomplished, or at least permitted.

Mr. Titus avers that Cteniza attaches moss, sticks, and fine pebbles to her door at times, and her object in so doing he believes to be to **The Ta-rantula Hawk.** hide it from the "Tarantula hawk." This insect attacks Trapdoor spiders within their nests if it can find them, and it is a most formidable adversary, for it grows to be about two inches long. Mr. Titus had the impression, which is the popular one, that the Tarantula hawk feeds upon the Trapdoor spider; but by reference to Vol. II. of this work it will be seen that the object is simply to procure food to place within a cell wherein the wasp bestows the egg of her future offspring. The insect referred to by Mr. Titus is no doubt the so called "Tarantula killer" of the southwestern States, the beautiful Pepsis formosa Say, which I have figured,[1] and whose habits in connection with the destruction of the large hairy Tarantula I have fully described.

This observation proves most interesting, and confirms in so far what I have said of the reactionary effect of hostile agents and environment **Enemies Influence Architec-ture.** upon the architectural habits of spiders. I have traced[2] the well known habit of the Tiger spider to cover its burrow with a mossy vestibule to which is attached a rude sort of door, to the purpose to protect itself from the attacks of an invading wasp, Elis 4-notata. I had no hesitation[3] in using the knowledge thus furnished by the habits of Tigrina as a key to interpret the motive power of Trapdoor spiders in their remarkable industry. I had no facts in my possession, and could only reason from analogy, but offered some conjectures[4] as to the character of the enemies whose assaults are thus met by this rare counteracting ingenuity. I also ventured to predict, from the various facts alluded to, that the "enemies they most dread may be reasonably looked for among diurnal creatures, and not among those of nocturnal habits." The observation of Mr. Titus, which is unfortunately

---

[1] Vol. II., Plate V., Fig. 2 ; see also full description, pages 384, 385.
[2] See Vol. II., pages 404, sq.   [3] Id., 409.   [4] Id., 414.

without details, confirms all that I had there suggested, and shows that the enemy which the Trapdoor spider is most concerned to evade belongs to the same remarkable family, if it be not identical with Pepsis formosa.

Mr. Titus adds that the moss and lichens placed by the Trapdoor spider on the outside of its door will keep green through the wet season, and dies **Mimicry.** when the dry time of the year comes. He once saw a nest made in a bank of beautiful green moss, the door of which was completely covered by the plant. The nest was discovered by the circumstances that the moss upon the lid was not quite so large as the surrounding moss, and it was only by this that the nest could have been detected. The fact showed that the moss had grown since the preparation of the lid, and this of course could only have been by the active or permissive agency of the spider architect.

Some observations made by Miss Thompson upon the maternal habits of a family of Trapdoor spiders are interesting. The young spiders have **Baby Spiders.** a pinkish white color when first hatched. The hatching occupied a period of about three weeks. A little later the color of the spiderlings is described as a silvery amber touched with pink. They remained in the nest with the mother, and at times perched upon her body and legs, where they looked like shining pink pearls against the glossy black. This is the custom of the young of all Citigrades, and, so far as is known, of Tunnelweavers.

Of the cocooning habit of this species Mr. Titus says that the female lays her eggs upon the inside of her nest, about three inches above the **Cocoon.** bottom thereof. The eggs are fastened to the sides of the burrow with glossy threads. It is thus seen that the cocooning habit of Cteniza Californica is precisely like that of the Trapdoor spiders of Venezuela, as described by Mr. Eugene Simon.[1]

---

[1] A copy of his figure showing the position of the cocoon within the nest will be found in Vol. II. of this work, page 140, Fig. 172, the burrow of Psalistops melanophygia.

CHAPTER IV.

WEATHER PROGNOSTICATIONS—SUNDRY SUPERSTITIONS— COMMERCIAL VALUE OF SPIDER SILK.

I.

THERE is perhaps no opinion concerning spiders that is more widely disseminated and popularly believed than that they have the power to prognosticate weather changes. As long ago as the days of Pliny the notion was entertained. That author affirmed that prognostications may be based upon the spider's behavior. For example, when a river is about to swell the spider will suspend its web higher than usual. In calm weather these creatures do not spin their ordinary webs, but when it is cloudy they do so, and therefore a great number of cobwebs is a sure sign of rainy weather.[1]

Weather Prognostication.

An interesting and romantic incident, based upon this supposed faculty, is associated with the wars succeeding the French Revolution. Quatremer Disjonval, a Frenchman by birth, was an adjutant general in Holland who took an active part on the side of the Dutch patriots when they revolted against the Stadtholder. It happened to him to be captured and condemned to twenty-five years imprisonment at Utrecht. Here for eight years he relieved the tedious confinement by many curious observations upon his cell companions, the spiders. Among other matters he discovered that they were in the highest degree sensitive to approaching changes in the atmosphere, and that their retirement and reappearance, their weaving and general habits were intimately connected with weather changes. He became wonderfully accurate in reading these living barometers, so much so that he could prognosticate the approach of clear weather from ten to fourteen days before it set in.

This ability served him a high advantage when the troops of the French republic overran Holland in the winter of 1794. They kept pushing forward over the ice with the assurance of ultimate victory, when a sudden thaw in early December threatened the destruction of the whole army unless it were instantly withdrawn. The French generals were seriously thinking of accepting a sum offered by the Dutch to withdraw their troops, when Disjonval sent them a message advising against such action. He had hoped that the success of the republican army might lead to his release, and therefore sent a letter to the

Quatremer Disjonval.

---

[1] Pliny, Natural History of Animals, Chapter XI., section xxiv.

French general wherein he pledged himself, from the peculiar actions of the spiders, of whose movements he was able to judge with perfect accuracy, that within fourteen days there would be a severe frost, which would make the French masters of all the rivers, and afford them sufficient time to complete and make sure the conquest which they had commenced before it would be followed by a thaw. The commander of the French forces, it is stated, believed these prognostications, and pushed on his armies. The cold weather, as Disjonval had predicted, made its appearance in twelve days, and with such intensity that the ice upon the rivers and canals was capable of bearing the heaviest artillery. On the 28th of January, 1795, the French army entered Utrecht in triumph, and Disjonval, as a reward of his ingenuity, was released from prison.[1]

One might think that, under the circumstances, the French captive would have been safe in predicting a change from a sudden thaw to severe cold in the Netherlands at that season of the year. No An Ex-planation great amount of prophetic skill would be required to make a success in like circumstances at least nine times out of ten. Nevertheless, a prediction based upon such commonplace affairs as the ordinary course of the weather would doubtless have produced no impression upon the mind of the commanding general; but when fortified by the strange and mysterious association with the behavior of spiders the prediction must have appealed powerfully to the imagination, and, supposing the truth of the story at all, have turned the balance in favor of the plan recommended by Disjonval. Nevertheless, the spiders obtained and have retained credit for the matter.

The popular notion that spiders are reliable barometers of weather changes is thus well expressed by the late Mr. Wood: Spiders are all very The Pop-ular No-tion. chary of using their silk, and never trouble themselves to make webs when a storm is impending. They are, therefore, excellent barometers, and if they all take to mending their nets or spinning new webs, fine weather is always at hand.[2] It has happened to me numbers of times to hear predictions of the weather based upon the condition of spider webs, made by American farmers in various parts of the country. From what I have heard I imagine that the notion is widespread, particularly throughout the Middle and part of the New England States.

I have tried to find whether there exists in Nature any sufficient basis for this opinion, and to that end have made numerous notes of my observations. It is undoubtedly true that spiders are sensitive to weather changes, that is to say, extremely cold weather or long protracted rains will keep them in or drive them to their nests or other retreats under

---

[1] Quarterly Review for January, 1844, quoted by Cowan.
[2] Rev. J. G. Wood, "Homes Without Hands," page 320.

leaves, branches, rocks, or whatever shelter may present. Here they will remain until the storm is over. I always know, after such a sudden change in the summer or autumn, that it will be useless to go into the fields for the purpose of studying spider webs until after an interval of a day or more, and until the weather has moderated. But I am constrained to say that this prognostication is after the facts instead of before them. Let us proceed to apply the test of scientific observation to the belief. Perhaps I may better enable the reader to decide upon the matter by quoting from my journal records of notes made during several years, which, as will be seen, concern summer showers and ordinary rains of the season.

II.

"Annisquam, Massachusetts, August 14th, 1888.—Severe rains yesterday. The Zillas have spun this morning and the day is bright, but a strong wind is blowing. . . . August 17th.—Orbwebs were freely spun this morning, but a shower with thunder came on about 11.30, and the rest of the day was overcast and showery. "Philadelphia, September 8th, 1888.—A warm, moist morning, soft showers falling freely, yet in spite of this a number of my colonized Argiopes have spun webs or parts of webs as though confident that the day would be clear. There were heavy rains all day, which the spiders did not appear to mind, hanging upon their webs during the showers, which seemed to be no inconvenience to them. They would stop spinning their snares while the rain fell, but take up the unfinished work during the intervals. . . . September 9th.—Rains this morning. At eight o'clock the sky was overcast, drizzling rains following a shower. The showers are soft like those of late spring. Many spiders are out upon their webs, and have so continued during the day. 3.20 P. M.—Epeira vertebrata spinning an orb; the former one evidently destroyed. . . . 5.30 P. M.—A large Domicile spider is making a new orb. No rain this evening, but the sky much overcast now and threatening.

Notes of Weather and Webs.

"September 10th.—All the spiders appear this morning in beautiful new webs. There has been no rain since yesterday afternoon. The sky is overcast this morning. The Government weather predictions for the twenty-four hours between 8 P. M. September 9th and 8 P. M. September 10th indicate cloudy weather, with rains and cooler temperature. Some of the spiders have moved their positions. The day kept clear though overcast, and a few drops of rain.

"September 11th.—This morning I found no activity among the spiders in the way of web making, but most of them were upon orbs which appeared to be new, and probably were made during the early morning. The Argiopes and Vertebratas were out in special force. Rain began about 10.30 A. M. and continued an almost steady downpour until eight or nine o'clock at night. The Argiopes hung to their webs and took it without flinching. So also did one of the Vertebratas.

"September 18th.—Cloudy. Spiders with full webs. The day clears up. No rains. . . . September 19th.—Spiders seen with full orbs everywhere. Evidently they were at work last night. At nine o'clock morning the weather cloudy. . . . September 20th.—Argiopes at work while raining. One spinning after a heavy shower, the drops of rain hanging like beads upon her legs. She is finishing the spirals, part of which had been made earlier in the morning. All the spiders are out on their webs. The weather showery, close, warm. There was a heavy rain at night and yesterday.

"July 11th, 1889.—Yesterday evening the Orbweavers on my vines were busy spinning webs, two young Epeira trivittata being especially observed; the heavens were clouded at the time and rain threatening. The day had been overcast and was slightly showery. A shower fell in the early part of the night, and between three and four o'clock one of the heaviest rains that I ever have known in this climate. The whole morning has been showery up to nine o'clock.

"September 11th, 1889.—This morning a number of Argiopes were seen spinning their webs; others had formed large perfect snares upon the vines. Several Epeira trifolium have also made snares; also Epeira labyrinthea. Last night and yesterday a very heavy gale blew. Some of the Argiopes remained upon their webs during all the storm until the shield, spirals, and almost everything except a few of the main radiating lines were absolutely melted away. They do not seem to care for the rain. Other spiders are hid in their nests, or covered under leaves and other places of refuge. This action of the Orbweavers would seem to prognosticate a fair day. 9 P. M.—On the contrary, the weather has been rainy during the greater part of the day, rain falling incessantly, wind blowing with more or less violence, a raw, disagreeable autumn day, and the night even worse. Again my spiders have failed to prove themselves true weather seers.

"September 12th.—The gale of yesterday and the preceding day continued during the night, with rain. This morning is again windy and rainy. A little cessation of the storm at nine o'clock permitted to visit the vines. A number of Argiope cophinaria had made perfect snares. Insularis, Trifolium, Labyrinthea, and Arabesca all were out upon completed snares, or had begun webs and proceeded as far as the spiral scaffolding. There is not the same activity nor the same number of webs that would have appeared had the day been bright, but undoubtedly many of the spiders have disregarded the weather. Two Argiope cophinaria have cocooned during the night or morning in the midst of the storm. . . . Eleven o'clock A. M.—The rain still continues, a heavy fall. . . . 2 P. M.— Wind and rain still continue. . . . 6 P. M.—Rain and wind continue. 12 midnight.—The storm raging with unabated violence.

"September 13th, 7 A. M.—Sky overcast. 9 A. M.—Showery. Argiopes working in the rain and between the showers. Some are hanging in the centre of their webs by a few lines with their webs melted around them. They begin to build, and finish beautiful webs. Other spiders are also at work. 6 P. M.—It has been cloudy and rainy all day. September 14th.—The severest storm known for many years has been raging all along the sea coast. The spiders are badly out in their predictions. . . . September 17th, 1889.—A beautiful display of webs on the vines in the manse yard. Argiope cophinaria especially, but other species also have made snares, which I have rarely seen equaled for beauty. . . . 1 P. M.—One of the heaviest rains of the season has fallen this morning. Evening, the whole afternoon, and night has been cloudy. Frequent violent rains. . . . September 18th.—The papers are full of particulars of the violent local rain storms of yesterday. . . . September 20th.—A cold and clear day. Spiders are out upon their webs, but the Argiopes are mostly engaged in cocooning.

"May 1st, 1890.—Warm bright day, many spiders out with new webs, Epeira strix and the young of Argyroepeira hortorum and Theridium tepidariorum. A storm came up about 5 P. M., and rain has fallen at short intervals up to 8 P. M."

It is useless to continue these quotations, as my notes are fairly represented by these given above. Any one who reads them must come to the conclusion that either spiders have no ability to prognosticate the ordinary weather changes of a summer season, or else they are so indifferent to those changes that they spin their webs regardless of them. My conclusion is that the persons who trust to the presence or absence of spider webs as an infallible prediction of the state of the weather will frequently be disappointed.

Conclusion.

III.

That spiders have in some way been associated with good or bad luck is a widespread superstition, which dates indeed from classical times. A very few of these superstitions may not be quite out of place here. It is an old English and Scotch notion that small spiders, termed "money spinners," are held to prognosticate good luck if they are not destroyed, or injured, or brushed off from the person on whom they are first observed. To destroy these money spinners is held "an equivalent to throwing stones at one's own head."[1] One might feel justified in encouraging this fancy both on the grounds of mercifulness to animals, and as covering at least the germinal truth that spiders do contribute roundly to man's good fortune by their faithful service.

Spider Superstitions.

It is a Southampton superstition, which appears to survive to this day,

---

[1] Jamieson's Scottish Dictionary.

that in order to propitiate these money spinners they are to be thrown over the left shoulder.[1] Old Fuller, who was a native of Northampton, thus quaintly moralizes upon this superstition: "When a spider is found upon your clothes, we used to say some money is coming towards us. The moral is this, those who imitate the industry of that contemptible creature may, by God's blessing, weave themselves into wealth, and procure a plentiful estate."

The superstition prevails that if a spider approaches either by crawling toward or descending from the ceiling upon a person, it forebodes good to such person; and on the contrary, if the spider runs hurriedly away it is an omen of bad luck. If one kill a spider crossing his path he will have bad luck. A spider should not be killed in one's house, but out of doors. If in the house, it is a saying with English common people, "that you are pulling down your own house." If a spider drops down from its web, or from a tree directly in front of a person, such person will see a dear friend before night. A variation of the superstition is, that if the spider be white it foretells a friend, and if black an enemy. In the Netherlands, a spider seen in the morning forebodes good luck, in the afternoon bad luck.[2] The same tradition prevails among Germans, for a German, now resident in Hartford, Connecticut, informed me (and I have since heard the same from others) that in the country parts of Germany the people are careful not to see a spider in the morning, under the belief that it will bring bad luck. If they have reason to suspect the presence of a spider they will most scrupulously look the other way. But in the evening the same parties want to see a spider, because they believe that then seen it will bring them good luck. He gave me the following German rhyme expressive of the above superstition, to which I add a rude metrical translation :—

Money Spinners.

Luck in Seeing Spiders.

"Spinne am Morgen	Spider in morning
Kummer und Sorgen ;	Brings trouble and care ;
Spinne am Abend	But spider at evening
Erquickend und labend."	Refreshing and cheer.

This notion has found its way into Ireland, for my cook, who came from County Kildare, the east of Ireland, has told me that the tradition prevailed in her section that if a wee spider would come upon the head it was a sign that the person would get a new bonnet. If it fell upon the dress or coat a new dress or coat would result. She further said that in a little reading book, in use in one of the government schools, there was a rhyme like the following :—

Traditions.

"If you want to live and thrive,
Let the spider run alive."

---

[1] Notes and Queries, Vol. II., 165.   [2] Thorpe's North. Antiq., III., page 329.

Another notion connected with spiders is that certain kinds of wood prevent their settling and spinning cobwebs. There is a common saying at Winchester, England, that no spider will hang its web on a roof of Irish oak, and the cicerone who shows the Cathedral Church at St. Davids points out to the visitor that the choir is roofed with Irish oak, which does not harbor spiders, though cobwebs are plentifully seen in other parts of the cathedral.[1]

The same faculty of repelling spiders is attributed also to chestnut and cedar woods,[2] and the old roof at Turner's Court, Gloucestershire, four miles from Bath, which is of chestnut, is said to be perfectly free from cobwebs.[3] Hence, it is said, the cloisters of New College and of Christ Church are roofed with chestnut.[4]

I have at least once met this superstition in Italy, and am free to say that there is no basis for it in fact. I do not remember ever to have visited a public building, particularly a church or chapel, in which I have not been able to trace somewhere the webs of spiders. No doubt, however, some public edifices are inhospitable sites for araneads, simply for the reason that they give little encouragement to the presence of those insects which form a necessary part of spider subsistence. Naturally enough, spiders will not resort to and cannot abide in places where they do not procure sufficient food. The spiders which are most frequently found inside our homes and public edifices are certain Lineweavers, mostly of the genus Theridium, and one or two species of Tubeweavers. To these spiders we are indebted for the common cobwebs of our ceilings and corners. The above are only a few of the curious beliefs that have grown up around the spider among all races of men. A number may be found in Cowan's work heretofore quoted.[5]

IV.

The possible commercial value of spider silk as an available textile in industrial art has often been considered. It is not surprising that one **Spider Silk in Industrial Art.** who sees the immense snares and bulky cocoons of tropical araneads or of some American species of Nephila, Argiope, and Epeira that inhabit Florida and Southern California, should think that such quantities of strong and beautiful spinningwork might be put to practical use. As early a traveler as De Azara[6] tells of a Paraguay spider whose spherical cocoon, an inch in length, was utilized for spinning by the inhabitants of that land, not only on account of its bulk, but its bright and fast orange color. It is probably a congener of Nephila clavipes, the "silk spider," of whose silk, according

---

[1] Notes and Queries, second edition, IV., 298, and Id., 377.
[2] Ibid., page 523.   [3] Ibid., page 421.
[4] Ibid., page 298.   [5] Curious History of Insects.
[6] Voyages dans L'Amer. Merid. Don Felix de Azara, 1809, I., 212.

to Mr. Jones,[1] the ladies of that island make use for sewing purposes. Mr. Jones succeeded in reeling from the spider a few yards of yellow silk.

From time to time, and from various quarters, one gets accounts more or less definite and trustworthy that various rude tribes, and more civilized nations, indeed, have availed themselves of spider web fabric for dress. Such is probably the story that found place years ago in an American literary magazine[2] of the Emperor Aurengzebe of Hindostan, who reproved his daughter for the indelicacy of her costume, although she wore as many as seven thicknesses of spider cloth! I remember reading somewhere an account, though the details have passed from memory, of a royal garment woven of spider silk for her Majesty the Empress Victoria by some of the loving subjects of her world wide empire.

That the silk of spiders can be reeled from their spinnerets in considerable quantities I long ago proved by experiment. That little spools of silk sufficient for show purposes can be gathered by winding off the thick foundation lines from the snares of indigenous Orbweavers, I also know; and further, that many spider cocoons can be collected, from which, by ordinary treatment, small quantities of silk thread may be prepared which are available for knitting petty objects. But that aranead spinningwork can be obtained in temperate regions, at least, by any practical process, in sufficient amount to justify business investments, I do not think at all likely for many ages yet to come. Until present industrial conditions shall be so far changed, and the present sources of raw silk so greatly modified as to warrant prolonged and costly experiments, and subsequently large outlays, men will adhere to the silk moth. For mere curios spider silk is available; for profitable commerce it is not practicable. However, the efforts to utilize this material in the domestic arts are entitled to some recognition in these pages.

As early as A. D. 1709 M. Bon, president of the Court of Accounts of Montpelier, communicated to the Royal Academy of that city a discovery which he had made of a new kind of silk obtained from the egg bags of several species of spiders, probably Orbweavers.[3] His method was as follows: Having collected a large number of cocoons he beat out the dust, then washed them carefully in water, and allowed them to boil for three hours in a pot containing water, soap, saltpetre, and a little gum arabic. The cocoons were then washed, dried, and carded with extremely fine combs. The result was a gray thread much finer than that of the silk worm, and capable of receiving all the

Sources of Spider Silk.

M. Bon's Pioneer Attempts

---

[1] A Naturalist in Bermuda, London, 1859, John Matthew Jones, page 126.

[2] Atlantic Monthly, June, 1858, page 92.

[3] Hist. and Mem. de l'Acad. Roy. des Sciences, 1710. Dissertation by M. Bon, Sur l'utilite de la Soye des Arraignées, Latin and French, 1748.

different dyes. From this product, in the natural color, M. Bon obtained two or three pairs of stockings and gloves of an elegant gray color, which were presented as samples to the Academy. The pamphlet in which these novel results were made known attracted much attention.

In 1710 the Academy of Sciences of Paris deemed the question sufficiently important to investigate thoroughly, and accordingly commissioned the eminent entomologist Reaumur to prepare a report upon the invention of M. Bon. Reaumur took up and prosecuted the inquiry with much intelligence and zeal, and came to the conclusion that the culture of spider silk could not be made a profitable industry in Europe, although he intimated that exotic species might repay further attempts. The difficulties which proved most formidable lay both in the maintenance of the animals and the nature of the silk product. He computed that more than half a million spiders (663,522) would be required to produce a pound of silk, and to procure natural insect food for this vast multitude appeared impossible. This obstacle, however, was partly overcome by the discovery that spiders would subsist upon chopped earthworms, and upon the soft ends or roots of feathers.

Reaumur's Results.

Then the solitary habit and indiscriminate voracity of the araneans presented serious difficulty. They could not be trusted together, or near one another, for unless separated by artificial barriers they waged ceaseless warfare, and great numbers were slain and eaten. This cannibalistic propensity immensely increased the difficulty of breeding and maintaining a spider plant. The supply of silk obtained from cocoons, moreover, is necessarily limited by the fact that they are not true cocoons, as spun by the larvæ of both sexes of insects, but egg bags woven by females alone.

Further, M. Reaumur decided that spider silk is greatly inferior in strength and substance, the silk worm producing a thread ninety times as strong proportionately. He also adjudged the advantage to be with the insect silk in lustre. In both these points the spider product seemed unavailable for weaving cloth.

A half century after Bon's attempt, A. D. 1762, the Abbe Raymond de Termeyer, a Spaniard, took up the matter, and for more than thirty years (1762–1796) pressed his investigations and experiments with admirable ingenuity and persistence, not only in Europe, but in South America with the large fauna of that continent. He invented a method of confining the spider while he reeled off the extruded silk; but his experiments brought the establishment of a profitable industry in spider silk no nearer solution than M. Bon had done. The whole amount of thread obtained in all his experiments did not exceed fifteen pounds. Perhaps had he not been called away from America, his most promising field, by what he terms "an unexpected command and an irresistible power," we might have chronicled

Abbe Termeyer's Experiments.

a more favorable issue.[1]   We are inclined to sigh with the disappointed
enthusiast: "What a pity, and what a loss!"   Termeyer's line of inquiry
differed from Bon's, in that he took his silk principally from the living
subject, while Bon wrought from cocoons.

Termeyer's attempt to reel silk was suggested by observing the manner
in which the spider extruded its spinningwork when swathing a fly.   His
contrivance consisted of three parts: first, a body rest (Fig. 51), consisting
of a piece of cork slightly hollowed in the centre,[2] and supported upon
a pedestal; second, a foil consisting of a bit of tinned iron (Fig. 51, b),
about an inch wide, having a curved notch in the bottom corresponding with the cavity in the cork, on either side of which were soldered
two iron pins or wires (c, c) which were introduced into the cork.   The
spider was placed upon the body rest, as shown at Fig. 53, so that the
foil falling between the corselet and the abdomen kept the spider in position, and withheld the legs
from interfering with the
threads.

Fig. 51.   Fig. 52.   Fig. 53.

Abbe Termeyer's apparatus for reeling silk from living spiders.

Fig. 51. Body rest and foil.   Fig. 52. Reel.   Fig. 53. Spider in
attitude for yielding to the reel.

When about to reel the
silk the Abbe gave his captive a fly, which was seized
with the feet and jaws,
and at the same time the
spinnerets were opened by
unconscious association of
ideas, and threads thrown
out as if to swathe the fly.
The end of this filament
was then attached to a
small reel four and a half
inches in diameter, with cylindrical arms of glass. (Fig. 52.) This was
slowly turned and the silk wound off, as with the silk moth's cocoon.
Indeed, Termeyer wound upon the same reel a band of spider's silk and
a similar band of silk worm's silk, of which he remarks that the comparison shows evidently how much more brilliant and beautiful the first is
than the second, so bright that it appears more like a polished metal or
mirror than like silk.

In more recent times the whole subject has been gone over by Professor
Wilder, now of Cornell University, New York.   For several years, in various

[1] Raimondo Maria de Termeyer, Ricerche e Sperimenti Sulla Seta de Ragni, Milan.
(Astor Library, New York.)   We are indebted to Dr. Bert G. Wilder for a translation of
Termeyer's report of his experiments, which he published in Proc. Essex Institute, Salem,
Mass., 1867.

[2] I have ventured here to insert this cavity, which is lacking in Termeyer's sketch, as
reproduced by Wilder.

scientific and popular publications,[1] he communicated many interesting observations, and advocated with much enthusiasm the possibility of establishing a new silk industry from the spinningwork of some of our American spider species. He was led into his investigations while serving as an army surgeon (August, 1863) in the war against the Southern rebellion, and with especial view to furnish suitable employment for the multitude of negro slaves who had been launched upon liberty by the rude force of war, and without the responsibility and occupations demanded for prosperous freedmen. While encamped in South Carolina his attention was arrested by the remarkable spinning qualities of a species of Nephila[2] which inhabits the Carolina sea islands and Florida. He invented an ingenious apparatus for reeling off silk from the spinnerets, and better adapted to the long cylindrical abdomen of

Wilder's Experiments.

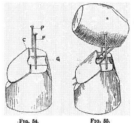

Fig. 54.   Fig. 55.
Professor Wilder's apparatus for reeling spider silk.
Fig. 54. The body rest.   Fig. 55. The foil cork in place.

Nephila than that of Abbe Termeyer, whose method he was quite ignorant of until three years later, but which he then studied and gave to the general public.

The first specimen from which Professor Wilder tried to reel silk remained quiet under the process for an hour and a quarter, and until he had obtained one hundred and fifty yards of thread; but its successors were less complaisant. He accordingly contrived an apparatus substantially like Termeyer's, which also served the double purpose of keeping the animal in an immovable position, and preventing her from cutting the extruding thread with her feet. The contrivance consisted of two large corks, a bent hairpin, two large toilet pins, a bit of card, and a bit of lead. One cork served as a body rest, and the bottom was loaded with the lead, one half its top beveled off at an angle of 45°, and the card (Fig. 54, c) fixed upon the oblique surface so that its upper edge projected an eighth of an inch. Into the horizontal half of the cork was cut a shallow groove (g), on either side of which were stuck two pins (p, p) about an inch apart.

The second cork served as a foil; it was rounded and smoothed at the smaller end, and a hairpin pushed obliquely through the lower corner of

[1] Proceed. Amer. Assoc. Advnct. Science, 1865; Proc. Boston Soc. Nat. Hist., Oct. 1865; How My New Acquaintances Spin, Atlantic Monthly, August, 1866; Two Hundred Thousand Spiders, Harper's Magazine, March, 1867; The Practical View of Spiders' Silk, The Galaxy (an extinct magazine), July, 1869.
[2] See Vol. I., page 146, and figures.

the larger end, so as to form an angle of 45° with the lower side. About a quarter or third of an inch from the cork both sides of the pin were bent outward so as to double the space between them. In use, the foil was so placed that the prongs of the hairpin passed underneath the card, on the beveled face of the basal cork, and the upper cork itself rested upon the heads of the common pins therein. Then the spider was laid upside down into the groove, so that the projecting anterior part of the abdomen brought up against the edge of the card, and the legs were in front of the pins. Next, the foil was pushed gently down, the prongs passing under the card until the narrow part near the cork embraced the spider's pedicle. The legs, being then set free, clasped the foil, which thus effectually withheld them from the spinnerets.

The natural tendency of the spider being to throw up its spinnerets against the top cork and make an attachment disk, a thread was thus

obtained from which to begin reeling. Dr. Wilder states that Nephila could retard the flow of silk by pressing the spinnerets against one another, but says that if the reeling is regular she cannot wholly prevent it. He suggests the use of some anæsthetic for the silk worm, to permit direct reeling of thread from the mouth tube, but apparently did not think of rendering Nephila complaisant and tractable by similar treatment. He suggests methods for reeling silk from several spiders at a time, but seems not to have tried the experiment. He computes that one spider will yield at successive reelings one grain of thread,[1]

FIG. 56. Nephila in position for reeling silk. (After Wilder.)

and that four hundred and fifty would be required to yield one yard of silk, or fifty-four hundred for an ordinary dress pattern of twelve yards; this is less than half the amount produced by the same number of silk moth larvæ, a comparison which corresponds substantially with that of Reaumur, although the two men greatly differ in their estimate of the number of spiders required to obtain a fixed amount of silk.

The most recent and apparently the most successful attempt to procure spider silk from culture I find described in a paper of M. Gautier on the habits of spiders.[2] He there informs us that an Englishman named

---

[1] This requires seven thousand spiders to the pound of avoirdupois of silk, which seems a much smaller estimate than Reaumur's. But no comparison can be made, since one estimates from reeling product, another from cocoons; one from Nephila, another from Epeira; and indeed one cannot always determine which method is referred to by the various writers.

[2] I have not the original, and quote from a translation printed in the New York *Sun*. In the summer of 1892 I tried to find some one in London who knew of this gentleman, but failed therein. The story, however, bears the appearance of authenticity.

Stillbers has made cloth of spider's silk which has been employed for purposes of surgery. He only uses tropical spiders, from which, thanks to a scientific culture, he has obtained a much greater return than **An Eng-** was foreseen by Reaumur. The spiders which he uses are large **lishman's** species from America and Africa. They are placed in octagonal **Attempts** cases, where a sufficiency of insects is served to them every day. In the room where the cases are kept a constant temperature of 60° Fahrenheit is maintained, and a liquid composed of chloroform, ether, and fusel oil is allowed slowly, to evaporate. That is to say, spiders spin best when they are under the influence of an anæsthetic, as Professor Wilder had suggested, the reason for which has heretofore been alluded to.

Mr. Stillbers is said to keep five thousand of these cases in a room forty yards long by twenty wide and five high. The spiders lay eggs of various colors, enclosed as usual with cocoons. These are gathered up and prepared by the same mechanical and chemical operations as the cocoons of the Bombyx moth. One cocoon yields one hundred and twenty to one hundred and fifty yards of thread by a process which is kept absolutely secret. The stuff obtained has a texture resembling ordinary silk, but thick, stiff, and a dirty color. It is all the more necessary to bleach it, because the color is by no means uniform. It is bleached by treatment with oxygenized water. Then it is tanned and softened, when it assumes a pretty yellow tint, and becomes brilliant and smooth.

To make a thread say a mile in length requires between forty and fifty cocoons. This is a great advance on Reaumur's calculations, but still falls far short of a practical industrial success. The stuff obtained must be sold at a very high price in order to obtain the merest compensation for all this trouble and expense. Thus, the last attempt at economizing the silk product of spiders returns to the method of Bon, to utilize the cocoon, and abandons the reeling process of Termeyer and Wilder.

CHAPTER V.

MOULTING HABITS OF SPIDERS.

I.

YOUNG spiders usually make the first moult within the cradle where the eggs have been laid. The young of Lycosa and Trochosa remain· in the cocoon until the second moult, after which they emerge and clamber upon the mother's back, where the third and fourth moults occur before the little fellows begin ·independent housekeeping in miniature burrows of their own. Wagner asserts[1] that the mother softens and partly tears the cocoon at its selvage, thus aiding the exit, and that without such help the little ones fail to escape, and die; a statement which I feel sure must be modified. Young Attoïds, having undergone the moult, shift their positions to the ·opposite end of the cocoon, and then moult a second and even third time before egress; as is shown by the fact that one finds within the same cocoon three separate heaps of skins cast at different ages.

Moulting of Young.

The subject of cannibalism within the cocoon has ·already been considered,[2] with the general conclusion that it is rare among spiderlings, but sometimes occurs. The Trochosas observed by Wagner would appear to be among the exceptions; for not only does the mother in captivity devour her broodlings, but the latter feed upon one another, a fact which is closely related to differences developed at the moulting period. At the end of two or three months a considerable difference in size appears among the young of the same brood; some are more vigorous and agile, others feeble, and those weaklings commonly fall a prey to their stronger fellows.

Cannibalism.

The cause of this inequality is traced to the fact that the eggs do not all hatch at one time, and that a whole day or more may intervene between the hatching of one division and another. In general, one remarks in a cocoon, at the second moulting period, one group whose individuals lack two or three days of the time, others ·on the eve' thereof, and still others in the act of moulting. This circumstance alone will explain why, after two moults, the stronger spiderlings are able to overcome and eat the feebler ones. It would thus seem that only the more vigorous enter upon independent life, while the feebler or those which come more tardily from the egg contribute to the perpetuation of the species by yielding

---

[1] La Mue, page 344.　　　　[2] Vol. II., page 200.

themselves to nourish the stronger. To this end also serve the infertile or undeveloped eggs, of which there are usually more or less in most cocoons.

II.

We may pursue in detail the habits of a few species during their first moult. A brood of Epeiras that much interested me appeared upon the honeysuckle vines in my manse yard on the morning of May 19th. **Young Epeiras.** They were then assembled beneath a large leaf which formed the roof of a little room of clustered foliage. (See Fig. 57, central group.) The assemblage was in a hemispherical mass an inch and a half in diameter and three-fourths inch to an inch thick. The entire outward opening of the cavity in which the spiderlings were gathered was filled with a rather closely spun tissue of silk lines, which extended downward for several inches, attached intermediately at several points to other leaves and forming a hollow cone of spinningwork. In the evening, May 19th, the assemblage was broken up into several distinct groups that hung like bunches of tiny grapes at various points of the cavity.

May 20th was a showery day, and at one time there was as severe a downpour of rain as falls in this climate. I feared the effect of such a torrent upon the baby spiders, but found that they stood the shower with no apparent inconvenience. Fortunately, there was but little wind, or the lashing of the vines might have been more disastrous than the rain, which, however, had the effect of causing the spiderlings to break up their separate groups and reassemble again into one ball. A few of the more adventurous spirits had separated themselves from the mass and were struggling with minute particles of moisture that beaded the defensive spinningwork, and appeared to be engaged in drinking. Indeed, the whole brood, consisting of several hundred individuals, seemed rather to enjoy than dislike the rain. No doubt at this stage water is necessary, or at least helpful, for their nourishment.

May 21st was a cold day for the season, and the spiderlings hung without any change in the assemblage above described. Towards evening, however, a few were engaged in shedding their skins, having detached themselves from the main mass for this purpose, and suspended themselves by lines that threaded the entire width of their dwelling.

May 22d, large numbers of the brood were engaged in moulting. By the use of an ordinary pocket lens the little fellows were seen pulling off their tiny coats, which they did in a few minutes without diffi- **Mode of Moulting.** culty, leaving the moults suspended upon the lines. They themselves came out looking bright and fresh, the abdomen a clear yellow, the fore part of the body transparent white. By noon the leafy domicile was filled with the grayish white skins which had been shed, giving the whole affair the appearance of the manse yard when the laundry has yielded its stock of white goods upon washing day. Observation of this brood shows that there is practically no difference between the habits of young Epeiroid spiders in the United States and those of the baby Argiopes so well described by Mr. Pollock.[1]

By five o'clock in the evening the entire cavity was filled with the little creatures who had gradually separated from the mass in order to cast their skins. Great numbers of grayish white moults occupied the lines directly in front of the opening, to which point many of the spiderlings preferred to come for their moulting. (See Fig. 57.) By standing upon a chair and using a lens I could see the entire process in various individuals. The feet were thrust out, upwards, grasping the supporting lines, the abdomen doubled up until it was almost at right angles with the cephalothorax, and sustained by threads outgoing from the spinnerets. The skin of the cephalothorax as it cracked open and escaped was so transparent that in the light of the setting sun it glistened like silver. The legs were gradually disengaged by slight regular movements, and issued white and transparent. At various points in the nest this process could be seen in divers stages of completion.

This interesting colony remained in or near the original place of assemblage for a week, during which time they migrated to nearby parts of the vine, forming thus several separate groups. From these they gradually, but rapidly at the last, spun themselves away and disappeared by aeronautic flight. I saw only one case of cannibalism in the entire brood; one individual was seen feeding on the carcass of a comrade, which it may or may not have slain.

Disper-
sion.

Mrs. Mary Treat[2] has related the moulting manners of a brood of young Turret spiders, Lycosa arenicola. When they were two weeks old these spiderlings strung innumerable lines of web across the mother's back, upon which they disposed of the castoff skins of their moults. Up to this time they had been massed upon her abdomen, as well as upon her cephalothorax, but then the little creatures, as if by common consent, entirely forsook the abdomen as a resting place and devoted it to the uses of a dressingroom. Sometimes two or three were divesting themselves at the same time. They fastened themselves by a short thread to one of the lines strung across the mother's back, and this held them firmly while they undressed. The skin cracked all around the cephalothorax and was held only by the front edge; next the abdomen was freed, and then came the struggle to free the legs. The little one worked and kicked vigorously and seemed to have no easy task, but came out of the old dress in about fifteen minutes, although exhausted and almost lifeless. However, it was soon as bright and active as before.

Young
Turret
Spiders.

[1] Ann. Mag. Nat. Hist., 1865, page 460. See also Vol. II., page 228.
[2] Home Studies in Nature.

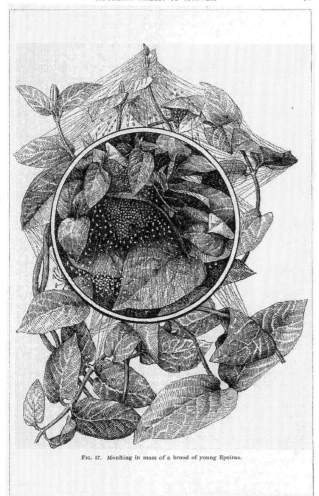

Fig. 57. Moulting in mass of a brood of young Epeiras.

It was fully a week before all the brood had moulted, and it was an odd appearance that the mother presented, with the innumerable little ragged castoff dresses hanging all over the lines upon her abdomen. In these broods Mrs. Treat never observed any tendency towards fratricide and intentional cannibalism, although she records that in the absence of other food the mother crushed some of her young and held them so that the little cannibals could suck the juices. The killing of the young may, however, have been simply accidental.

A somewhat similar phenomenon may be observed in the moulting of young Dolomede spiders. Within the large nest of woven and thatched leaves made by the mother,[1] the young spend the early period **Young Dolomedes.** of their life after issuing from the white cocoon, which is swung in the midst of the little leafy wigwam. Herein they string innumerable lines from wall to wall and from roof to floor, on which they sport and hang in groups, and in due time suspend themselves for the act of disrobing. One who peeps within the nest after this process, or after it has been abandoned by the brood, will see great numbers of castoff skins hanging to the network of interior lines.

The same fact may be observed in the case of the Theridioid young who are in the habit of remaining within the parental snare, or maze **Therid-ioids.** of crossed lines, for a period after hatching. They usually take a position in mass at the upper part of the snare, and thereon, when the first moults are made, they hang their rejected skins. In this habit there appears, indeed, to be little difference among broods of young spiders. The place in which they happen to be when Nature compels a moult is the place in which the phenomenon occurs. No doubt, the spiders that remain for the first moults in the nest provided by parental instinct must have a better chance for life, as against the exigencies of weather, than those which, like the Orbweavers, seek their own moulting domicile and shed their skins in any available locality.

III.

Blackwall expresses the opinion, after having frequently witnessed the moulting of spiders in their natural haunts as well as in captivity, and **Manner of Moulting.** having examined the cast skins of numerous species belonging to more than sixteen genera, ranging through all the tribes except the Territelariæ, that the process of moulting is substantially uniform among all kinds of spiders.[2] Wagner's observations have led him to the same conclusion, which I am able also to confirm from observation of the act and study of castoff skins, including therein the Territelariæ, whose moulting I have seen in several individuals and found to follow the general rule.

---

[1] See Vol. II., page 145, Fig. 177.   [2] Researches in Zoology, page 308.

Orbweavers moult as follows :[1] Preparatory to casting its skin the spider spins several strong lines in the vicinity of its snare, from which it suspends itself by the feet and a thread held in the spinners. After remaining for a short time in this situation the covering of the cephalothorax gives way laterally, disuniting immediately above the insertion of the falces and legs, so that the head and thorax are the first parts liberated.

The line of separation pursues the same direction, passing midway through the pedicle until it extends to the abdomen, which is next disengaged. As the thread held by the spinnerets is usually shorter Orbweav- than the legs and undergoes little alteration in length, the ab-
er's domen is gradually deflected from its horizontal to a vertical Moulting. position, nearly at right angles with the cephalothorax. By this change, attended with numerous contortions or undulatory movements of the body, the spider frees the abdomen, which falls back in a wrinkled saclike mass united to the dorsum of the cephalothorax by the upper half of the tegument of the pedicle.

The legs are the last and most difficult to detach, but are drawn out downward and usually entire by successive muscular contractions and strains, with brief intervals of rest. Blackwall thinks that the spines with which the legs are provided facilitate the operation; for, as they are directed down the limbs and are movable at the will of the animal, when it has partially withdrawn the legs from their sheaths by contracting them, it can prevent them from reëntering by slightly erecting the spines and thus bringing their extremities in contact with the inner surface of the integument. As the spines also and simultaneously with the legs undergo moulting, it may be doubted if their service in this respect is very great.

When the spider has completely disengaged itself from the sheath it remains for a short period relaxed and exhausted, suspended solely by a After thread from the spinnerets. The entire process, as above de-
Moult. scribed, may be completed in about twenty minutes under normal conditions, but varies in length of time according to circumstances. After a short rest the spider adjusts its position, making itself more secure upon the suspensory lines by seizing them with the feet; stretches its legs, bends and unbends them, passes them through the mouth, and hangs in repose until its strength is sufficiently restored and its limbs have acquired the requisite firmness, when it ascends its filaments and seeks its nest or retreat, or takes position upon its snare.[2]

---

[1] This detailed description is that of an Epeiroid, and is made from my own observations combined with those of Blackwall, Wagner, and others.

[2] Blackwall, Researches in Zoology, pages 306, 307; British Spiders, Introduction, page 7; Id., Ann. and Mag. Nat. Hist., XV., 1845, page 230; Id., Trans. Linn. Soc., XVI., pages 482–484; Wagner, La Mue des Araignées, page 284.

An example or two of moulting as seen in special individuals will serve to define more clearly the above general description. A nearly mature female of Argiope cophinaria was observed (August 19th) in the final stage of moulting. When first seen she was suspended head downward to the central shield of her snare, as represented in Fig. 58. The cephalothorax had already escaped from the shell, and the dorsal part of the moult still clung to the pedicle and stood straight out at right angles to the body. The abdomen was just ready to escape, and, indeed, slipped out of the shell as I approached, and the skin lay in a rumpled mass at the end of the thread by which the creature was suspended.

Argiope Moulting.

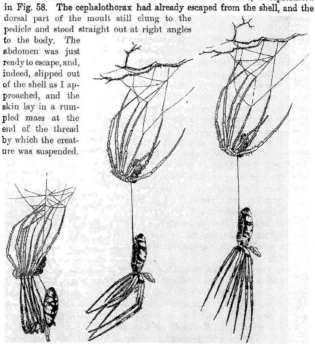

Fig. 58.   Fig. 59.   Fig. 60.
Fig. 58. Argiope in the last act of moulting. Figs. 59 and 60. Argiope stretching her legs just after moulting.

The body was bent upwards in a horseshoe shape, and the legs were partly freed from their moult. A few paroxysms occurred by which the legs were forced further and yet further out of the skin; then, first escaped the first pair, then, in a very brief space thereafter, the two second legs; immediately the third pair followed, and in brief succession the fourth pair. The spider's body dropped downward, and she stretched herself as though finding the sort of relief that a human being does when he yawns. (Fig. 59.) The limbs were finally extended to their utmost tension, the respective legs of each pair being precisely opposite to and a little separated from each other. (Fig. 60.) Shortly thereafter, still maintaining this parallelism of the several pairs, the hind legs were elevated, and then successively the others, until they were all a little more widely separated than represented in the figure. After a few minutes repose in this position the legs were doubled up, and the feet placed in a little circle upon the mouth organs, as represented at Fig. 61. The colors of the body were almost the same after as before moulting, only fresher and brighter, with the exception of the palps, which were nearly destitute of color and almost transparent.

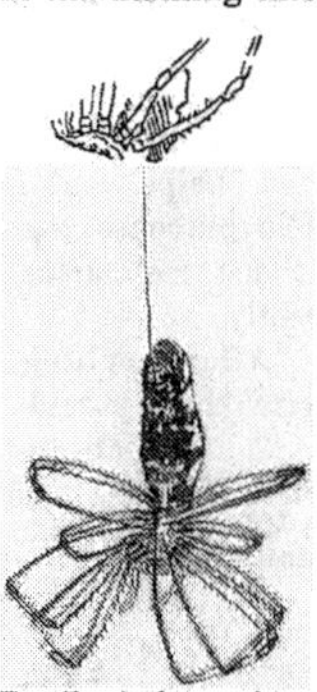

A female of the same species was found (September 6th) just after moulting. A rudimentary web had been constructed consisting simply of the characteristic central space, although the silken shield was but slightly marked, and an irregular line of straggling thick white silk represented the usual zigzag ribbon beneath. This was suspended among

FIG. 61. Argiope moistening her feet after moulting.

surrounding grasses and weeds by several radii so that it remained quite firm, the whole structure being about three inches long and two wide. The cast skin was attached to the upper part of this moulting frame, the feet being turned upward, the claws holding to the upper lines of the notched zone. The corselet skin swung backwards, showing that the spider had come out in that way by pulling downward. She herself was hanging to the lower portion of the moulting frame in the usual manner, her feet attached to the shield below the moult. The rings upon the legs

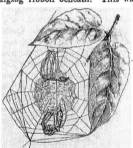

FIG. 62. Argiope resting after moulting, suspended to her silken shield.

showed white as they did in the moult. The animal when fully mature does not show these rings, but the tarsus and metatarsus are generally a uniform black color.

IV.

A female Linyphia communis moulted as follows: The spider was trussed upon threads stretched across a paper box in which she was confined, her body at an angle of about 45°, the abdomen apparently resting upon the bottom of the box. The fore legs were extended forward so that the feet came well together, each leg being fastened upon a thread. The cephalothorax and head were bolstered against several cross lines. When the observation began the skin of the corselet was loose and cast back above the abdomen. The spider was in the act of moulting the face below the eyes and falces. This was done by a succession of regular motions by which the body was pressed backward against the moulting frame, and the skin of the legs, palps, and falces at each swell was pushed a little more forward. The purchase was less as the skin was more and more rejected, and at the metatarsus and tarsus the legs were extricated by pulling them gently.

Other Tribes— Linyphia.

After moulting the legs were folded together under the sternum, and were then passed through the mouth in the manner of spiders when cleansing the hairy armature of the legs. The new skin looked fresh and bright; the black rings upon the legs retained their hue; the flesh colored and brownish parts were whitish; the abdomen was little changed as to color. Shortly after moulting the spider turned over and assumed an upright position upon her moulting frame. When touched she kept quite still, decidedly in contrast with the normal habit of the species.

Behavior After Moulting.

The Medicinal spider (Tegenaria medicinalis) suspends itself beneath its web in order to moult. The skin divides at the edges of the cephalo-thorax, leaving the casts of the sternum and the three first pairs[1] of legs, together with the mouth parts, on one side; the shield, abdomen, and last pair of legs on the other. In the cast skin the corselet is thrown backward and downward, and is surrounded by the hind pair of legs. It is united to the abdomen, which is represented by an irregular mass of black skin, the softness of that organ preventing it from maintaining the firm outlines of the other parts, which more resemble the shell of true insects. The skins of the first three pairs of legs are thrown forward, nearly or quite touching above the face, as when one throws his arms over his head.[1] This represents the position maintained during the act of casting the skin.

Medicinal Spider's Moult.

With Trochosa singoriensis the rejected dorsum of the cephalothorax is held to the abdomen by the skin of the pedicle, which is rent longitudi-nally into two nearly equal parts. The upper part unites the corselet to the abdomen, and the lower part ties the sternum thereto. The cephalothorax parts along the edge above the inser-tion of the legs; the last pair of legs first escape from the old skin, then the third, and the others in order. In unsheathing the legs the spider finds a point of support in the legs themselves; that is, she supports

Lycosids. Tarentula.

---

[1] This may not be the rule, but is true of the case described.

herself upon the third pair in unsheathing the fourth, upon the second to free the third, and so on.

The abdomen does not commence to moult until after the cephalothorax. It is disengaged from the skin, without the aid of the legs, by means of contractions of the abdominal muscles, which produce undulatory movements of the skin in the direction from the cephalothorax toward the spinnerets. The cast skin of the abdomen is always much wrinkled, owing to its extreme softness and fineness, that permits it to fold up under pressure. In this saclike abdominal moult one finds the moulted lungs and glands, and fragments of the moult of the intestine and muscles. (Wagner.)

The skin of Mygale when cast is sometimes so little broken, as shown by Fig. 63, that by placing the corselet shell upon the sternum and pasting **Mygale Moult.** it down to the falces, a casual observer might think it a living creature. It will be seen from the cut that the abdomen has been withdrawn from the old teguemnt forward through the circular rent at the base, where it was united to the pedicle. Even the long spinnerets retain their habitual position curled upward along the apex. The abdominal skin of this spider is so much thicker than that of ordinary araneads, and withal is so heavily covered with strong hairs, that it more readily retains its usual form, instead of shrinking up in a wrinkled mass, as with most species. Sometimes, however, the cast

FIG. 63. A cast skin of Mygale, showing slight rupture of parts.

is not so complete as here shown. In this figure the line of rupture along the sides of the cephalothorax is well shown; also the usual mode in which the pedicle is parted, uniting the abdomen on the one hand and the corselet on the other to the sternum. The mandibles have evidently been withdrawn backward by a motion the reverse of the abdomen, as shown by the unbroken moult fallen forward upon the moult of the mouth parts which adhere to the sternum.

V.

Some spiders issue from the eggs with their feet free; others, as Epeïroids and Theridioids, come out having their feet adhering under the abdomen. They remain thus for six days, more or less, when they cast the first tegument and quit the cocoon.[1] The skin thus enclosing the legs is

---

[1] Simon, Histoire Naturelle des Araignées.

in fact the vitelline membrane, and deliverance therefrom constitutes the first moult.

The number of times that spiders change their skin before they become adult is not uniformly the same as regards every species, not even perhaps the same within the individuals of any one species. **Periodicity of Moulting.** For example, Blackwall made some careful observations upon the frequency of moulting in a young female of Zilla x-notata[1] and Epeira diademata, which I arrange for sake of comparison in the following tabulated form:—

	ZILLA X-NOTATA.	Days Interval.	EPEIRA DIADEMATA.	Days Interval.
Disengaged from egg	March 30th,		April 14th,	
1. Moulted in cocoon	April 8th,	9	April 24th,	10
Quitted cocoon	May 1st,		May 3d,	
2. Moulted	June 4th,	54[2]		
3. Moulted	June 22d,	18	June 21st,	50[2]
4. Moulted	July 12th,	20	July 10th,	19
5. Moulted	August 4th,	23	August 3d,	24

There were five moults in each case; the intervals between disengagement from the egg and the first moult were about the same; the Epeira remained much longer in the cocoon than the Zilla, perhaps for some local reason; its next moult, the first after emerging, was not recorded. But there was evidently an irregularity during the interval from moult 1 to moult 3 where the record can again be compared. However, if we sum up the comparison from the dates of leaving the cocoons, the moulting intervals of the two species are of nearly equal length. Perhaps this table represents fairly enough the normal periodicity of moulting with Orbweavers.

With Lycosids (Trochosa), according to Wagner, the second skin is rejected in about six days, at a time when the younglings dwell in part within the cocoon and partly on the mother's back. The third **Lycosids. Trochosa.** moulting occurs six to seven days thereafter upon the mother's body. The fourth moult occurs seven to eight days after the third, partly on the mother's body and partly in the maternal burrow. After this moult the younglings leave the burrow and begin independent life. At this time the spiderlings have attained more than one-tenth their normal size, and have before them a series of moults amounting to ten in all.[3] The time required for full development and for completion of all the moults is from one hundred and sixty to one hundred and ninety-five days, excluding winter months.

The periodicity and the safety of moulting are modified by various

---

[1] Epeira calophylla Blackw.

[2] There is probably a break here in the observation, and a loss of several moults.

[3] Wagner, Note on the Tarentula, Comtes rendus de la Sect. Zool. de la Soc. Imp. des Sc. Nat. de Moscow, 1866.

conditions. Blackwall had already noticed that food and temperature exercise a decided influence,[1] and Wagner has confirmed the fact. Moulting is suspended in winter in natural site; but if spiders be housed in warm rooms the moult may be artificially stimulated, but the moulting intervals are then longer than usual in natural condition. If a young brood of the same age be exposed to different degrees of heat they will shed their skins at intervals corresponding thereto, the warmer ones earlier, the colder later. The lack of sufficient nourishment retards the moulting epoch, and tends to make the act more difficult and dangerous, so that many spiders die in or after the act from inanition.

Modifying Agents.

Causes affecting the normal health of the organism modify the moult. Blackwall discovered that young spiders infested by the larva of Polysphincta carbonaria, an insect belonging to the Ichneumonidæ which feeds upon their fluids, never moult.[2] Wagner notes the effect of the prick of a Pompilus sting upon two male Trochosas; one stung July 8th remained sick and languid until August 7th, an unusually long period, and then moulted. During the act, probably for lack of vigor, the legs were contorted and deprived of motion. Another male, stung at the same time, passed an interval of a month and ten days before moulting, a great retardation as compared with subjects of his own age who had long before that shed their skins. This spider began moulting August 17th, and on September 2d, when it was moribund, it had only achieved the moult of the abdomen and corselet. The legs were withdrawn from the old skin in the morning and appeared to be normal, but in the evening they were bent up and flattened, doubtless the result of imperfect alimentation during the two months succeeding the sting, and of the imperfection of the interior moult.

Pompilus Sting.

Neither of these wounded spiders made any preparation for moulting by stretching supporting frames of silk lines. The normal conduct and periodicity of moulting appears thus to depend upon three kinds of agents: first, the interior conditions of the animal's development; second, the expenditure of the reserve for the maintenance of internal heat and locomotion; and, third, external conditions, such as heat and food.[3]

VI.

After moulting, all spiders, old as well as young, are in a state of greater or less feebleness, proportionate to the difficulty and length of the act. They hang in a relaxed and helpless condition upon their proper snares, if sedentary, or upon temporary scaffolds woven as supports; one

[1] Zoological Researches, page 309.
[2] Brit. Assn. Advanct. Sci., 14th meeting, pages 70, 71.
[3] Wagner, La Mue, page 357.

may even touch them at that time without responsive signs of life. Of course they are then at the mercy of their enemies, to whom, in point of fact, they do fall an easy prey in great numbers.

One would naturally expect from what has elsewhere[1] been written of protective habits, that Nature, intent upon the preservation of the species, would provide some recourse against this peril. Accordingly we find that, upon the approach of the moulting period, many spiders prepare for the emergency; some creep away into crevices of rocks, crannies of walls and fallen wood, hollows of trees and stumps, underneath loose bark, stones, and like sheltered positions. Many species overlap and join together the edges of leaves, and moult within the tent thus formed; some preëmpt the cocoons or lodges of other spiders, and some appropriate the nests of sundry insects.

The Attoïds moult within the silken cells which are the characteristic dwellings of the family and make no other provision therefor. Curiously, they have a fancy for cells other than their own, and for old rather than new. They freely avail themselves of strange cells, and the ones which appear most to please them are the abandoned nests within which the females have laid their eggs, after the young have quitted them. Hence, one will sometimes find the shed skin of one or even two vagabond males inside such sites.

Protective Habits.

Tubeweavers seek the funneled part of their snare and moult beneath the outspread curtain, as do also the Linyphiæ. Lycosids moult within their burrows, which they previously close, showing thus the same sense of need that leads them to cover their nests in winter and the cocooning season.

Orbweavers do not have the same degree of secretiveness at this period; at least many of them are found moulting upon their snares, as shown in the various accompanying figures, with no special provision for concealment or protection. Often they do not even get behind their webs or seek the shelter of adjacent foliage. The Lineweavers also moult upon their webs, but then their position underneath and within their maze of crossed lines, as with Theridioids and with Pholcus, would in itself seem to be a good protection. As a general but not invariable rule it may be said that spiders having a fixed abode, as all the sedentary groups of Lycosids that live in burrows, many Attoïds, etc., cast their skins on or in their snare or lodge. But those species which have no fixed dwelling seek divers shelters for moulting.

All the burrowing spiders observed by Mrs. Treat closed their dwellings just before they moulted, and before making their cocoons. When this work was over they cut the threads and threw the covers back, sometimes entirely severing them. At other times a sort of hinge was left on one

----

[1] Vol. II., page 407.

side and the door fell back, keeping an attachment to the wall.[1] This is especially the case with Lycosa tigrina.

Wagner confirms this observation in the case of the European tarantula, which closes its burrow with a sort of pent house above the door, and suspending itself to the sides thereof, head downward, passes its moult. In captivity these preparations change in their details, but their aim remains the same, and is always attempted in one way or another.[2]

It seems probable, as suggested by Wagner, that the nature and extent of these precautions depend upon the facility with which the moult is accomplished; the greater the facility the less the precaution.

Facility in Moulting. For example, young spiders appear to experience little or no difficulty in shedding their coats, which they do in a few minutes—young Orbweavers in from three to ten minutes, young Trochosas in two minutes. As the spiders advance in age the succeeding moults are passed with increasing difficulty, the last moult being often the hardest to achieve. Now, young spiders make no preparations and take no precautions in moulting, and drop their skins wherever they chance to be, a carelessness which disappears with the approach of adult life.

With Young Spiders. Thus Wagner records that the young of Trochosa singoriensis make no protective defenses during the early period of life, when they moult easily; but as the act is made more difficult and protracted by advancing age, they cover their burrows when they feel the moulting period coming on.

The Thomisoïds quite generally shed their skins easily and rapidly, and, accordingly, they do it openly, only spinning a supporting thread over the petals of a flower or the surface of a leaf. Even in the case of some young spiders one may see evidence of the same sensitiveness to danger, for if the young of Attus terrebratus, who are in the habit of moulting en masse within the maternal cocoon, be removed after the second moult and put in a suitable place, every one will spin a little cell within which the third moult is separately made. This quick perception of the change of condition and ready adaptation thereto is justly noted by Wagner, who relates it as a fine example of instinctive wisdom.

My tarantula "Leidy," distinguished by having reached the greatest age of any spider known to science, finally died in the act of moulting when more than seven years old. Its death is another example of a fact which I had previously observed, that the act of moulting is frequently attended by dangers of one kind or another to spiders. It is common to find

---

[1] My Garden Pets, page 82.
[2] One needs to distinguish between the word tarantula, which is the popular name for the huge Mygale of our southwestern States, and the genus Tarentula of the Lycosids. The Turret spider (L. arenicola) and my Tarentula (Lycosa) tigrina are closely related in habits and structure to the famous "tarentula" of Italy, and the well known Tarentula (Lycosa) Narbonensis.

specimens without one or more limbs, also with distorted and abbreviated limbs. I have frequently found males lacking several legs. The theory commonly adopted is that in most of these cases the loss has

Moulting Dangers. resulted from conflicts, perhaps among rival lovers in attendance upon the same female. Something of loss may be attributed to this cause, but I am satisfied that in a much larger degree losses and malformations are due to the accidents of moulting.

One example I may cite, the loss of two limbs experienced by a large tarantula which I had kept under observation. This spider lay upon its back in the araneary during part of the time of moulting, and on its side during the remainder thereof. The skin was cast by a succession of movements of the body or parts of the body recurring at reg-

Limbs Lost in Moulting. ular intervals, reminding one of labor pains among mammals. For some reason two of the legs refused to separate from the skin, and after a prolonged struggle they were broken off at the coxæ, and remained within the moult. (See Fig. 64.) One foot of another

FIG. 64. Cast skin of Mygale, showing stumps of legs broken during moulting.

leg shared the same fate. This moult occurred in the spring; during the latter part of August of the same year the spider again moulted. The moult was a perfect cast of the animal, the skin, spines, claws, and the most delicate hairs showing, and their corresponding originals appeared bright and clean upon the spider. When the castoff skin

was removed the dissevered members were lacking thereon, but on the spider itself new limbs had appeared, perfect in shape but smaller than the corresponding ones on the opposite side of the body. The dissevered foot was also restored. The rudimentary legs had evidently been folded up within the coxæ, and appeared at once after the moult, rapidly filling out in a manner somewhat analogous to the expansion of wings of insects after emerging.[1]

It is possible that the tarantula " Leidy " was too much exhausted by long previous fasting to endure the severe strain upon the organism in the act of moulting, although judging from the disjecta membra of the

[1] See Proceedings Academy of Natural Sciences, Philadelphia, 1883, page 196.

skin recovered from the burrow it had succeeded in casting them all off without any mutilation. The spring of 1887 was a backward one, and I **Effect of** experienced great difficulty in procuring insects for food from **Nourish-** the immediate neighborhood. The annual supply of grasshop-**ment.** pers and locusts upon which I had relied came very late. Perhaps had the spider been strengthened by a few weeks generous feeding previous to its last moult it might still have been alive.

VII.

With each moult spiders undergo a change in color and patterns more or less decided. Some species have such neutral colors and are so uni-**Color** formly marked that the differences are not decided; but some **Changes.** undergo such decided changes that different species have been established for the same spider upon specimens taken after different moulting periods. In some species the colors and markings of the youngling, after the first moult or two, fairly represent the markings at maturity; in others the difference is so great between the two stages of life that it is quite impossible to identify young individuals, or distinguish the young of several species with accuracy.

Among the young of Lycosa and Attus, according to Wagner, these modifications are effected with the female and male so equally and uniformly during the first four or five moults, and with Trochosa during the first six or seven moults, that one is scarcely able to distinguish the sex. With the final moults these distinctions become more and more marked, though not always to the same degree. The differences in relative length of legs and in the shape of the palps also begin to appear; for example, the male Trochosa singoriensis at the seventh moult equals in body size and relative length of the legs those of the female at the sixth moult. The same is true of Attus.

Among Orbweavers generally, and in spiders of various tribes observed, the change in color (and in form also) is most decided in the males; that **Change in** is, the young male carries the typical colors and general shape **Males.** of the adult female; the younglings of both sexes after the initial moults resemble each other perfectly, and tend to resemble the adult female. Thus the young male of Dictyna philoteichous bears a close likeness in color and pattern to the adult female; but after final moult the difference in color is quite marked, as well as in shape of palps and contour of body.

Professor Peckham[1] finds a close resemblance between birds and spiders in their moulting changes, and his special studies of the Attidæ are

---

[1] Occasional Papers, Nat. Hist. Soc., Wisconsin, Vol. I., 1889, page 16, George W. and Elizabeth G. Peckham.

valuable and interesting. He concludes that when the adult male is more conspicuous than the adult female, the young of both sexes closely resemble the latter in form and color. On the contrary, when the female is more conspicuous the young follow the more modest colors of the male, especially in the earlier moults. When the adult sexes resemble one another the young of both sexes favor the common type.

Peck-ham's Studies. Attidæ.

As examples of the above, Phidippus johnsonii female has the abdomen red and black, with a white base and some white dots; the male is bright vermilion red, with sometimes a white band at the base. The young of both sexes resemble the less showy mother until the last moult, when the males assume their bright livery.

In another species, Habrocestum splendens, which the Peckhams illustrate with a good plate, the young during the first moults more closely resemble the female, which is the less showy sex. The male is a brilliant fellow, who dons his gorgeous livery at the last moult just as he becomes mature, though in some species the nuptial robe is acquired one moult before maturity.

Among the Laterigrades the same rule obtains, several species of Thomisids showing greater brilliancy of color among the adult males, while the young males resemble the female until the last moult.

Lateri-grades.

In Sparassus smaragdalus[1] the female has a deep green body and legs of somewhat lighter shade; the male[2] has green corselet and legs, but the entire dorsum of the abdomen yellow, with a wide herring-bone median stripe of red and the folium margined on each side with the same color.

FIG. 65.   FIG. 66.

FIG. 65. Immature male palp of Zilla atrica. FIG. 66. The same when mature.

The young at the first moult are a dull whity-yellow color and grayish legs, but in subsequent moults are said to resemble the mother.

Figs. 65 and 66 will illustrate the difference resulting from the final moult of male spiders generally. The drawings are made from a male Zilla atrica, Fig. 65 being the form shortly before the last moult and Fig. 66 that of the mature male. In some species the difference between mature and immature palps is much more striking.

It is not correct to say that these modifications are effected in the interval of the last moult alone. In point of fact the distinctions begin to appear earlier, but they are commonly so difficult to detect, and the apparent change effected during

[1] A fine female of this species with her cocoon and young was taken by Rev. W. F. Anderson, of Fordham, N. Y., on the mountains of Switzerland and brought to America safely. It was sent to me alive and lived several weeks. The young were all a dull yellowish color with livid legs, but I could not preserve them beyond the first moult.

[2] Blackwall, Spiders Gt. B. & I., Vol. II., pl. v., Fig. 61.

the last moult is relatively so much greater, that one reasonably comes to locate the transformation within the last moulting period. In some cases, however, the coloration of the sexes is more strongly differenced during the stages previous to final moult. Thus the color of male Trochosas during the two or three last moults is notably clearer than that of females. With Attus, in the corresponding periods, one may observe the sexual peculiarities in color distinctly appearing, and that they deviate more and more intensely with each moult, until the last fixes the distinction. Like the male, the female awaits the final moult for perfect development, at which time the genital cleft is freely opened and the hood and scapus assume those various forms which serve as valuable specific characters.

"Leidy," one of my captive tarantulas, shed its skin several times. The first moult occurred some time in August (1882). I had been absent on my usual summer vacation, and returning August 31st saw the animal lying on the soil about the middle of its araneary, with its feet gathered together, looking dull, gray, and faded out, apparently dead. I shook the globe. No responsive motion followed, and I left without more careful observation, concluding that the spider was dead. I was not able to visit it again until the fifth day of September following. I threw off the cover of the globe and put my hand in to take out the dead body, which lay apparently in the same position, in order to preserve it in alcohol. At my touch the animal leaped to its feet, and as I hastily withdrew my hand it presented itself quite changed in appearance. The body was a fresh bright color, the cephalothorax a clean whitish gray, the head and fangs dark brown. The abdomen was black, with brown hairs covering it. The legs were black, with yellowish brown hairs and spines. I at once understood that the spider when first seen was in the torpid condition which usually immediately precedes the act of moulting. In the interval between my visits it had cast off its skin, which I found lying in a tolerably complete condition on one side of the glass. Another tarantula, a male, which I received when quite young, came to me a dull reddish brown, but during successive moults at last appeared a bright black brown, almost black.

Change in Tarantulas.

VIII.

We may thus summarize the most important moulting phenomena, as above disclosed. 1. The first two moults of spiderlings occur within the cocoon or on the mother's back; several occur before entrance upon independent life, sometimes as many as four. 2. The individuals of a brood do not all moult at once, and those moulting first, having the greater strength, in some species feed upon the younger and weaker individuals; in these cases survival depends upon priority of moult. In other species cannibalism is absent or rare, and

Biological Summary.

early moulting gives no such advantage. 3. Young spiders moult rapidly and easily, and with little loss of vitality; therefore precautions against dangers incident to that period are scant or wanting. 4. The method of moulting is substantially the same in all species. 5. The head and thorax moult first; the old skin cleaves horizontally about midway of the latter, the shells escaping backward and upward and downward, the mouth parts adhering to the sternum moult; the pedicle splits longitudinally, holding the above parts severally to the upper and lower front of the abdomen. The abdomen next follows, the shell escaping backward entire; then follow the legs, which are withdrawn from the old sheaths downward by an interrupted series of muscular contractions and strains, usually escaping entire and in succession, beginning with the first pair.

Methods.

6. The number of moults varies according to species, from seven to ten being the most common limit; these are made at intervals more or less regular, and at corresponding periods of time in individuals of the same species or even genera. 7. The periodicity of moulting is modified by amount of food, by temperature, and by causes affecting the spider's normal health; it is wholly suspended by the presence within the body of a larval parasite and by the prick of a wasp's sting. 8. A period of relaxation and exhaustion, more extended and severe as age increases, follows the moult. 9. Many species protect themselves against this after-moult weakness by various precautions in accord with their social nesting habits, such as creeping into crannies and leaf-rolled dens, covering over burrows, et cetera. Sedentary spiders usually spin a moulting frame. These precautions are commonly cotemporaneous with the increasing difficulty of moulting.

Periodicity.

10. Changes in color and pattern, more or less decided, occur with the successive moults. 11. The young males and females are scarcely distinguishable just after the first moult, but the difference grows more distinct with successive moults until the last moult, when the animals are mature and the sexual characters distinctly marked; at this time the male form most widely diverges from that of the typical young. 12. The changes of skin are often attended with loss of limbs or parts thereof, and death sometimes results from inanition; the attendant weakness exposes the subjects to assaults of stronger congeners and alien enemies, so that moulting thus becomes a factor of danger to individual life, and so to the perpetuation of the species.

Changes.

IX.

We pass now from the biological phenomena of the moulting period to consider the physiological and histological processes and changes connected therewith. The most complete study of these has been made by Mr.

Waldemar Wagner, the Russian savant, whose skillful and patient researches have heretofore been quoted. I shall draw chiefly upon his work "La. Mue des Araignées" for the material which it serves me to present here. For full information the student is referred to Mr. Wagner's paper. It will sufficiently answer the purpose to give a general notice of the development of the process, and to illustrate these by quoting somewhat in detail, in the case of several organs, the genesis of the new skin and the rejection of the old.

Physiological Moulting Changes.

Ordinarily the matrix of spiders presents a layer of protoplasm with numberless cells. This is colored equally throughout, the cell, as usual, being more, the plasma less, intensely colored. As the moulting time approaches one ceases to observe this equality of coloration; the superior layers next the old cuticle are colored more and more feebly; and finally, at the moment when the contingent layer ceases to be colored and loses its granulation, and when, consequently, we may rightly consider it as transformed into a chitinogenous layer, then the future tegument begins slowly to separate from the old.

As the new skin is retracted from the old cuticle the interval thus formed is filled with liquid, in small quantity at first, but gradually increased with the enlargement of the cavity. The new tegument increases rapidly; within the cephalothorax, not being able to have full extension, it forms many folds (Fig. 67, f.n.t), destroying thereby the layer of the old

Forming New Skin

FIG. 67. f.n.t, folds of new tegument; Mt, matrix; section showing folds of new tegument under the old (o.t).

cuticle, which is drawn from it, and from which it is disengaged little by little. On the abdomen one does not observe these foldings of the new skin, because the old is there so pliable that it does not impede the growth of the new. At the moment when the new hairs are completely formed there remain of the old skin, separated from the new, only the tubes which serve as sheaths for the new hairs; and, on the day of the moult or the day before, these sheaths are destroyed, and the liquid underneath the tegument disappears.

The moult of the eyes takes place simultaneously with the other organs; the matrix in process of growth insinuates itself between the vitreous body and the preretinal envelope, the cells of which are in that way forced upward and lose their regular form. The moulting of all the eight eyes does not take place at once, but probably at different times. With Attus terrebratus, at least, the lateral eyes are the first to end their moult. For a period longer or shorter before the rejection of the old skin, according to the stage of development, spiders lose their sight, and after the moult vision is not restored all at once.

Eyes and Lungs.

During the moulting of the lungs breathing is difficult, but the time occupied therein is short. Two of the three layers which compose the tracheæ are shed, the broken parts of the old tube remaining among the silk threads of the moulting frame, and all glands formed by ectodermic invagination also lose their linings.

The pharynx, œsophagus, and rectum take part in the moult, as do also the tendons, especially the muscles of the limbs, the matrix growing around the old tendon and forming a new one, while the old one atrophies and is cast away with the tegument.

During moulting the number of spherical blood corpuscles, which usually is only three to four per cent. of the total number of corpuscles,
The Blood. increases to ten per cent., almost all the red corpuscles being transformed into spheres. Want of movement during the process seems to be one but not the sole cause of this change in the condition of the blood; and it must be remembered that a development of all the internal parts of the body takes place at the moulting period, so that the casting off of the integuments, etc., is really but a secondary act.

In the last stages of development, when the embryonic cuticle is secreted, according to Schimkevitch certain cells of the chitinogenous layer are augmented in dimensions, taking a spherical form and dispersing themselves under the chitinogenous layer, which forms above them something like a raised vault above the cuticle. These are the future hairs. The nuclei of these cells are of large size, but they are not very distinctly separated in contour from environing cells. Some time thereafter these cells increase in size, but their nuclei deviate proportionately less. These cells are already developed immediately under the cuticle; it is evident that in growing
Moulting of Hairs. they have pushed the chitinogenous cells against the sides. The contours of the trichogenous cells deviate more apparently, and the adjoining chitinogenous cells, from a contour ordinarily distinct, stretch out, taking a semilunar form and investing by their sides the trichogenous cells. The hair of spiders is a unicellular formation.[1]

Mr. Wagner describes the origin of a new hair in detail. At a certain epoch in the life of spiders more or less preceding the moult, according
Origin of New Hairs. to the various species, the inferior layer of the old skin, as we have already seen, is retracted from the others, and the interval thus formed is filled with liquid. The retraction is effected slowly. The sections made during that stage present the appearance of Fig. 70. The inferior layer of the cuticle (in.lr.ct.o, Fig. 70) is removed far from the other layers (ct.o), and the intervening space is filled with liquid (lq), which is so formed that it floats the tube which at the time has served to form the primitive hairs. The position in which one sees it in the figure is not the most common, but is quite rare; it is often seen bent up in one fashion or another next the seat of the hair.

---

[1] W. Schimkevitch, "Materiaux pour la connaisance du developpment des Araignées," Memoirs Acad. Sci., St. Petersburg, 1886, Suppl. ant. LII., No. 5.

While these things take place, one of the inferior trichogenous cells of the matrix (nw.hcl) augments in size and stretches itself in the direction of the hair tube (tu.oh) in the form of a light papilla (nw.hcl), which originates the new hair. Some time thereafter this papilla ensheaths itself within the cavity of the old hair tube, perforating the layer of the matrix (mtx) situated above it.

Sometimes, near the extremity of the new hair, as is shown in Fig. 68 (tu.oh), one sees also the unsheathed tip of the new hair (nw.hr) standing up freely, and the remainder

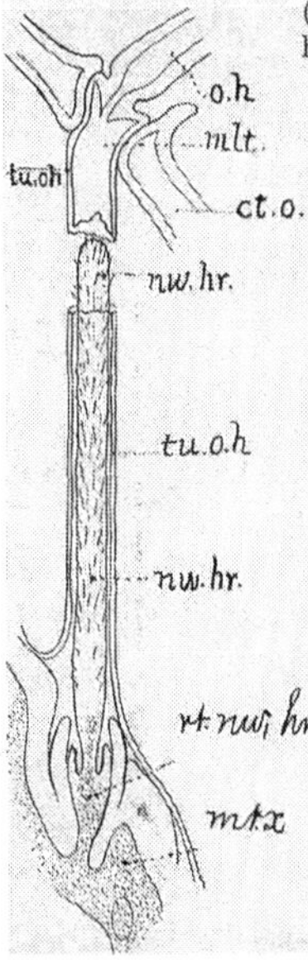

Fig. 68. Fig. 70.

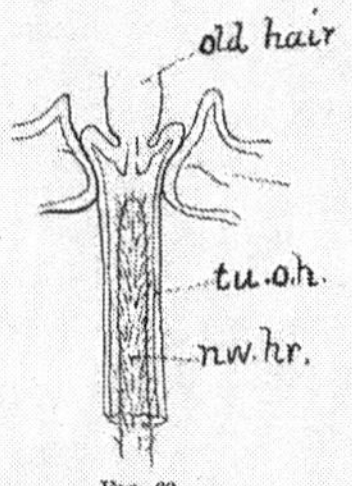

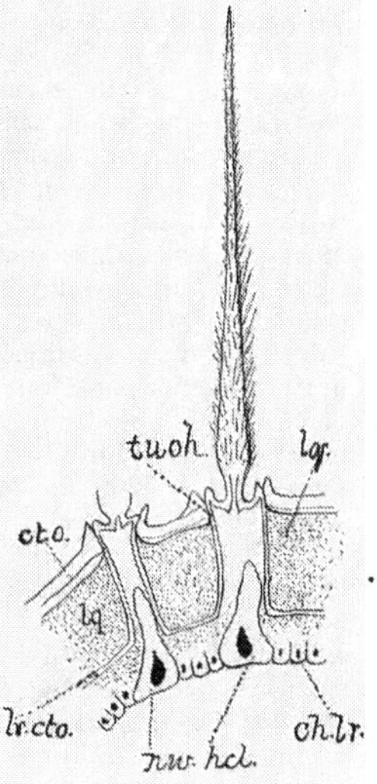

Fig. 68. New hair, nw.hr, shown within its old sheath, tu.oh, the tip exposed, and the base of the moult, mlt, pushed off. Fig. 69. Tip of new hair shown within the old sheath before moulting. Fig. 70. Section showing the origin of a new hair (after Wagner); in.lr.cto, inferior layer of old cuticle, ct.o; lq, liquid filling interval; tu.oh, tube or sheath of the old hair; ch.lr, chitinous layer; nw.hcl, new hair cell; mtx, matrix; rt.nw.hr, root of new hair. References uniform in the figures.

thereof enclosed within the sheath. At the same time is seen the pubescence upon the new hair, and its root (rt.nw.hr, Fig. 68). The relations of the tip of the hair to the old tegument are shown at Fig. 69. At this epoch the liquid (lq, Fig. 70), which filled the space between the old and new skin, disappears; afterward the sheath of the new hair breaks and is seen entirely formed under the old skin.

X.

The poison glands after the first moult are situated with Attus in the mandibles, and do not pass beyond those limits. Further on, as with the Lycosid (Trochosa), the inferior end of the gland is placed rather below the median line of the mandible. (Fig. 71, gl.v.)

Poison Gland.

Afterwards the superior parts of the gland issue beyond the mandible for a third of their length (Fig. 72), and their inferior extremities do not reach the median line of the mandible. Finally, after a series of moults, the gland is no longer found within the mandible, but extends wholly beyond it (Fig. 73), and is withdrawn farther and farther with every moult. At the same time the dimensions of the gland increase as well as the number of muscular fibres; a fact which gives us a measure of the phylogenetic development of any spider thus tested.

With the genus Epeira the poison glands are relatively large. They are placed in the anterior part of the cephalothorax at some distance from the mandibles, with which they are joined by quite a long conduit. With the Clubionidæ these glands are of a much smaller size than with Epeira, and only a part thereof is found within the cephalothorax, the remainder of the gland being situated within the basal joint of the mandible. With the Mygalidæ the whole gland is located within the basal joint. It will thus be seen that the modifications in the poison gland accomplished by successive moults in advancing age are, first, the increase in size; and, second, the change of position.

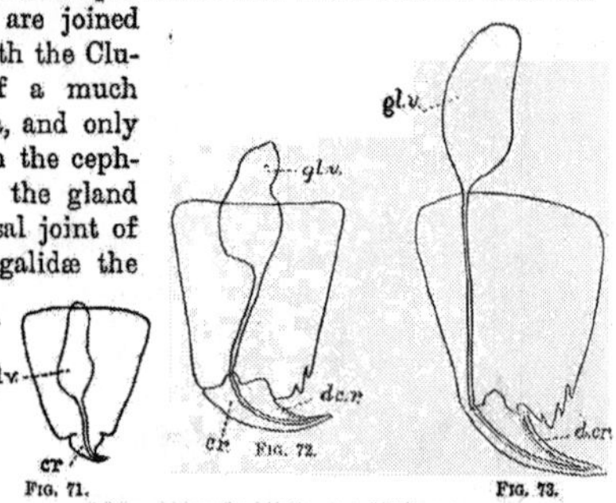

FIG. 71.   FIG. 72.   FIG. 73.

MOULTING OF POISON GLAND IN TROCHOSA.
FIG. 71. Venom gland, gl.v, at an early stage, inside the mandible; cr, the duct. FIG. 72. Gland at further stage. FIG. 73. Gland after final moult, withdrawn beyond the mandible.

A study of the rejected moult of a spider shows well the thickened points of insertion of the muscles of the abdomen, which in that organ are, characteristically, immediately upon the cuticle. Upon the inferior or ventral part of the abdominal moult, according to Wagner, there are two median rows of thickenings, consisting of sixteen pairs, a little removed from one another (Fig. 74, Nos. 1-16); then two rows of lateral thickenings. (Fig. 74, Nos. 17-39.) These rows consist of twenty-three pairs of thickenings disposed first (near to the lungs) irregularly, and then in lines almost straight and parallel to the two medians. In all, there are on the inferior face of the abdomen thirty-nine pairs of thickenings, representing the points of insertion. Besides

Abdominal Muscles.

these there is a large number of points of immediate insertion of the muscles upon the cuticle without thickening of the chitine at that point.

Like conformations are observed upon the dorsal face of the abdomen. These are near the local thickenings of the chitine at the points of inser-

tion of the dorso-ventral muscles, and in the interval comprised between the lines of the local thickenings upon the dorsal face and the lateral line of thickenings. (Fig. 74, Nos. 17 and 39.)

An examination of rejected teguments shows the existence of two forms of points of insertion, which differ in structure and size. One presents a thickening, more or less considerable, of a sculpture tuberculous upon the exterior (Fig. 75), and having a row of cells and walls more or less thick upon the interior. (Fig. 76.) Each of these cells serves for a point of immediate insertion upon a fasciculus of muscles, so that the number of cells indicates the number of fasciculi. The other mode of immediate insertion of the muscles shows no thickenings of the chitine; but one observes at these points upon the tegument certain

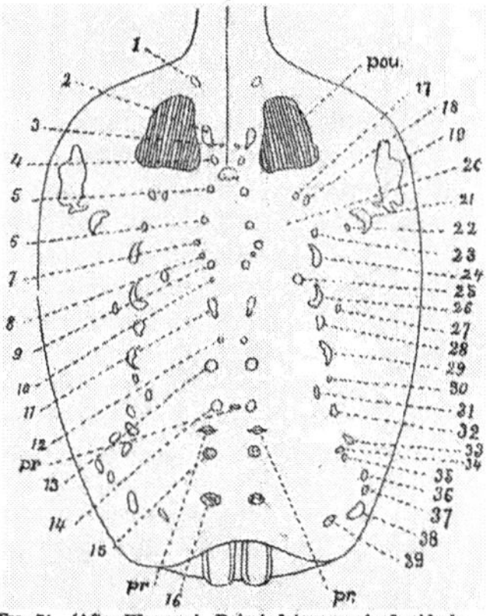

Fig. 74. (After Wagner.) Rejected tegument of spider's abdomen, ventral part, showing points for muscular attachment; o, genital orifice; pr, pores; 1-16, sixteen pairs of points of insertion of two median series of muscles; 17-39, twenty-three pairs of points of insertion for two lateral series; pou, gills.

linear thickenings, hardly distinct, which indicate the limits of the point of insertion of the muscles and of each of the fibres taking part.

One can hardly fail to note that the arrangement of these points of attachment indicates in a general way, and indeed with tolerable accuracy,

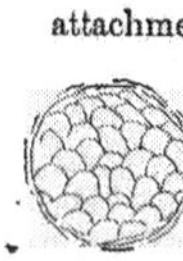

FIG. 75. FIG. 76.
POINTS OF INSERTION OF DORSO-VENTRAL MUSCLES.
FIG. 75. Exterior tuberculous sculpture, with thickening. FIG. 76. Interior cells, without thickening. (After Wagner.)

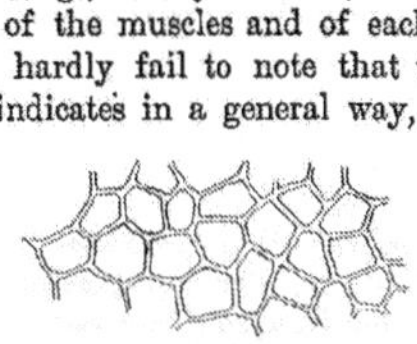

the outlines of the markings which distinguish these animals. For the most part these are so grouped as to form a folium or rude figure of a leaf with various irregularities, as scallops and dentations, upon the edge. The thickenings first described are always of a yellow color, more or less intense; a color

which is the dominant one in nearly all families of spiders showing decided hues. In the second form of attachment the cuticle is colorless and transparent. In both cases the skin above the points of insertion is without hairs, and the linear thickenings of the teguments, in approaching these points, surround them in concentric circles. I have heretofore suggested that all action of the muscles upon the abdominal teguments may have much influence in the distribution of color for the formation of various patterns which mark spiders.[1]

Fig. 77 represents the tarsus and metatarsus of a Lycosid foot (Tarentula), taken from the rejected skin. The tendons by which the tarsus and claw are moved are shown within the leg, and one sees the thickening of the tendons (e.t) at their free ends upon which the muscles are inserted. The claw and the entire dentition thereof, as may be seen, have left a perfect cast in the moulted teguments.

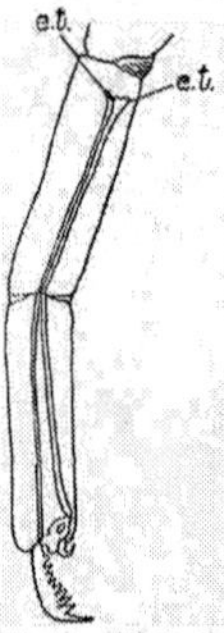

Tendons of Foot.

The tendon passes from the claw in the form of a thick double cord, traversing the tarsus apart and uniting at the articulation with the metatarsus, which joint is traversed nearly to the articulation with the tibia, where the cords join with the short stout muscles inserted into the cuticle at that point. The silk glands in moulting undergo changes both in form and number, as with the tubuliform glands of tarentula shown at Fig. 78, after the second and third moults. Three different forms of glands appear (gl 1, gl 2, gl 3), corresponding possibly with the ampullate, tubuliform, and pyriform glands of Epeira.

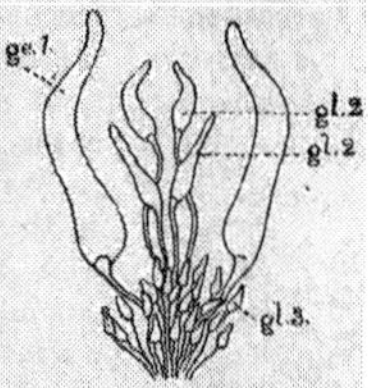

Silk Glands.

FIG. 77. Moult of final joints of a Lycosid foot, showing the tendons (e.t) that move the tarsus and claw.

FIG. 78. Silk glands of a tarentula at the periods of second and third moults.

Adult spiders of the two sexes, according to Wagner, do not always possess the same silk glands. The females have glands not observed in the males, which this author believes serve to supply the cocooning silk. However, in the earliest stages of life the silk glands of the male and female are alike.

XI.

These physiological facts in the moulting processes of spiders Mr. Wagner has himself thus summarized: 1. The rejection of the old skin constitutes only a part of the moulting process, and that a secondary one. 2. The processes of a moult, in some of their features, commence a comparatively long time before, and end after the rejection of the skin and in connection therewith. 3. The spider, partly before casting off the old skin, partly at the moment of the act, and even for a brief period afterward, is deprived of some of its faculties: of sight,

Summary.

---

[1] Proceed. Acad. Nat. Sci., Philadelphia, 1888, page 173.

hearing, and touch, of movement, and even of respiration for a short time. 4. To the moult are subject most of the ectodermic and part of the mesodermic products. 5. The blood corpuscles, which with spiders are formed at the expense of the endoderm, are subject at each moult to periodic modifications, the final result of which is their proliferation. The number of red blood corpuscles increases from being three or four per cent. to become ten per cent. of the whole. 6. Cotemporaneously with the above named periodic processes of moulting are wrought certain constant processes of interior and exterior modifications, which are chiefly accomplished at the moulting period with which they are found in more or less direct connection. 7. The modifications to which spiders are subject during their post embryonic development are by no means limited to shape and to the final development of the genital organs. 8. With the moulting of spiders are connected certain special faculties, which are proper to the animal only during the moulting period, such, for example, is the faculty of renewing lost organs. Thus, if a spider's foot be lost during that period of life within which it is subject to moulting changes, it will be renewed after every moult; but if a limb be lost after the same period it will never be restored. 9. In connection with one or another of its various moults, the spider is found in possession of certain provisional organs, some of which soon disappear, others only with sexual maturity. 10. Finally, it may be stated that the moulting processes of spiders are almost exactly similar to those of the larvæ of insects which undergo an incomplete metamorphosis, as the Orthoptera, Pseudoneuroptera, and Hemiptera.

CHAPTER VI.

REGENERATION OF LOST ORGANS AND ANATOMICAL NOMENCLATURE.

I.

THE regeneration of a spider's lost legs and palps is a fact well known, and has been a subject of much observation, experiment, and speculation.

Renewal of Lost Organs. In brief, it may be said that the action is the result of two processes, the atrophy of the old tissues and the formation of new, which are effected simultaneously and in the same period of time. If a spider's leg be severed by accident or amputated, all the tissues which fill the stump of the leg gradually disappear, and within the cavity so formed a new organ originates and is completely developed.

Among early writers upon this phenomenon is Dr. Heineken, who reached the following conclusions:[1] (1) Spiders can not only reject a mutilated extreme joint, but reproduce it; (2) as the period of moulting ceases reproduction ceases also, even from the suture; **Heine-ken's Ob-serva-tions.** (3) the power of reproducing the limbs is restricted to certain periods of spiders' lives, but as soon as the animal ceases to moult its skin, in other words, becomes adult, its limbs cease to be reproduced; (4) until the growth of the limb was perfected, it appeared to Heineken that between the different moults little increase took place, and the act of moulting two or three times seemed to accomplish the full formation of the limb; (5) up to the period of the next moult the stump or suture, whichever it might be, remained externally unchanged; (6) the animal retired into a covering which it had woven for a day or two and then came forth with the limb or joint renovated. The above conclusions have been since shown to be substantially correct.

Some of his inferences, however, are not yet verified, for example, his experiments and observations which seemed to indicate that spiders possess the power of throwing off their joints at will, at least under certain circumstances. A large Lycosa, dropped by him into boiling water, instantly parted with six legs at the sutures. Another Lycosa on being held by one hind leg instantly threw it off. I have met nothing that confirms these statements, and know no similar records; and do not believe

---

[1] Experiments and Observations on the Casting Off and Reproduction of Legs in Crabs and Spiders by C. Heineken, M. D., Zoological Journal, Vol. IV., 1828-9, pages 284-294.

that spiders have the power to cast off limbs at will. The first example may have been a coincidence of the experiment with the full time for moulting, when the old tegument was just ready to be cast and was at once rejected through the sudden shock of the hot water plunge and the violent death struggles. The second case may have been simply an actual loss of a leg by handling.

Dr. Heineken, however, seemed to have no doubt as to the power of the animal to reject a leg. Moreover, he noted that the spiders which cast off crushed limbs were " hunters;" those which retained them the webmakers; a difference for which he accounts by supposing that the former, perhaps, have the strongest inducement to the act, as an inert and powerless joint would be a greater inconvenience to them than the loss of the whole limb. Furthermore, a webmaker, being of stationary habit, is less liable to accidents than the hunter, which is constantly on the move, and generally exposed. On this point I may remark that I have often met Orbweavers with one, two, four, and even five legs wanting, the result either of moulting mishaps, or of adventures and battles with assailants of various sorts. It is not uncommon to find males in this condition, a consequence of the unfavorable attitude of females in courtship. One also occasionally finds spiders with contorted legs which we would think might better be off than on, did the aranead have the power of self amputation. Certainly, these lost and wounded limbs did not prevent the spinning of snares, for I have seen in two cases, at least, an Orbweaver with all the legs wanting on one side weaving an efficient web.

Self Amputation.

Mr. Francis R. Welsh writes me that he saw an Orbweaver, which was probably Epeira insularis, that had lost seven of its feet (not legs) grasp with its spinnerets a spiral of its web, underneath which it hung, and hang thereto by the spinnerets only. It did not attach a dragline. It afterwards hung and moved by bowing its legs over the spirals of its web. A loss of this peculiar nature would probably have been occasioned by impeded moulting, and illustrates not only the perils of this act, but also the spider's power of adapting itself to extraordinary disadvantages.

Blackwall has published several important papers on this subject.[1] But for the most thorough and satisfactory studies of the regeneration of excised members we are indebted to Mr. Waldemar Wagner. This enthusiastic araneologist has pursued the entire histological development of certain organs, especially the legs, from the moment of amputation until the appearance of the new limb, and I shall undertake to interpret, substantially, the facts as recorded by him.[2]

Wagner's Work.

[1] See, as already quoted, Trans. Linn. Soc., Vol. XVI., pages 482-84 ; Proceed. Brit. Assn. Advc. Sci., Vol. XIV., pages 70-74, and Spid. Gt. Brit. and Ir., Introduction, page 7.

[2] " La Regeneration des Organes Perdus chez les Araignées." Voldemar Wagner. Bull. d. l. Soc. Impér. des Naturalistes de Moscow, 1887, No. 4.

II.

According to this observer, if a spider (Trochosa) lose a foot while very young, it will be restored at the last moult with such perfection that it cannot be distinguished from the others. If the member be lost at a more advanced age, after the eighth or ninth moult, it will be renewed imperfectly, and although the number of joints will be complete the restored limb can easily be distinguished. This is illustrated in Fig. 79, a drawing of the Huntsman spider, Heteropota venatoria, a specimen which I obtained in Florida, one of whose hind legs is seen to be shorter than the other, a moulting defect. In such cases the defective limb is usually not only shorter but smaller, of paler color and with less numerous hairs. Such a fact indicates that Nature has provided a certain amount of vital force and substance, for the exigencies of a spider's life, which cannot wholly answer the draughts made by the regeneration of adult limbs, although responding invariably to the

Lost Limbs.

Fig. 79. Huntsman spider with one leg (4) shortened in moulting.

recuperative demands of early life. I have already referred to another case in point, a large tarantula, when speaking of the dangers of the moulting period.[1]

As to the relative perfection with which lost limbs are reproduced, Blackwall considered it to be in inverse ratio to the extent of injury. Thus he found that palps and legs detached at the coxa were usually reproduced symmetrical but diminutive; while those amputated at the articulation of the digital with the radial joint, and near the middle of the tibia or of the metatarsus, were always much larger and unsymmetrical when restored. In point of fact, therefore, the development of the new limb depends upon the vital capacity of the undetached part. Thus, if a leg be amputated near the middle of

Imperfect Reproduction.

---

[1] See Chapter V., above.

the metatarsus the whole portion between that joint and the body will be reproduced of the same dimensions as the corresponding parts on the opposite leg; but the severed metatarsus and the tarsus will reappear much diminished. Precisely corresponding results follow similar excisions of any other joint.[1]

In order to define the time necessary for regeneration of a lost limb Wagner cropped the feet of a number of subjects of different ages. He found that if the foot was amputated a little before the moulting period it was not renewed after the act, but instead a whitish papilla was seen in the stump where the future organ originates. After the next succeeding moult the member appeared small, pale, short, but after the moult next following that it was thicker and more like the normal. If the leg were amputated immediately after a moult it would be restored during the interval preceding the next moult.

As a general rule it may be announced that a lost organ is restored in a period of time equal to that which separates two successive moults at that

Period-icity of Regener-ation.

stage in the development of the spider during which the limb is lost. If the foot of a spider is removed only two or three days after a moult, a new limb is formed in the period which remains until the following moult. For example, if the leg of Trochosa be removed during the period of the second moult, the forma-tion of a new member requires only five days, as that number is just the period which separates the second moult from the third. If the leg be cut away in the period of the sixth moult, a new one is formed in the space of ten days, because between the sixth and seventh moult there is an interval of from ten to twelve days. Thus, if the leg is clipped one or two days after a moult, a new member will ordinarily have time to form before the following moult.

On the contrary, should a leg or a part thereof be removed at a period, before the moult next to follow, shorter than that naturally required at that stage of development for complete renewal, then the appearance of a new member will be deferred until another moult shall occur;[2] that is to say, two moults must intervene before the lost part is made good. For example, Mr. Wagner cut off the leg of a Trochosa four days after the sixth moult; the seventh moult took place ten days thereafter, but a new limb was not then formed, and did not appear until the next following moult, viz., the eighth, which occurred eighteen days after the seventh. In such a case, and all others, the stump of the severed limb, healed and overclosed by its chitinous cicatrix, retains the same appearance until replaced by the new member.

This process of regeneration will be continued, as often as losses occur, during that period of life when the spider is subject to moult, that is to

---

[1] Blackwall, Brit. Spid., Int., page 8.   [2] Blackwall also observed this fact.

say, until complete sexual maturity. After that period no lost organ or part thereof will be renewed, no matter how long thereafter the spider may live. The function is thus evidently designed to favor the maturity of the species and so insure its continuance.

III.

As has been said, the regeneration of lost organs is the result of two processes, equally important and interesting, the atrophy of the old tissues and the formation of new. If one cuts off a spider's foot all the Atrophy tissues which fill up the remainder of the limb disappear, and of Old simultaneously a new organ is originated and completely devel-Tissues. oped within the cavity of the joint from which the old tissues have been atrophied.

Immediately after the operation Nature begins to cover up the wound, the blood cells form a thick cellular mass, which in the course of three

days is formed into a hard, dark scab, which serves as a stopple to the open wound. This is shown at Fig. 80, and a longitudinal sec-tion at Fig. 81. In the latter, one sees the old cuticle (ct) united with the scab, and the chitinous membrane

FIG. 80. View of healed stump of a spider's amputated limb. FIG. 81. Lon-gitudinal section of same; ctx, the cicatrix; ct, cuticle; mss, granular mass next the cicatrix; ch.c, chitinized cells; b.c.a., amœboid blood cells; b.c.r., red blood cells.

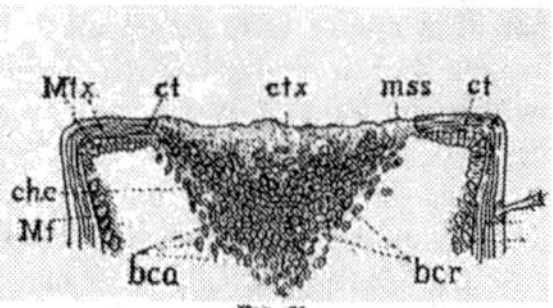

under which is the row of deep cells of the matrix (Mtx). Immediately under the surface of the cicatrix lies a mass (mss) of unstratified Forma- granular matter, next to which is a nest of cells arranged in tion of rows one above another, gradually diminishing in length and Cicatrix. receding into the cavity beneath. Of these the upper tier are "chitinized" cells (ch.c), and the remainder blood cor-puscles, both red (b.c.r) and amœboid (b.c.a). The transi-tion or destruction of these cells, and their metamorphosis into the structureless mass of the covering cicatrix, is accomplished gradually.

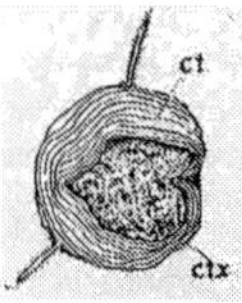

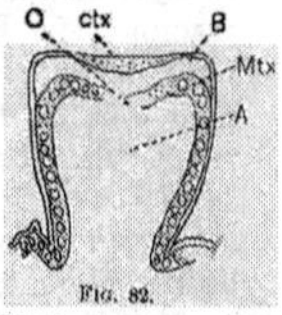

Now the matrix, which alone of all the tissues does not undergo entire degeneration, begins to retract little by little from the cuticle, thus parting from the Retrac- cicatrix. It commences at the top of the joint tion of within whose cavity its ends are approximated. Matrix. It is withdrawn more and more, forming above the old tissues a sort of cupola, to use Wagner's word.

FIG. 82. Appearance of stump when matrix, Mtx, begins to re-tract; A, B, lower and upper cavities; O, ori-fice between them; cicatrix, ctx.

This may be seen in the series of figures 82, 83, 85. At first the cupola and the cavity beneath are large (Figs. 82, 83), but as the tissues that fill the cavity decay and disappear they diminish more and more (Figs. 85, 86), shrinking gradually towards the base of the joint.

The tissue of the matrix is only atrophied in part; really, it is in those parts which converge above the cavity of the joint at the summit of the cupola, that is to say, where the amputation has produced a rent, that one observes the regeneration of the fatty tissue. The deep regular cells seen at the junction (Mtx) entirely disappear, and also afterward the part contiguous to the matrix.

The red blood cells, after part of them have been regenerated and given birth to the cicatrix, dispose themselves in different parts of the joint. They glide between the fibres of muscles and into the clefts which are formed by their gradual bending.

Red Blood Cells. The orifice formed after the operation at the summit of the joint between the fragments of the matrix (Fig. 84, O), in spite of the gradual approximation of the extremities of the matrix (Figs. 85, 86, O), remains open. In the degree that the summit of the cupola sinks the orifice becomes narrower, which, however, does not hinder the blood cells from penetrating thereby from the lower part of the cavity of the joint (Figs. 85, 86, A) into the upper part situated above the cupola (Figs.

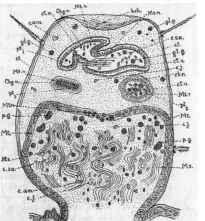

FIG. 83. A transverse section of the stump of a foot eighteen days after amputation; Mt.r, place in the matrix where the regeneration of grease is seen; Ms, muscles; gl.g, globules of grease; p.g, portions of grease; pt. 1, part of the new formed foot next the base; pt. 3, extremity of the same; pt. 2, part between the above two; Ms.n, muscles of the new formation; ct.n, cuticle of same; chg.n, chitinous layer of same; c.am, free nuclei of red blood cells; c.sa, amœboid blood cells; cj, color cells.

85, 86, B), to contribute to the work of nourishing the new limb. It is only when the new organ (Fig. 86) is completely formed that those disappear. At that epoch the extremities of the old matrix, then sunken almost to the base of the stump, are joined, and the orifice between the cavities below and above the cupola is closed. Shortly thereafter the process of the disappearance of the old tissues ends.

IV.

The atrophy of the muscles begins by their regeneration into globules of grease (gl.g, Fig. 83), which, as the process progresses, are conjoined, embracing one another without interblending, forming large drops **Atrophy** or portions of grease (p.g), whose ultimate role is to undergo **of Mus-** decomposition and serve as nutritive matter. Three days after **cles.** amputation these fatty globules may be seen among the fibres of the muscles, and along with them blood corpuscles; the latter undergo modification in size, increasing to three or four times their original volume. It appears that the regeneration of the grease does not continue until the final disappearance of the muscles, but extends to the epoch at which the transverse striation disappears.

Eighteen or nineteen days after amputation the muscles that had decayed from the joint show fine fibrilles, without transverse striation, which generally disappear very gradually and slowly. A few days thereafter there remain only a few fragments of muscular tissue floating within a structureless mass, wherein the blood cells, which they there meet under their special forms, put an end to their activity, and destroy the final remainders of mortification.

Fig. 83 presents a view of the process as above described, which is represented in the lower part of the drawing beneath the cupola formed by the **Origin** retracted matrix. Here is indicated the point in the matrix where **of New** the regeneration of fatty globules has begun (Mt.r); the globules **Limbs.** and drops of grease (gl.g, p.g) are seen floating in the structureless mass of nutritive material which is being recruited by the decay of the muscles.

One sees the distribution of this material through the orifice in the cupola, into the upper cavity of the stump, and therein the new limb is seen in process of formation. The part next the base of the new joint is shown at pt 1; that which is to form the tip at pt 3, and the middle part of the neoformation at pt 2.

V.

Soon after amputation, three to five days with spiders of middle age, in that part of the matrix which at this epoch is retracted from the **Origin** cuticle, one sees appearing upon the superior or dorsal part of **of New** the cupola an excrescence or papilla. (Fig. 84, pt.) The matrix **Leg.** and the conjunctive tissue beneath it are bent, and thus form this excrescence. The cavity of this excrescence contains blood cells. This papilla grows rapidly in length and soon bends at the summit. (Fig. 85, pt.) At this epoch its basal part begins to articulate, presenting thus the first joint of the foot, the basal joint, pending which the others do not exist at all. This joint is the thicker part of the neoformation.

(Fig. 85, ar.) It afterwards increases from one end to the other more equally in length than in width, and nineteen or twenty days afterward, with a tarentula of middle age, it presents an organ long, almost completely formed, gradually diminishing from the base towards the extremity, and much twisted. The articulations of this organ are formed successively from the base towards the periphera, the second forms after the first, the third after the second, etc. Consequently, this process is accomplished nearly as that described by Clarapede[1] for Lycosids in their embryonic period. Thus is accomplished successively the isolation of the tissues between the base of the joint and its extremity. Pending this; near the base of the foot, one readily observes muscular fibres (Fig. 83, pt 1, and Ms.n), and on the surface a chitinous membrane (ct.n), a little elongated at the base. (Fig. 83, pt 2.) Here one does not notice that the muscles are isolated, or anything but chitinous tegument (ct.n); the tip of the foot (pt 3) only the chitinogenous layer is isolated.

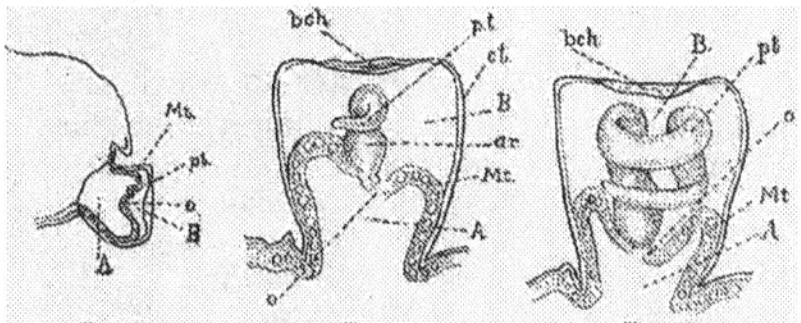

Sections of this new organ at different stages of its development show that it first originates the matrix of the papilla; its cells at its periphera are in immediate contact with the blood, environing the neoformation. In the meantime, as the neoformation grows, the exterior layer of the matrix is more and more feebly colored; at this epoch the superficial part may be regarded as chitinous tegument. The tegument is very fine, and does not form simultaneously upon all the neoformation, but gradually, its growth extending from the base towards the periphera. Thus the neoformation presents, from the exterior inwardly, first, a chitinous tegument; second, a chitinogenous layer beneath it; third, the matrix. Within the latter the hairs originate from elongated trichogenous cells, in no way differing from that already described during moulting,[2] except, of course, that no sheath of the old skin appears in the process.

SERIES SHOWING THE GRADUAL DEVELOPMENT OF THE NEW LIMB WITHIN THE OLD STUMP.

FIG. 84. Side view, showing on the retracted matrix, Mt, a small papilla, pt, the beginning of the new leg. FIG. 85. A further stage of growth; ar, base of the new organ. FIG. 86. New limb just before the moult, folded up within the stump. Other letters as above. (After Wagner.)

Order of Origin:
Hairs.

The above statements present the general features of the processes from which results the regeneration in spiders of a lost organ. Mr. Wagner adds the following deductions: 1. The blood corpuscles, under the influence of certain conditions, are subject to metamorphosis, the final result of which is the formation of a tissue which resembles chitine. 2. The

---

[1] Recherches sur l'evolution des Araignées. [2] See Chapter V., Fig. 13.

matter of those tissues which are atrophied and undergo the regeneration of the fat benefits the organism in three ways: First, in transmitting
Deductions. that matter by means of the amœboid blood cells, which assimilate and plasmatically digest it,[1] essentially after the regeneration
of grease becomes less intense, and even appears to have touched its limit. Second, in transmitting the matter by means of the red blood cells which assimilate it by absorption. Third, by means of globules of grease, which in themselves present certain forms whose role is to transport the nutritive matter into the whole body. 3. It is doubtful whether, without coöperation of blood cells, regeneration of fatty tissues would proceed to complete disappearance; if it could occur it would be after a considerable and indefinite time. 4. The process of degeneration of muscular tissue in spiders, in its general traits, suggests that of vertebrate animals.

VI.

In the descriptive matter to which the second part of this volume is devoted certain terms and abbreviations are used that require explanation

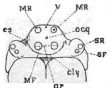

FIG. 87. Face of orbweaver; cly, clypeus; es, eye space; MF, MR, midfront and midrear eyes; SF, SR, sidefront and siderear eyes; oc.q, ocular quad; v, vertex.

Nomenclature of Description. and illustration, which it seems best to insert together at this point for convenience of reference. The quadrilateral described by the middle group of four eyes (MF and MR) is called the ocular quadrilateral, or more commonly "ocular quad" (Fig. 87, oc.q), as at once a brief and definitive term. The ocular area or eye space (es) is that part of the face over which the eight eyes are distributed. The curvature of the eye rows forms an important characteristic in determining species. The rows are said to be procurved when the concavity is directed forward toward the mandibles, and recurved when the concavity is directed backward. When the row is straight, or nearly so, it is said to be "aligned," or nearly so. In deter-

The Eyes. mining the curvature the eye rows have been looked at from the front, and a little below the horizon of the front row. It is often difficult to determine the exact curvature, especially as it differs with the point of vision. In

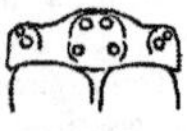

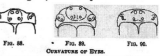

FIG. 88.   FIG. 89.   FIG. 90.
CURVATURE OF EYES.
FIG. 88. Front row recurved, rear row procurved. FIG. 89, Both rows procurved. FIG. 90. Front row nearly aligned, slightly recurved, rear row procurved.

Fig. 88 the front row of eyes is recurved, the rear row slightly procurved; in Fig. 89 the front as well as the rear row is procurved; in Fig. 90 the front row is nearly aligned, the rear procurved.

---

[1] See Wagner, "Du Sang des Araignées."

In measuring the eyes, the unit for determining the distance between any two, as MF or MR, is the diameter of the pair under consideration. In measuring the distance between the middle group of eyes, whether of the front or rear row, and the side group, the unit of measure is usually the linear space occupied by the same, that is, the distance from outer margin to outer margin; in other words, the width of the front or rear sides of the middle quadrilateral. This line is spoken of technically as the "area" or the "alignment." (Fig. 87, ar.) Thus the area of MF is the line from the outside of one midfront eye to the outside of the other. Sometimes the space between the midfront eyes is also used as a unit.

The clypeus (cly, Fig. 87) is that part of the face lying between the front row of eyes and the anterior lower margin thereof, where it meets the mandibles. The clypeus is spoken of as "high" or "low," as

Clypeus
Epigy-
num.

the space above named is greater or less; or sometimes the words "wide" and "narrow" are used as equivalent to the above. The height or width of the clypeus is measured by the distance of its margin from the two midfront eyes, and the diameter of one of these is the unit for such measurement.

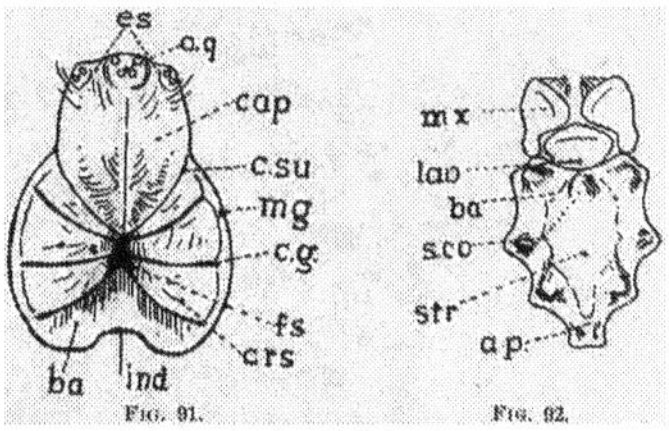

The views of the epigynum or external part of the female genital, in the various plates, are taken with the spider lying upon the back and the observer looking down upon her from the direction of the spinnerets. The drawings therefore show the lower side of the organ unless otherwise specified.

The following terms are used in

CEPHALOTHORAX AND PARTS THEREOF.

Fig. 91. Crs, corselet; c.g, corselet grooves; fs, median fosse; ba, base of corselet; ind, indentation thereof; cap, caput; mg, margin of corselet; c.su, cephalic suture; es, eye space; o.q, ocular quad. Fig. 92. Sternum and mouth parts. Str, sternum; s.co, sternal cones; ba, base of sternum; ap, apex of same; lab, labium; mx, maxillæ.

describing the cephalothorax: The dorsal part is known as the "corselet" (Fig. 91, crs), of which the posterior part is the base (ba), and this is

Cephalo-
thorax.

usually smooth, truncated, and indented where overhung by the abdomen. The depression on the summit is the median fosse, or simply "fosse" (fs); the slight furrows extending from the fosse to the insertion of the legs are "corselet grooves" (c.g) or furrows. The line of union between caput and corselet is the "cephalic suture" (c.su); that part of the head extending from the suture to the vertex I have called the caput (cap), as distinguished from the vertex or forehead and the face. The under part of the cephalothorax is the sternum (Fig. 92, str), which is a more or less oval or cordate or subtriangular shield, sometimes much indented or scalloped at the coxal insertions and pointed at the base (ba). Opposite the coxæ, in some species, are slight eminences technically referred

to as "sternal cones," or simply "cones" (s.co). The labium (lab) and
maxillæ (mx) are also shown in Fig. 92.

VII.

The parts of the abdomen referred to in description are as follows:
The dorsum (Fig. 93, do) is the rounded top of the back, of which the
Abdomen anterior part is called the "base" (ba), and the posterior the
"apex" (ap). The "folium" (fol) is a leaflike figure which usually
occupies one-third or more of the middle part, extending longitudinally
from base to apex. The median line (mn) of the dorsum often has a
deeper shade of color than the edges of the folium. The blackish "pits,"
or round dark spots (pt), symmetrically arranged in pairs on each side of
the median line of the folium, indicate points of attachment for internal
muscles. The venter
(vn, Fig. 94) is the part
lying between the epi-
gynum (ep) and gills (gi)
and the spinnerets (sp);
pits or dimples may also
be seen on the venter
symmetrically arranged
as on the dorsum. The
median ventral band is
usually bordered by a
strip of lighter color
more or less broken.
When the abdomen is
thickened at the apex,

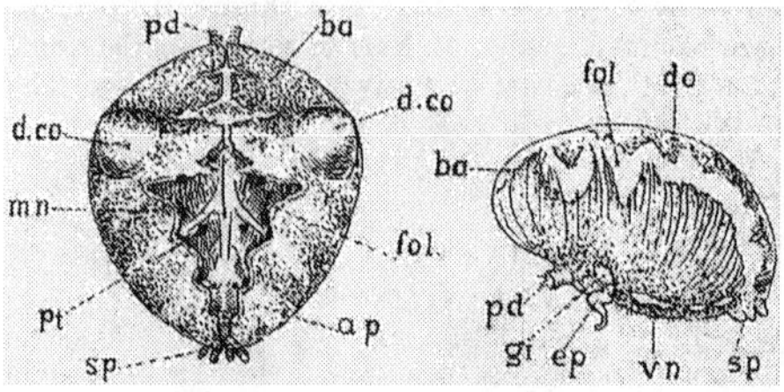

Fig. 93.   Fig. 94.
ABDOMEN AND PARTS.

Fig. 93. Dorsal view; pd, pedicle uniting to cephalothorax; ba, base or
fore part; d.co, dorsal cones or shoulder humps; fol, folium or dorsal
pattern; ap, apex; sp, spinnerets; pt, pit marking insertion of muscles;
mn, median line of dorsum. Fig. 94. Side view; vn, venter; ep, epigynum; gi, breathing gills; other references as above.

the space between the dorsal and ventral apex is known as the apical
wall of the abdomen. On either side or shoulder of the dorsal base some
species have conical prominences known as dorsal cones or tubercles (d.co).
These are sometimes placed at several other points, usually along the margin of the dorsum. In some species these are hard and spinous.

The exterior parts of the epigynum referred to in the description are
illustrated at Figs. 95, 96. The atriolum vulvæ, or vulval porch, is a
Epigy- vaulted porch or hood that curves over the genital opening, in
num. front of which (anterior) it is located. It is usually chitinous,
and covered at the base with hairs. The middle part in many
species is prolonged into a shaft or scapus atrioli, which is sometimes a
mere flap, and again is much lengthened. Sometimes it is smooth, sometimes wrinkled or rugose, often with a row of stout hairs thereon. Frequently it is grooved or hollowed along the under surface. The tip is
more or less attenuated and rounded, or sometimes pointed; in many
species it is widened into a bowl or spoon, usually oval, but sometimes

nearly circular. This part of the scapus I have termed the cochlear or spoon (cch, Fig. 95). I believe that the scapus serves as an ovipositor, to assist in the direction and arrangement of the eggs as they are extruded.

It has occurred to me to inquire whether there exists any relation between the peculiar shape of the scapus and the character of the cocoon laid by the female, especially in the position and arrangement of the eggs; but I have reached no generalization on this point. It is probable that this organ al-

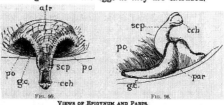

VIEWS OF EPIGYNUM AND PARTS.

FIG. 95. Lower part, that is, the part next the ground when the spider is walking; at, atriolum; scp, scapus; cch, the cochlear or spoon of the scapus; po, portulæ; gc, genital cleft. FIG. 96. Side view, to show the parm or buckler; other references as above.

so aids to clasp and direct the palpus of the male, or at least to regulate the flow of the sperm, in the act of fertilization.

In some species, as Epeira labyrinthea, Fig. 96, there is thrown up a sort of wall in front of the scapus and on the anterior margin of the genital cleft, which I have termed the parmula, or simply the parm or buckler (par, Fig. 96). In some species at least, as in Fig. 96, there appears a groove or channel in the median part, indicating that this may also aid in directing the eggs during extrusion, or again in guiding the inflow of the sperm.

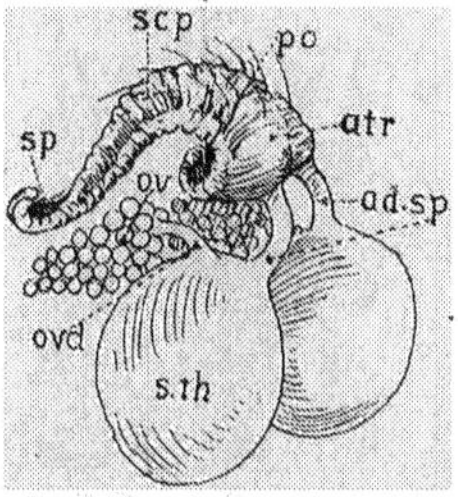

Underneath the vulval porch are (usually) two concave openings or little gates, portulæ spermathecæ (po), which receive the sperm from the male palp, whence it flows from Oviducts: each portula through a duct, aditus Sperma- spermathecæ (ad.sp), sometimes simple, thecæ. sometimes much convoluted, into the spermathecæ (s.th). Thence another smaller duct, the oviduct (ovd), admits the sperm to the eggs (ov). The ovary also has an opening into the portulæ, by which the eggs are discharged when cocooning. The epigynum is subject to immense variations in form, and these variations are nearly constant for every species; hence they are good specific characters. I have not been able to find in them any fixed generic value, although there is often a noticeable tendency to a general likeness of form in any closely related group of species.[1]

FIG. 97. View of epigynum in connection with ovaries (partly diagrammatic); atr, atriolum; po, portulæ; scp, scapus; sp, cochlear or spoon; ad.sp, aditus spermathecæ, sperm duct; s.th, spermathecæ; ov, the eggs; ovd, oviduct.

---

[1] I hesitate somewhat to propose the above anatomical terms, as the field is one which I have not specially cultivated.

VIII.

It is now generally believed by araneologists that the digital joint or bulbus of the palp is the copulative organ of male spiders. We are indebted to the careful observations of Menge[1] for the discovery of the method by which the male transfers the sperma from his sexual aperture to his palps, and so, through the palps, to the vulva of the female. Blackwall, however, with that keenness of observation and judgment which characterized this distinguished English naturalist, independently observed the process.[2] A male Agalena labyrinthica confined in a phial spun a small web, and among the lines of which it was composed he perceived that a drop of milklike fluid was suspended. He had not observed how it was deposited, but saw that the spider by the alternate application of his palpal organs speedily imbibed the whole of it. The conclusion which he derived from this circumstance was to confirm the acute suspicion of M. Dugés that the palpal bulb alternately performs the office of an absorbing syphon and an organ of ejection.[3] The fact first made known by Menge has been abundantly confirmed by Ausserer[4] and Professor Bertkau,[5] and later by Mr. F. M. Campbell,[6] and in part by the author.[7]

Use of Male Palps.

The former belief that the testes have their outlet into the palp has now very few, if any, advocates. Those who have most carefully studied the anatomy of spiders, from the earliest to the latest students,[8] have found that no connection whatever can be traced between the organs which prepare the spermatic fluid and the palps, and that the testes are far from these latter organs, on the under side of the abdomen, near its anterior extremity, in a position corresponding to the female vulva. The little slit there, on which the efferent ducts of the testes have their orifices, may sometimes be seen with the naked eye or a simple magnifying lens. Under such circumstances one may well ask with Thorell,[9] what is more natural than to suppose that the sperma, previous to coition, is in some way or other transferred to the intromittent organs, the palps? That this transmission always takes place in the manner directly observed by Menge in the case of Agalena and Linyphia is indeed

Location of Testes.

---

[1] Ueber die Lebenweise d. Arachn. Naturf. Gessellsch. in Danzig, IV., I., pages 39–41 (1843); Preuss. Spinn. f. inst. I., page 106 (1866).

[2] Journal Proceedings Linn. Soc. Zool., VII., pages 157, 158.

[3] Observ. sur les Aran., Ann. d. Soc. Nat. Ser. Zool., VI., pages 189, 190.

[4] "Beobactungen über die Lebenweisse," etc., Zeittchr. Ferdinandeums, 1867.

[5] "Ueber den Generations apparat der Arachniden," Arch. Nat. Gesch., 1875, page 254.

[6] "Pairing of Teg. guyonii," Linn. Soc. Jour. Zool., XVI., 103.

[7] See Vol. II., page 41 and page 73.

[8] TREVIRANUS, Ueber d. inneren Bau d. Arachn., page 77, tab. iv., Fig. 33 (1812); MENGE Preuss. Spinn., I., pages 32, 33 (1806).

[9] Synonyms Europ. Spid., page 503.

not proved, and it is possible that the male sometimes applies the genital
bulbs to the sexual aperture, and thus charges them with the
fertilizing fluid, a fact which Menge seems to suspect may be
the case with Lycosa rurestris.[1] That this transfer is so seldom
observed, and by so few persons, will not excite wonder on the part of one
who has devoted much time to the study of the habits of these reticent
creatures, and who knows the difficulty of obtaining a complete observation
of even the most common of its habits. Possibly the extrusion of the
sperma by the male upon the little silken receptacle from which it is
absorbed into the palps takes place very rapidly; or it may be done long
before the act of fertilization; perhaps, as Thorell suggests, immediately
after the last change of the skin.

Transfer of Sperm.

The Cymbium is that part of the modified digital joint of the male
spider's palpus upon which is placed the copulatory apparatus which it
sustains. In many species of Orbweavers it covers one side of
the digital, having the appearance of the
half of a seed husk, or shell of
grain, and is covered more or
less thickly with hairs and
bristles. (See Fig. 98, cym.)

Cymbium.

The alveolus (alv) is the
concavity in the cymbium
within which is located the
copulatory apparatus proper.
Its form depends upon the
structure of the cymbium, with
certain spiders occu-
pying half of the sur-
face, in which case the cymbium has the form of a canoe.
With others it is smaller, as
with Segestria, for example;
with others again larger and

Alveolus.

Fig. 98. Digital joint of palp of a male spider (after Wagner),
schematic longitudinal section of the cymbium with the copulatory apparatus drawn out; emb, embolus; teg, tegulum;
mea.san, meati sanguinis, minute ducts for conducting blood
from the hæmatodocha (hæm) into the receptaculum seminis, rec.sem; the arrows indicate the course of the blood;
gla.cl, glandular cells; alv, the alveolus or hollow of the cymbium in which the apparatus rests.

occupying the whole surface of the cymbium. The simpler the structure
of the apparatus the less is the alveolus in circumference and depth, and
vice versa. The more complicated is the apparatus the more space it
embraces in every sense.

The alveolus serves as a seat for the Hæmatodocha (haem), a follicule
placed in the form of a spiral, and intended with some spiders to serve
as a seat for the copulatory apparatus itself. The alveolus has no independent significance; its form and size in its development are dependent

---

[1] Lebensweise d. Arach., page 43.

upon other parts of the apparatus. The Glandular Cells (gl.c) are situated at the margin of the alveolus. The chitinous teguments above these in the cymbium are pierced by a number of ducts, which eventually serve as conduits for the secretions of the cells.

The Hæmatodocha is a delicate chitinous saclike organ, of which the bottom is inserted into the alveolus, and the superior margin into the inferior face of the tegulum. Thus the entire basal part (receptaculum seminis), which is situated under the tegulum, is within the cavity of the sac. The walls of the hæmatodocha are very thin and plaited, so that if the sac be full it is able to attain large dimensions. In its ordinary condition with most spiders it is folded in a spiral and situated within the alveolus.

Hæmatodocha.

The bottom part, by which the sac is attached to the concavity of the alveolus, has a round orifice (orificium hæmatodochæ, or.hæm), which unites the cavity with the lacuna of the cymbium situated beneath. Mr. Wagner says that during fecundation the hæmatodocha is not filled with sperm, as certain observers suppose, but at that moment is filled with blood. Its role is to transmit, under pressure of its elastic walls, the blood which has come to the lacuna into the cavity of the receptaculum seminis, through certain fine ducts (meati sanguinis), by which the cavity of the receptaculum is united with the cavity of the sac. It follows that the function of the hæmatodocha is that by its means the blood takes part in fecundation by penetrating, although in small quantity, along with the sperm into the genital cavity of the female.

Upon the superior part of the copulatory apparatus, and in part within the cavity of the hæmatodocha under the tegulum, is situated a tube with thick, chitinous walls, which in different species of spiders differs in shape, length, and position. This is the receptaculum seminis (rec.sem). The extremity of that part situated under the tegulum, the basal part of the tube, is always closed. The opposite end terminates in an orifice at the summit of the embolus (emb). At the moment of fecundation the receptaculum is filled with sperm. Its role is as follows: Shortly after the moult it is filled with sperm, which the male has forced from his genital orifice upon a woven thread or tissue previously prepared. He plunges his palp into his drop of sperm, which, by the law of capillarity, mounts into the receptaculum through the orifice (or) of the embolus, the only organ of the copulatory apparatus that receives and conserves the sperm. The sperm can only be discharged by the orifice of the embolus through which it entered. The slender ducts (meati sanguinis), which serve to unite the receptaculum with the cavity of the hæmatodocha, are of such extreme fineness that they cannot serve as a conduit for the sperm into the hæmatodocha, and it is only the blood plasm which, under the pressure of the walls of the hæmatodocha, can be made to penetrate into the cavity of the receptaculum. Consequently, the role of the latter is

Receptaculum Seminis.

that of a passive organ, receiving the sperm and transmitting it into the genital cleft of the female under the pressure of the blood which enters into the hæmatodocha through the meati sanguinis.

The receptaculum, in the region of the embolus, is throughout its whole length fine, smooth, and without pores. Further toward its closed extremity **Blood Ducts.** is a mass of minute ducts which pierce the walls, and which Wagner has named meati sanguinis, blood ducts. All spiders are provided with these ducts, and their role is to serve as conduits to the blood from the hæmatodocha into the receptaculum.

The Tegulum is a quite thick plate of chitine serving to cover in from above and to protect the receptaculum. Many spiders, as the Attids, Thomisids, and others, have here no chitinous conformation except the tegulum; but spiders which have a more complicated organism of this apparatus are provided with many other auxiliary organs, in the form of laminæ, dentations, and excrescences of the most unique and varied forms. The embolus is an organ of a chitinous nature, for the most part subiliform or having the form of a switch. At its extremity there is a small orifice (Fig. 98, or), by which the sperm enters and issues. The articulation of the embolus with the tegulum may be mobile or immobile.

The action of the above parts is as follows: The male applies to the genital cleft of the female the exterior face of his palp, and by numerous **The Role.** contractions of the abdomen, in which the subcutaneous muscles take part, forces the blood through the orifice into the cavity of the hæmatodocha, which it expands, pushes out the copulatory apparatus, and having by way of the blood ducts penetrated into the cavity of the receptaculum seminis, impels the sperm through the embolus into the genital cleft of the female. When the blood begins to abate, returning into the body of the male it fills anew the sac as full as at first, an operation which is repeated until fecundation is terminated; then the palp is withdrawn from the genital cleft, the hæmatodocha contracts, and the tegulum resumes its position.[1]

---

[1] La Mue des Araignées, M. Waldemar Wagner, Ann. Sc. Nat. Zool., 1888, 367–371.

PART II.

DESCRIPTION OF GENERA AND SPECIES.

AMONG Orbitelariæ I include, with Thorell, all spiders that spin a so called geometrical web. This may be arranged in a more or less circular plane, perpendicular or horizontal to the horizon, as the case may be, which is the characteristic web of Epeira and many other genera. Or it may be arranged in a circular plane, which lacks one segment of greater or less size, usually in the upper part of the snare, as in the case of Zilla and some species of Epeira. Or again, it may consist simply of a single sector of a circle, as in the case of Hyptiotes, the well known Triangle spider. I include among the families of Orbitelariæ, Uloborus, which makes a circular snare, suspended horizontally, but without the usual Epeïroid armature of viscid beads upon the spiral lines.

Orbitelariæ Defined.

It does not seem that any spider which spins a snare of the general character here described can be properly placed in any suborder other than Orbitelariæ. Yet it may well be that there are Orbitelariæ, even Epeïroids in the strictest sense, which spin either no web at all, or an irregular one; just as (to quote Thorell's comparison) there are many Tubitelariæ that do not fabricate webs of the form characteristic of that group.[1] Pachygnatha, for example, I include with the Orbitelariæ, as is now done by the best araneologists; but, so far as is known, it makes no web, and appears to live underneath stones and capture its prey after the fashion of the wandering spiders. Of course, such a conclusion as this cannot be wondered at; for no one will claim that a natural classification of animals can be based upon habits alone, although in the case of spiders it certainly is true that there is a quite constant relation between the natural habits and the natural order of systematic life.

Habits and Classification.

Some of the older arachnologists still cling to the term Araneïdea to denominate the order of true spiders; but for the most part the word Araneæ is now thus used. This term was first proposed by Sundevall, in his Conspectus Arachnidum, in 1833. It was

Name of the Order.

---

adopted by Thorell, in his splendid paper on European Spiders, continued in his "Synonyms," and has been retained by him ever since adopted.
Count Keyserling used it in his latest works, as does the French araneologist, Eugene Simon (1892), in the first volume of his revised Natural
History of Spiders, which promises to be a monumental work. In America Dr. George Marx, the editor of Keyserling's Epeïridæ, and Professor
and Mrs. Peckham agree with myself in the use of the word. On the
contrary, Mr. Cambridge, in his beautiful work on the spiders, in Biologia
Centrali-Americana, retains the earlier term, Araneïdea. This concensus of
araneologists seems to be justified. Aranea was the original generic name
of all spiders, but there is now no genus of true spiders thus called, and
therefore there can be no objection on that account to call the order
Araneæ, a word which seems to be eminently proper, since it signifies
spiders, just as, for example, Aves signifies birds, and Serpentes signifies
snakes.

As to the use of the terms Araneïdes and Araneïdea, Thorell urges two
objections: first, that it is altogether illogical, since the addition of the
Araneæ terminal syllable indicates an enlargement of the conception that
Approved lies in the word to which such termination is appended. For
instance, by Carabidæ we mean all Carabi, and besides a number of animals more or less closely allied to them. By Araneïdea, therefore, properly speaking, we would mean spiders and animals closely related
to them. Moreover, the form is regarded by Thorell as liable to the
objection that it is a hybrid, being a compound of a Greek and a Latin
word, and should on that account, if for no other reason, be rejected,[1] an
opinion in which the purists, at least, will concur.

Before passing to the description of species it will be well to state the
author's reasons for certain changes therein made in the Hentzian nomenclature of certain well known species. In the year 1887, while
Changes visiting the Zoological Library of the Kensington (London)
in Nomen- Museum of Natural History, my attention was called to a series
clature. of manuscript drawings of American spiders. These proved to
be the original notes and figures of John Abbot, beautifully drawn and
colored from nature, upon which Baron Walckenaer had based the descriptions in his Natural History of Apterous Insects. This find seemed to me
most important. It was my first knowledge that the drawings were extant,
and I know of no printed statement of what had become of them after
Walckenaer's death. No English or American araneologist had made any
use of them, and if known at all their value was not appreciated. How
they came into the possession of the Kensington Museum I could not learn,
but no doubt they were bought by the authorities soon after Walckenaer's
death.

---

[1] Syn. Europ. Spid., page 507.

I spent a portion of one day in study of the figured Orbweavers alone, and took copious notes of those species which were entirely familiar to me. After my return to Philadelphia I made careful studies of Walckenaer's published descriptions, comparing the same with my notes, and thereupon published a paper in the Proceedings of the Academy of Natural Sciences of Philadelphia (1888). Therein was considered the necessity for revising the nomenclature of various Hentzian species of Orbweavers, of which a brief tabulated list was given.

The paper caused an animated discussion in the Academy upon the limits within which the law of priority should obtain. The Hentzian names had so long prevailed, and were so widely inwoven with **Law of** our aranead literature, that it was held by a few that they ought **Priority.** to be retained, since a change would cause embarrassment to naturalists, confusion in popular literature, and thus detriment to science. On the contrary, the majority present, among whom were the eminent President of the Academy, the late Professor Joseph Leidy, and Professor Dall of Washington, held that the earlier names should in all cases be adopted, no matter how much inconvenience might be entailed thereby.

This position has since received general approval. Professor T. Thorell, who is justly regarded as the most eminent of living araneologists, and **Approval:** whose authority on such a point is of special value, thus wrote:[1] **Professor** " The discovery of Abbot's drawings of American spiders is **Thorell;** indeed a fact of the greatest interest, not only to Americans, but **Dr. Marx.** to all arachnologists, and I congratulate you on having had the luck to make this discovery. Of course, I have read with great attention what you have said on the subject. As to me, I do not entertain the least doubt that you and Professors Leidy, Lewis, and Dall are right, and that the earlier names should in all cases be adopted. The law of priority must be respected, and is the only one that prevents arbitrariness, and that gives stability to nomenclature. I think, then, that in all such cases, in which Walckenaer's species can, with tolerable certainty, be recognized, his names should be preferred to those more lately published, even if such names are more commonly used, or the species better described or figured under the newer names." To the approval of this most distinguished authority may be added that of Dr. George Marx, who, in his Catalogue of Described Araneæ, fully accepts the conclusions, and adopts the revised nomenclature suggested by me.

To this general concensus, however, there was one notable exception. Mr. J. H. Emerton[2] published a criticism upon my paper, in which **Emerton** he rejected the conclusions, and depreciated the value thereof, **Excepts.** as well as of the Abbot manuscripts, which he further claimed to have seen in 1875, although he had never in any way made known

---

[1] Proceed. Acad. Nat. Sci., Phila., 1888, page 430.
[2] Psyche, Vol. V., No. 149-150, Sept.–Oct., 1888: Cambridge, Mass.

his information, nor anywhere alluded thereto in his published descriptions of spiders. I replied to this criticism,[1] vindicating the value of the Abbot drawings, and strengthening the ground upon which my judgment rested. Nevertheless, in view of the above challenge, I resolved to make more thorough study of the manuscripts before publishing the final volume of this work, and meanwhile allowed, for the most part, the Hentzian names to stand in Vols. I., II., which are chiefly concerned with the habits of spiders.

Accordingly, in the summer of 1892 I visited London, and gave a week to the study of Abbot's drawings, confining my attention to the Orbweavers.

Second Visit to London. I verified my former notes, carefully compared Abbot's figures and brief memoranda with Walckenaer's published descriptions, and made colored copies from tracings of most of the Orbitelariæ.[2]

This work done, I submitted the whole to Mr. R. I. Pocock of the Kensington Museum, who kindly went over the same, comparing Abbot's figures with Hentz's, and with Walckenaer's descriptions. In every instance he was able independently to reach a conclusion as to identity that almost exactly tallied with my own. These studies confirmed all that I had previously published, and added several species to the list of Hentzian names that must yield priority to Walckenaer. Accordingly, in the following pages I have felt compelled to revert to the earlier nomenclature.

Result Confirmed. This is done with sincere regret, as the well known names of Hentz are embodied in all my previous publications, and the use of the index alone can disentangle the resulting confusion. Nevertheless, the rectification of nomenclature must sometime inevitably be made, and postponement could only add to the confusion. It seemed better, therefore, to face the difficulty at once with an honest effort to bring in final order by the just sovereignty of the law of priority.

I now regret that I permitted confidence in my first judgment of the value of the Abbot drawings to be so far shaken as to lead me to retain the Hentzian synonyms in the first two volumes of this work, instead of at once eliminating them. For this I can only excuse myself by the fact that the books were going through press while the matter was still under discussion, and before my own conclusions had been assured by the judgment of leading araneologists. I have done the best, under the circumstances, to lighten the inconvenience which the reader may thus have been caused, by full references in the synonoma, and liberal references and cross-references in the Index.

---

[1] Proceed. Acad. Nat. Sci., Phila., 1888, page 428: "The Value of Abbot's Manuscript Drawings of American Spiders."

[2] My thanks are especially due to Mr. William Caruthers, Keeper of Botany, for kind services during these studies. He gave me the use of his office, procured for me all required books and material, assisted me with his extended knowledge and experience as a biologist, and added thereto the charm of a hospitable host.

GENUS EPEIRA WALCKENAER.

THE original genus Epeira of Walckenaer has been divided and subdivided at various times, yet still contains by far the greatest number of orbweaving species, not only in the United States, but other parts of the world. That it must further be subdivided is apparent to any one who has given much attention to the species which are grouped thereunder; but equally plain that it will be difficult to distinguish sharply the characteristics upon which good generic differences may be based. The following are the principal character-istics herein held to mark the typical Epeira.[1]

The cephalothorax is moderately high and oval, more or less shortened or cordate the corselet for the most part being well rounded. It is moderately high, the fosse placed near the centre of the summit, from whence the corselet slopes more or less sharply to the base, which is truncated and often indented. The corselet grooves are sufficiently distinct; the cephalic suture well marked. The margin of the corselet is often a narrow belt free from pubescence more or less shelving to the articulation with the legs. The head is lowly arched, somewhat depressed at the face, which is wide, though narrower than the corselet, and somewhat quadrate. The sternum is shield shape or cordate, somewhat longer than wide; the labium longer than wide, strong, half as high as the maxillæ, which are as wide, or nearly as wide, as long.

The eyes are placed in three groups, of which the side eyes are upon tubercles or elevated bases, the four middle eyes upon a rounded eminence, the ocular quad being in the form of a quadrilateral whose greatest width nearly equals the length. The front side of the quad is usually a little wider than the rear. The space between the sidefront and midfront eyes is equal to 1.5 times the area of the latter, or from 2 to 2.5 the intervening space thereof. The space between the siderear and the midrear eyes is usually greater than the above. The clypeus is low, rarely exceeding 2 to 2.5 times the diameter of the midfront eyes. The eye rows are not widely separated at their greatest point of divergence and closely approximate at the sides; the front row is slightly recurved, the rear row slightly procurved.

The legs are in order of length 1, 2, 4, 3; stout in all the joints; the tarsus and meta-tarsus gradually diminishing in size, and not noticeably thinner than the other joints. They are clothed heavily with hairs and bristles, and abundantly with strong spines. The palps in the female are armed as the legs; the digital joint with a strong claw; longer than the radial joint; the cubital joint being about half the length of the latter, and the humeral joint approximating the length of the digital, but usually longer. The abdomen is usually subglobose or oval, rounded at the base and diminished at the apex, or is a triangular ovate. The spinnerets are distal and the base overhangs the cephalothorax sometimes for half its length. The skin is soft and pubescent. The epigynum has generally a prolonged scapus.

The male in general form and markings resembles the female. Tibia–II is frequently widened at the tip, and in some species provided with strong denticulate clasping spines. The coxæ are often marked by spurs at the articulation with the trochanter, or upon the base. He is commonly much smaller than the female, but sometimes nearly equals her in size.

It will probably be observed in the following descriptions that the above specifications of the typical Epeira are not strictly adhered to, and in some points indeed are widely divergent therefrom. But a strong indisposition to multiply new genera, together with the confusingly interblended characteristics above referred to, have prompted to, if not justified such a course. No doubt future students, who may have a greater amount of material in hand, will be able not only to indicate necessary divisions, but to unite some of the genera already created from this overflowing group.

---

[1] For the technical terms used in description of species see Chapter VI, page 124 sq.

No. 1. Epeira sclopetaria (CLERCK).   Plate I., Figs. 9, 9a; Pl. II., Figs. 10, 11.

1757.	*Araneus sclopetarius*, CLERCK	. . Aranei Svecici (Svenska Spindlar), p. 43, pl. 2, tab. 3, Fig. 1.
1757.	*Araneus sericatus*,[1] CLERCK	. . . Aran. Svec., p. 40, pl. 2, tab. 1.
1789.	*Aranea undata*, OLIVIER	. . . . . Encycl. Méthod., iv., p. 200.
1833.	*Epeira sericata*, KOCH, C.	. . . . Herr.-Schæff. Deutschl. Ins., pp. 120, 1.
1834.	*Epeira virgata*, HAHN	. . . . . . Die Arach., ii., p. 26, tab. 46, Fig. 113.
1837.	*Epeira frondosa*, WALCKENAER	. . Ins. Apt., ii., p. 66; ABBOT, G. S.,[2] No. 326.
1847.	*Epeira vulgaris*, HENTZ	. . . . . J. B. S.,[3] v., pl. 30, p. 469; Id., Sp. U. S., p. 108, pl. 12, Fig. 6.
1850.	*Epeira sericata*, KOCH	. . . . . . Die Arachn., xi., p. 110, pl. 385, Figs. 914, 915.
1851.	*Epeira sclopetaria*, WESTRING	. . Förteckning, etc., p. 34.
1855.	*Epeira sclopetaria*, THORELL	. . . Recensio Critica, p. 22.
1861.	*Epeira sclopetaria*, WESTRING	. . , Araneæ Svecicæ, p. 33.
1864.	*Epeira sericata*, BLACKWALL	. . . Sp. G. B. & I.,[4] p. 328, pl. 23, Fig. 238.
1864.	*Epeira sericea*, SIMON	. . . . . . Hist. Nat. d'Araign., p. 492.
1866.	*Epeira sclopetaria*, MENGE	. . . . Preuss. Spinn., p. 57, pl. 7, tab. 7.
1871.	*Epeira sclopetaria*, BLACKWALL	. . Spiders from Montreal, Ann. & Mag. Nat. Hist., viii., p. 429.
1884.	*Epeira sclopetaria*, EMERTON	. . . N. E. Ep.,[5] p. 303, pl. 33, p. 4; pl. 35, p. 10.
1889.	*Epeira sclopetaria*, MCCOOK	. . . American Spiders, Vols. i., ii.
1889.	*Epeira sclopetaria*, MARX	. . . . Catalogue,[6] in loc.

FEMALE: Total length, 12.5 mm.; cephalothorax, 5 mm. long by 5 mm. wide; abdomen, 8 mm. long by 6.5 mm. wide. The general color is grayish brown, which in some specimens is deepened into iron gray upon the abdomen, cephalothorax, and terminal joints of legs. The cephalothorax and abdomen are usually uniform in color.

CEPHALOTHORAX: A rounded oval, truncated at the base; the median fosse is a deep lateral pit placed back of the middle point, which is the highest; the sides of the corselet are rather steeply sloped, the head depressed, the caput inclined from its insertion to the eye space; color glossy black brown, covered sparsely along the sides with gray hairs, which form a thicker band at the margin, and are spread out more thickly along the sides of the caput forming heavy eyebrows, and covering the clypeus. Sternum shield shaped, longer than wide, but of nearly equal width to the third coxa; glossy black, covered particularly at the sides with gray hair; labium wider than long, the tip triangular; the maxillæ as wide as, or wider than each other, rounded on the sides, and, like the labium, dark glossy brown, light yellow at tips.

EYES: The ocular quad upon a rounded prominence most decided in front, which is decidedly wider than the rear, and about as wide as the sides. MF are separated by about two diameters, and are larger than MR, which are separated by more than one diameter. Side eyes upon tubercles, separated by about the radius of SR, which is slightly smaller than SF, and situated almost upon a line with it upon the side face, thus, as in the case of

---

[1] Strictly adhered to, the law of priority would give this name to the species, as it appears first in order of paging in Clerck's book. This seems to have led Koch and Blackwall, as below, to adopt the name Epeira sericata, the latter, however, subsequently abandoning that for the more generally accepted one. As the change would not affect the credit due to the first author, and would cause much confusion, the better known title is here retained.

[2] Georgia spiders, manuscript drawings.

[3] These abbreviations will be used throughout for the paper of Hentz in the Journal of the Boston Society Natural History (J. B. S.), and the collection of same by the late Mr. Edward Burgess, entitled Spiders of the United States (Sp. U. S.).

[4] Spiders of Great Britain and Ireland, so throughout.

[5] New England Spiders of the Family Epeiridæ, Trans. Conn. Academy, Vol. VI., 1884. Abbreviated throughout: N. E. Ep.

[6] By this single word I refer throughout to Dr. George Marx's "Catalogue of the Described Araneæ of Temperate North America," Proceed. U. S. Nat. Museum, Vol. XII., 1889.

E. strix, bringing it into the front row. MF are separated from SF by about 1.3 their area, and from the margin of the clypeus by about two diameters. Front row is recurved, the rear row, which is much the longer, procurved.

LEGS: Strong and stout; covered with gray hairs, rather sparsely with bristles, and with numerous yellow or blackish spines. The femora are yellow, or orange yellow, with broad dark brown or blackish terminal bands, which also encompass the patella. The tibia and metatarsus have dark brown annuli at the ends, and a broad one of similar color in the middle. Feet black, with a yellow band at the articulation with the metatarsus; palps heavily armed with gray bristles and long yellow spines, are colored as the legs, but lighter. Mandibles conical, arched at the base, where they slightly project beyond the clypeus; dark glossy brown or blackish brown in color; not so much contracted at the tips, nor so greatly arched at the base as E. strix.

ABDOMEN: A long oval, narrowing at the apex to the spinnerets, which are distal. The dorsal folium (Plate II., Fig. 10) is sharply outlined by a narrow undulating border of gray hairs, broadest at the base, and gradually diminishing in width to the spinnerets. A lance head point projects forward along the front from the basal part of this marginal line. The line is interrupted about one-third the distance from the base, giving it in many examples the appearance of two separate figures, the apical portion thereof being a triangle with scalloped edges, and a lance headed figure projecting from the middle. The general colors of the dorsum are blackish brown, the sides are mottled with gray waving longitudinal lines formed, like the dorsal figure, of long gray hairs. The venter is a wide trapezoidal figure, blackish brown in color, with yellowish gray lunettes on either side. The spinnerets are surrounded with black, and are themselves blackish or dark brown in color. The epigynum (Plate I., Fig. 9) has a narrow scapus but little widened at the base, and narrowing down to the portulæ on either side, which are prominent and present at times the appearance of the figure, but at others more compacted, and rather resembling the portulæ of Epeira patagiata. (Fig. 11a.)

MALE: Length, 7 mm.; in markings and general color closely resembles the female. The cephalothorax is more rounded and of a uniform bright brown color, apparently not so heavily haired, pubescent upon the top of the corselet, but with a marked ring of gray hairs encompassing the margin, and also along the edges of the caput to the eye space. The median fosse is a longitudinal slit. The sternum is rather more cordate in shape; the legs much longer and relatively thinner than in the female. The second leg is not specialized in any way, and there appears to be no special clasping spines or armatures at any point; the femora, especially of the first two pairs, are mottled beneath with dark brown spots, which sometimes may also be observed in the female. The character of the palp is shown at Plate I., 9a. It is easily distinguished from the male of E. strix, not only by the general appearance and character of the palpal bulb, but more easily at once by the absence of the curved metatarsus and the series of strong, black clasping spines upon the inside of the second tibia which characterize Strix. The first leg of Strix, also, is more heavily armed at the thickened tibia with numerous black spines.

DISTRIBUTION: Epeira sclopetaria is a common spider in many sections of the country. It is abundant along the seashore of New England and New Jersey. It is also found around the outhouses, stables, etc., in the neighborhood of Philadelphia. Hentz described it from South Carolina. Specimens are common in collections from the West, and it is probably distributed over the entire United States. It is also a European species of general distribution from Sweden southward, and probably shares with E. patagiata a world wide distribution through the northern temperate zone.

No. 2. Epeira patagiata (CLERCK).   Plate I., Fig. 11; Pl. III., Figs. 8, 9.

1757. *Araneus patagiatus,* CLERCK . . . Aranei Svecici, p. 38, pl. 1, tab. 10.
1834. *Epeira dumetorum,* HAHN . . . . Die Arachn., ii., p. 31, tab. 48, Fig. 117.
1837. *Epeira dumetorum,* KOCH . . . . Uebersicht des Arachniden-Systems, heft 1., p. 2.

1845. *Epeira patagiata*, Koch . . . . . Die Arachnid, xi., p. 115, tab. 386, Figs. 916, 919.
1861. *Epeira patagiata*, Westring . . . Araneæ Svecicæ, p. 36.
1864. *Epeira patagiata*, Blackwall . . Sp. G. B. & I., p. 320, pl. 24, Fig. 229.
1865. *Epeira formosa*, Keyserling . . . Beitrg. z. k. d. Orbitel. Verh. d. z. b. Ges. Wien,
          p. 828, pl. 19, Figs. 17, 18.
1866. *Epeira patagiata*, Menge . ∵ . . . Preuss. Spinn., i., p. 60, pl. 8, tab. 9.
1867. *Epeira patagiata*, Ohlert . . . . Aran. d. Prov. Preuss., p. 24.
1870. *Epeira patagiata*, Thorell . . . . Synon. Europ. Spid., p. 16.
1881. *Epeira hilaris*, Cambridge . . . . Spid. from Newfoundland, Proc. Roy. Phys. Soc.
          Edinbg., p. 112.
1884. *Epeira patagiata*, Emerton . . . N. E. Ep., p. 305, pl. 33, Fig. 3.
1889. *Epeira patagiata*, McCook . . . . Amer. Spid. and their Spinningwork.
1889. *Epeira patagiata*, Marx . . . . . Catalogue, in loc.

FEMALE: This species closely resembles E. sclopetaria and E. strix. The specimens differ much in their dorsal markings, as well as in their general colorings, as illustrated in Plate III., Figs. 8a-8f. Sometimes the specimens are quite light, at other times dark brown or black, and among northern individuals yellow bands are mingled with the dark colors. The specimens collected in the North appear to be colored more brilliantly than those found in the South. Total length of adults, from 13 to 8 mm.

CEPHALOTHORAX: 5 mm. long by 4 mm. wide, 2 mm. at the face; a rounded oval, not high, flattened at the summit; the fossa a deep semicircular pit; skin glossy, thickly covered with grayish white hairs upon the summit and the entire surface of the caput, and a line of similar hairs around the margin of the corselet. The head is somewhat depressed, the face with heavy gray eyebrows. The sternum shield shaped, longer than wide; dark or dark brown, covered with grayish yellow hair. The labium is semicircular; dark brown at the base and yellow at the tips, as are also the maxillæ.

LEGS: Order, 1, 2, 4, 3; femora, orange yellow to brown at the base, blackish at the tip; the patella and tibia similarly colored, the latter with a darker ring at the tip and a median annulus. The metatarsus and tarsus are yellow, the former with dark annuli at the tip and midway. The feet are blackish brown; strongly armored with yellowish gray hairs and bristles and dark spines, which are intermingled with shorter yellow ones, dark at the base. The palps are armed as the feet. The maxillæ are parallel, conical, glossy, blackish brown, and hairy.

EYES: Ocular quad upon an eminence, the rear eyes set at the base thereof; length and width about equal; wider in front than behind; MF separated by about two diameters, MR by about one; middle eyes not greatly differing in size. Side eyes not contingent; set upon tubercles; SF the larger, and separated from MF by about 1.3 the area of the latter. Front row recurved; rear row, which is the longer, procurved. The clypeus removed from MF by 1.5 the diameter of the latter.

ABDOMEN: A long oval; 8.5 mm. long, 6 mm. wide, across the base; narrowing towards the spinnerets; not greatly arched; narrowing to the apex, which is about as thick as the base. The dorsal folium resembles that of Sclopetaria and Strix, but lacks the strong marginal lines of long hairs on the former. The color is dark brown, with a yellow herring bone pattern in the centre. A lance head figure of brown, with yellow margin, divides the base, passing along the front. The margins of the folium are deeply scalloped; on either side a broad undulating band of yellow, followed along the sides by a band of dark brown; the venter a dark brown trapezoidal band, surrounded by yellow lunettes resembling Strix and Sclopetaria. The epigynum (Plate I., Fig. 11a) has a short spoon shaped scapus, which in mature specimens separates it from Strix and Sclopetaria.

MALE: Total length, 9 mm.; abdomen, 6 mm. long by 4 mm. wide; cephalothorax, 5 mm. long by 4 mm. wide. In appearance (Plate III., Fig. 9) it does not widely differ from the female. The cephalothorax is brown shading off into yellow, pubescent, with a narrow, hairy ring around the margin of the corselet; the caput covered with hairs longer and more numerous on the edges, forming a decided eyebrow. The legs are yellowish brown, annulated, strongly armed with hairs, bristles, and spines. Tibia-II is not specially thickened or provided with the clasping spines, but is well armored with long black spines. The palpal digit is shown at Plate I., Fig. 11b. It may be readily distinguished from those of E. cornuta-strix and E. sclopetaria by the absence of the long bifid crablike claw near the base.

DISTRIBUTION: Epeira patagiata is one of the earliest known species, having been described in the last century by Clerck, the Swedish pioneer in Arachnology. It is widely extended over Europe, and is one of the Syrian spiders collected in the Holy Land by Mr. Cambridge. Its distribution in our own country is quite general. I have collected it throughout the Eastern United States as far to the northeast as Massachusetts, and in the northwest at Portland, Oregon. The collection of Dr. Marx notes it as far south as Virginia, and to the northeast at Fort Simmo, Labrador. I have not found it abundant around Philadelphia. Along the northern boundary of the United States it has been collected on Lake Superior, in Michigan, in Montana, at Fort Yukon and Sitka in Alaska, and Fort Kavanah, in the Aleutian Islands. I have numerous specimens from Utah, collected by Professor Orson Howard, and numbers from California by Mr. Curtis. It is probable, therefore, that the spider has made the entire circuit of the world, and may be found in almost every country of the northern hemisphere. In this respect it is certainly entitled to a remarkable position among our spider fauna, but perhaps other species might share this distinction of cosmopolitan distribution had they been collected as diligently as Patagiata.

No. 3. Epeira cornuta (CLERCK). Plate I., Figs. 8, 9, 10, 11.

1757.	Araneus cornutus, CLERCK	Sv. Spindl., p. 39, pl. 1, tab. 2.
1805.	Epeira apoclisa, WALCKENAER . .	Tab. d. Araign., p. 16 (in part).
1835.	Epeira arundinacea, KOCH	Herr.-Schaeff., Deutchl. Ins., 131, 18–20.
1837.	Epeira apoclisa, HAHN	Die Arach., ii., p. 30, pl. 48, Fig. 116.
1845.	Epeira arundinacea, KOCH	Id., xi., p. 109, pl. 385, Fig. 913.
1845.	Epeira foliata, KOCH	Die Arach., xi., p. 119, Figs. 920, 921.
1851.	Epeira cornuta, WESTRING	Enum. Aran., p. 34, p. 21.
1855.	Epeira cornuta, THORELL	Recensio Critica, p. 21.
1864.	Epeira foliata, KEYSERLING . . .	Beschr. n. Orbitel. Sitz. d. Dresden, p. 92, pl. 7, Figs. 10, 11.
1866.	Epeira cornuta, MENGE	Preuss. Spinn., i., p. 58, pl. 8, tab. 8.
1870.	Epeira cornuta, THORELL	Syn. Europ. Spid., p. 15.

VARIETY: Epeira cornuta-strix HENTZ.

1837.	Epeira apoclisa-americana, WALCK.,	Ins. Apt., ii., p. 61.
1837.	Epeira foliosa, WALCKENAER . . .	Ins. Apt., ii., p. 68; ABBOT'S Ga. Sp. Mss., No. 39.
1846.	Epeira affinis, BLACKWALL	Spid. from Canada, Ann. & Mag. Nat. Hist., xviii., 77.
1847.	Epeira strix, HENTZ	J. B. S., v., p. 473; Sp. U. S., 112, xiii., 5.
1860.	Epeira apoclisa, GIEBEL	Spinn. a Illinois. Zeitschr. f. Ges. Naturwiss, xxxiii., p. 249.
1884.	Epeira strix, EMERTON	N. Engl. Ep., 305; xxxiii., 5; xxxv., 12.
1889.	Epeira strix, McCOOK	Am. Spid. and their Spinningwork.
1889.	Epeira cornuta, MARX	Catalogue, p. 544.

FEMALE: Total length, 12 mm.; cephalothorax, 6 mm. long, 5 mm. wide; abdomen, 7 mm. long by 5.5 mm. wide. I have classified Hentz's Eperia strix as a variety of E. cornuta, after comparing the former with specimens of the latter. Of these, one was furnished by Professor Waldemar Wagner, from Moscow, Russia, and the other by Mr. Thomas Workman, from Ireland. Both of these examples differ from Strix in that the more roundly arched abdomen is diminished backward more decidedly than Strix. Strix is slightly flattened upon the dorsum of the abdomen, and the sides are carried from the base to the apex with very little diminution in width, making an almost even oval in outline. The genital organs of the two species are alike. I have no male of E. cornuta to compare with that of E. strix. The following description is of the American specimen.

CEPHALOTHORAX: Oval, quadrate in front, rounded at the sides; truncate behind, where it is slightly notched; the slope is not steep to the median fosse, which is tolerably deep; caput arched, slightly elevated at the middle above the corselet. Color brown or yellowish brown, glossy, a thick row of gray hairs along the margin, which are more lightly distributed upon the sides and the deep cephalic suture, and more heavily upon the caput back of the eyes, and again upon the face in front. Sternum shield shaped, longer than wide, the width not greatly unequal throughout; dark glossy brown, with a lighter median band and hairy; sternal cones wanting. The lip as wide as long, subtriangular. The maxillæ obtusely triangular at the tips, as wide as long, and, like the lip, dark glossy brown, lighter at the tips.

EYES: The ocular quad on a rounded prominence, much more decided in front; MF, indeed, may be said to be placed upon large separate tubercles. The quadrilateral wider in front than behind, and the length about equal to the greatest width; MF are separated by about 1.5 to 2 diameters, and are larger than MR, which are set close together, not being separated by more than a radius. The side eyes are upon low tubercles, are separated by about a radius of SR, which is somewhat smaller than SF, and situated behind it low upon the sides, thus bringing it almost into the front row of eyes. MF are separated from SF by about 1.3 their area, or 2.5 times their intervening distance. The front row is but little recurved, the rear row, which is much the longer, is procurved. The space between SR and MR is at least three times the area of the latter; height of clypeus about twice diameter MF.

LEGS: Stout, rather short for the size of the species, leg–I measuring 16 mm. Order, 1, 2, 4, 3. They are heavily clothed with yellowish gray hairs, and with numerous blackish or blackish brown spines. Color yellow, with dark brown annuli at the ends of the joints; no median annuli; palps similarly colored and heavily armed at the tips. Mandibles conical, much arched at the base, where they slightly project beyond the plane of the face, are sparsely covered on the insides with gray hairs, and have a decided cog at the articulation with the side face; they are dark glossy brown, almost black.

ABDOMEN: A long oval, very little diminishing from base to apex; dorsum not highly arched, and, except in gravid females, rather flat; base overhangs cephalothorax; the apex rounded and high above spinnerets; dorsal folium wide, diminishing somewhat towards the spinnerets; the margin, which is undulating, is dark brown or blackish, enclosing a herring bone pattern of yellow, having a broken median band of darker color. A narrow ribbon of whitish gray borders the scalloped edge of the folium, merging into broad ribbons of yellowish color along the sides, the color of the yellow bands, particularly within the folium, being sometimes cretaceous. These are followed farther down by a band of darker color extending to the venter. The skin is glossy. The venter has a broad band of dark brown bordered by yellowish lunettes. The epigynum (Plate I., Fig. 10) has a short thin scapus, and the portulæ of the atriolum are separated by a broad, oval frontal plate, over which the tip of the scapus extends. In some specimens the abdomen is quite dark, even blackish, and this is not infrequent with the young.

MALE: The male differs little in color and marking from the female, and is not greatly less in length (9 mm.), and specimens may sometimes be found equal to the female. The palp is marked by a curved lanceolate hook at the extermity, terminating in a quite decided point. The second tibia is curved and armed upon the inside with strong, black clasping spines.

DISTRIBUTION: This species is one of the most common in the Eastern United States, and is distributed along the Atlantic seaboard and Middle States from Canada to the south, where I have collected it and observed its habits. Its extension westward I have not been able to trace beyond Wisconsin (Professor Peckham), but it is probably widely distributed. As E. cornuta, the principal form, it is widely dispersed over Europe. I have captured it in Great Britain and Norway, and a young specimen which I take to be this species on the upper railings of the dome of St. Peter's in Rome. Most students of spiders have been greatly troubled in determining the foregoing three species, namely, Epeira strix, E. sclopetaria, and E. patagiata. There are certainly strong resemblances between the three; and it is easy with immature specimens to confound the one with the other, indeed, it is often difficult to separate them; but in the case of well matured specimens this difficulty is not so great. Strix is marked from Sclopetaria by the more even oval of its abdomen, which also tends to be flat upon the dorsum; by the shorter and stouter legs, which also have no median annuli; by the general glossy appearance upon all parts of the body, in which even the abdomen shares; by the shorter pubescence; by the less distinctly marked line of whitish gray hairs around the margin of the caput and along the front of the face. The head of Strix is also a little more rounded than that of Sclopetaria and Patagiata, and the face is not quite so narrow. The rear eyes of the lateral pairs are set a little farther up upon the head than in Sclopetaria, but do not greatly differ in this respect from Patagiata. The dorsal folium of the abdomen is usually unbroken from the base to the apex, whereas in Sclopetaria there is a distinct separation into two figures, or parts of one figure, at a point about one-third the distance between the base and the apex. A decided difference also appears in the epigyna, the scapus of Strix being much shorter and smaller than that of Sclopetaria, and the general structure in other respects differing. The scapus on the epigynum of Strix more closely resembles that of Sclopetaria than that of Patagiata. (See and compare Plate I., Figs. 9, 10, 11.) All the above marks which distinguish Strix from Sclopetaria equally separate it from Patagiata. The latter is distinguished from Sclopetaria in the geographical province of Philadelphia by its smaller size, but specimens of Patagiata from the Pacific Coast are quite as large; see also the characteristics referred to in the above description. But it may be most readily distinguished in mature species by the shape of the epigynum.

No. 4. Epeira pratensis HENTZ.   Plate I., Figs. 6, 6a, 6b.

1847. *Epeira pratensis*, HENTZ . . . . . J. B. S., v., p. 475, pl. 31, Fig. 11.
1875. *Epeira pratensis*, HENTZ . . . . . Sp. U. S., p. 115, pl. 13, Fig. 11.
1884. *Epeira pratensis*, EMERTON . . . N. E. Ep., p. 310, pl. 33, 15 ; 36, 9.
1802. *Epeira pratensis*, KEYSERLING . . Spinn. Amerk., p. 184, tab. ix., 136.

FEMALE: Total length, 10 mm.; cephalothorax, 5 mm. long, 3 mm. wide; abdomen, 8 mm. long, 5 mm. wide; facial width, 1.4 mm. This species varies much in size, the above being one of the largest specimens.

CEPHALOTHORAX: A long oval, highest in front of the fosse, which is a deep longitudinal slit, and thence shelving to the base, which is truncated; head depressed, quadrate; cephalic suture well marked; corselet grooves distinct but not deep; colors yellowish brown, with darker brown band along margins of corselet, and a yet darker one on middle of caput; skin glossy, slightly pubescent, but the head is well covered with stout gray hairs, with a few long white bristles over the side eyes. The sternum is black, glossy, with pronounced cones in front of the coxæ, a few black hairs, longer than broad, the edges scalloped and the point attenuated. Labium longer than broad, dark brown at the base, lighter upon the subtriangular tip. Maxillæ dark brown, rather longer than wide, inclined towards each other, and obtusely triangular.

EYES: Ocular quad on a rounded prominence much more decided in front, the rear eyes, indeed, lying at the base thereof; the front wider than rear, and the sides longer than the width; MF separated a little more than one diameter, and slightly larger than MR, which are separated by about a radius; side eyes contingent on a slight tubercle; SF larger than SR; MF removed from SF by about 1.3 their area, and from the clypeus about two diameters; front row slightly recurved, the rear row longer and procurved.

LEGS: 1, 2, 4, 3, as follows: 14.3, 13.1, 12.5, 8.3 mm. They are stout, brownish yellow, with dark annuli at tips of joints, strongly armed with brownish spines, gray bristles and hairs; palps colored and armed as legs. Mandibles conical, dark brown, arched at the base, where they are on a plane with the face, smooth and glossy.

ABDOMEN: Ovate, widest at base, arched upon dorsum to spinnerets, which are underneath the apical wall; covered with yellowish gray pubescence. Color yellowish brown.

which is lighter upon the base and darker upon the apical half. Two interrupted lines of yellow color, on either side of the median, traverse longitudinally the dorsal field, frequently obliterated toward the apex; these encompass a long, lanceolate band of brown color. Beyond this, at about equal distance on each side, are six interrupted lines of black spots, approximating toward the apex, and which mark out the dorsal folium. The sides are a rather uniform mouse color. The venter has a black, subtriangular patch, with two smaller yellow patches at the end and a round yellow spot on each side of the black spinnerets. The epigynum (6a, 6b) has a rather narrow hood, with a strong, trowel shaped scapus, dark brown, hard, glossy, spooned at the tip, furrowed almost to the base, and pubescent.

MALE: Resembles the female, with abdominal markings somewhat plainer. The second femur is thicker than the others, the tibia somewhat thickened, curved, and with a series of short, stout clasping spines upon the inner side.

DISTRIBUTION: This species is widely distributed, especially along the Atlantic slope, having been collected from Massachusetts, where it is adult in the latter part of July or early August, and southward to Virginia. Westward, specimens have been obtained from Missouri and Utah (Marx). It is doubtless distributed over the whole United States along the parallels above indicated, and probably is also found in the southern group of States.

No. 5. Epeira marmorea (CLERCK). Plate I., Figs. 1, 2; Pl. II., 1, 2; Pl. III., 7, 8.

1757.	*Araneus marmoreus*, CLERCK	. .	Svenska Spindlar, p. 29, pl. 1, tab. 3.
1757.	*Araneus Babel*, CLERCK	. . . . .	Ibid., pl. 1, tab. 6.
1775.	*Arenea marmorea*, FABRICIUS	. .	Syst. Entom., ii., p. 415, 31.
1778.	*Aranea aurantio-maculata*, DE GEER	. . . .	Mem., viii., p. 222, pl. 12, Figs. 10, 17.
1797.	*Aranea regalis*, PANZER	. . . . .	Faun. Ins. Germ., 40, 21.
1802.	*Aranea melittagria*, WALCKENAER,	Fauna Parisienne, ii., 191.
1805.	*Epeira marmorea*, WALCKENAER	.	Tableau d'Araneides, p. 61.
1832.	*Epeira marmorea*, SUNDEVAL	. .	Svenski Spind., p. 241.
1839.	*Epeira marmorea*, KOCH	. . . .	Die Arach., v., p. 63, pl. 162, Figs. 379, 380.
1861.	*Epeira marmorea*, WESTRING	. .	Aran. Svec., p. 30.
1864.	*Epeira scalaris*, BLACKWALL.	. .	Sp. Gt. B. & I., ii., p. 331, pl. 24, 240.
1866.	*Epeira pyramidata*, MENGE	. .	Preuss. Spinn., p. 51, pl. 3, tab. 3.
1884.	*Epeira marmorea*, EMERTON	. . .	N. E. Ep., p. 307, pl. 33, Fig. 2.

VARIETY: Epeira conspicellata WALCKENAER.

1842.	*Epeira conspicellata*,WALCKENAER[1]	Ins. Apt., ii., p. 58; ABBOT, G. S., No. 126.
1847.	*Epeira insularis*, HENTZ	. . . . .	J. B. S., v., 470, xxxi.
1847.	*Epeira obesa*, HENTZ	. . . . . .	Ibid., Fig. 2.
1863.	*Epeira insularis*, KEYSERLING	. .	Neuer Orb., Sitzung. Isis, p. 91, tab. v., 3.
1860.	*Epeira annulipes*, GIEBEL	. . . .	Spinn. aus Illinois, Zeitschr. f. Gesammten Naturwiss., xxxiii., 250.
1870.	*Epeira marmorea*, THORELL	. . .	Syn. Eu. Spid., p. 9, vars. *intermedia*, *pyramidata*.
1881.	*Epeira obesa*, CAMBRIDGE	. . . .	Spid. Newfind., Proc. Roy. Phys. Soc. Edin., p. 112.
1884.	*Epeira insularis*, EMERTON	. . .	N. E. Ep., p. 309, pl. 32, Fig. 1.
1888.	*Epeira conspicellata*, McCOOK	. .	Proceed. Acad. Nat. Sci., Phila., p. 5.
1889.	*Epeira insularis*, McCOOK	. . . .	Amer. Spid. and their Spinningwork.
1889.	*Epeira marmorea*, McCOOK	. . .	Amer. Spid. and their Spinningwork, Vol. I., p. 77.
1889.	*Epeira marmorea*, MARX	. . . .	Catalogue, p. 546.
1892.	*Epeira insularis*, KEYSERLING	. .	Spinnen Amerikas, Epeir., p. 170, pl. 8, 126.

[1] Walckenaer has "conspicillata," perhaps a misprint; he named the species from the fancied resemblance of the dorsal pattern to a pair of spectacles. The identity with Hentz's E. insularis admits of no doubt, and if our American form be held as a true species it must bear the prior name of Walckenaer, as I first pointed out in 1888.

FEMALE: Body length, 15 mm.; cephalothorax, 7 mm. long, 6 mm. wide; abdomen, 12 mm. long, 7 mm. wide. This species varies greatly in size, ranging from the above to 10 mm. long, and an abdominal width of 5 mm. The description following is of the American variety, E. conspicellata (E. insularis), as the one most prevalent and generally known. The two varieties are so closely interblended that it is often difficult and indeed impossible, to positively distinguish them, but the figures in the plates will sufficiently show the features of the most divergent forms. E. marmorea (Plate III., 6) more closely resembles the European specimens, which I have compared with examples sent me from Russia by Professor W. Wagner. I have found it in northern New York and New England, and not elsewhere. The variety E. conspicellata is the one which appears to prevail west and south of the above limits. It differs from the *forma principalis* chiefly by the brighter orange hues, the more sharply outlined folium, and by somewhat more distinctly annulated legs. In general size, form, and habits the species are alike.

CEPHALOTHORAX: Almost as wide as long; truncated and indented at the base, shelving upward to the fosse, which is deep; cephalic suture well marked, corselet grooves sufficiently distinct; flat upon top, rounded at sides; head slightly arched, and a very little elevated above the level of the corselet; color varying from orange to yellow, with darker stripes upon the sides and middle of caput; skin glossy and lightly pubescent. Sternum heart shaped, indented at the edges, with decided cones in front of the coxæ and labium and upon the apex; flattened in the middle, lightly pubescent, orange brown, with a lighter yellowish median band; lip longer than wide, rounded at sides, tips subtriangular, more than half the height of the maxillæ. Maxillæ rounded at the sides, tips obtusely triangular and inclined towards each other, length and width about equal. Color of maxillæ and lip orange brown, with light yellow tips.

EYES: Ocular quad on a rounded prominence, wider in front than rear, sides about equal in length to front; MF separated by about 1.3 diameter, and somewhat larger than MR, which are separated by about one diameter. Side eyes on tubercles, SF larger than SR, separated by about the radius of SR; MF removed from SF by about 2.5 times their intervening space, or 1.3 their area. Front row recurved, rear row longer and procurved. Height of clypeus about 1.5 diameter MF or less; a line of bristles extends along its margin.

LEGS: Order, 1, 2, 4, 3; color usually a bright orange both above and beneath as far as the tibia, whence the color is yellow, with dark brown or blackish tips at the joints; provided with long gray hairs and stout bristles, yellowish upon the yellow parts and black upon the dark parts; numerous spines, long, yellowish white, with brown bases. Palps lighter in color than legs, tips black, and terminal joints heavily covered with bristles and spines.

ABDOMEN: Ovate, widest about the middle, highly arched on the dorsum, tapering to the distal spinnerets; color usually bright yellow, a broad folium narrowing to the apex with at least six indentations, which are distinctly marked out with broad margins of brown. The median design is composed of a series of cruciform figures and triangles, which vary more or less according to specimens, but are tolerably persistent. The figure upon the base usually resembles a Maltese cross, though sometimes the upper arm is obliterated, giving the appearance of a trefoil. The next figure is sometimes cruciform, but the central part is occasionally obliterated, leaving the points grouped as a quatrefoil. Waving lines of yellow brown pass along the sides from dorsum to spinnerets. The ventral figure is a subtriangular patch of velvety brown or black, bordered by a horseshoe band of yellow. Six compressed spots are arranged on either side of the median line. The spinnerets are usually orange color, though sometimes brown or blackish, with a yellow spot at either side of the base. The epigynum (Plate I., Figs. 1a, 1b; Figs. 2, 2a) has, in some examples, at least, the receptacles thrown up prominently above the surface; the scapus long, wrinkled, of nearly equal length throughout, but somewhat narrowing toward the tip, which is spooned, and the entire under surface furrowed.

MALE: Individuals vary much in size, one specimen before me being 9 mm. long; cephalothorax, 5 mm. long by 4 mm. wide across the corselet, diminishing to about 2 mm.

at the face. The color of the cephalothorax is uniform orange brown, with a light band, heavily pubescent, on the extreme margin. The skin is glossy, but little pubescent. The legs are orange to orange yellow, with orange brown annuli at the tips of the joints, and median annuli along femurs-I and II. The color of the tibia is rather lighter than that of the femur, and the metatarsus and tarsus still lighter yellow. The tibia of leg-II is curved, thickened about the middle, and with a triple row of short, black, toothed clasping spines upon the inner side. The spines upon the femora are shorter; those upon the tibia of leg-I are long, yellowish brown, with dark bases. The coxæ of leg-I has upon the side next leg-II a short, curved brown spur, curved toward the face; and coxæ-II has at the base near the articulation of the sternum a slight conical process. The palps are as in Plate I., Fig. 1c, 2b. The abdomen is a long oval, colored and marked upon the dorsum with a folium not greatly differing from that of the female.

DISTRIBUTION: As Marmorea is widely distributed throughout Europe and the American specimen is found in most parts of the United States, the species is thus seen to have an extended distribution. I have collected it from New England southward to Florida, westward through Pennsylvania, Ohio, Illinois, Wisconsin. It is found in Arizona, but I have as yet received no specimens from the Pacific Coast.

No. 6. Epeira trifolium Hentz.   Plate I., 3a–6; Pl. II., 3; Vol. II., Pl. IV.

1837. *Epeira jaspidata*, WALCKENAER . Ins. Apt., ii., p. 59; ABBOT, G. S. No. 111.[1]
1847. *Epeira trifolium*, HENTZ . . . . . J. B. S., v., 471; Sp. U. S., 110, xiii., 1
1847. *Epeira aureola*, HENTZ . . . . . Ibid., pl. xxi. (xlii.), 2.
1884. *Epeira trifolium*, EMERTON . . . N. E. Ep., p. 306, pl. 33, Fig. 8.
1889. *Epeira trifolium*, McCOOK . . . . Amer.. Spid. and their Spinningwork.
1889. *Epeira trifolium*, MARX . . . . . Catalogue, p. 548.

FEMALE: Total length, 15 mm.; cephalothorax, 7.5 mm. long, 6 mm. wide; abdomen, 12 mm. long, 8 mm. wide. This is one of our largest indigenous species, and varies much in size from the above length and upward to adult females of 10 mm. long. It is especially distinguished by great differences in color at different stages and among different individuals, as illustrated Vol. II., Plate I., p. 48.

CEPHALOTHORAX: A rounded oval or cordate, truncated and indented at the base; corselet flat on top, the fosse a deep transverse slit; cephalic suture sufficiently marked, corselet grooves rather indistinct; color brown, with a yellow marginal band on the dorsum, passing upward to the sides of the caput; head lowly arched; face depressed, wide, quadrate, with gray pubescence. The sternum is shield shaped, dark brown, the median band of lighter color; sternal cones marked; labium blackish brown, lighter at the tip, as are also the maxillæ, which are longer than or as long as wide.

EYES: Ocular quad on a well rounded prominence, the front wider than rear and narrower than sides; MF separated by 1.5 to 2 diameters, somewhat larger than MR, which are separated by one diameter or more; side eyes' on tubercles, separated by about the diameter of SR, which are smaller than SF; MF divided from SF by about 1.5 the area of MF, or say 2.5 times their interspace; clypeus height about two diameters or more of MF, and with marginal row of strong yellow hairs. The front row recurved and shorter than the procurved rear row.

LEGS: 1, 2, 4, 3; strong, stout, not long for the size of the species, abundantly provided with grayish yellow hairs and bristles and yellow spines, which are dark brown on the basal half; color yellow, with bright brown apical tips, patellæ entirely brown, feet brownish black. Palps light yellow, without annuli, thickly provided with strong yellow spines and bristles. Mandibles strong, conical, arched at the base, and swollen on the outside at the articulation with the clypeus.

---

[1] Abbot's drawing of this species in the MSS. in Kensington Library of the Brit. Mus. Nat. Hist. appears. to be E. trifolium Hentz; but there is enough doubt as between this and E. insularis Hz. to justify leaving the name thus.

ABDOMEN: The dorsal markings are difficult to describe, and vary considerably. The dorsum is well arched, rounding from base to distal spinnerets; covered sparsely with white hairs. Near the base three white patches of circular or irregular shape are grouped together in a form somewhat like clover or shamrock leaves, which doubtless suggested the specific name. These are attached to the yellowish stem which proceeds along the median line forward and backward. In some examples this trefoil pattern is repeated about the middle of the dorsum, and white or whitish patches of varying sizes are distributed in semicircular lines, with the convex part towards the apex, along the dorsal field. In some specimens the median band is continuous; in others it is broken up into groups of whitish patches resembling clover leaves. Plate II., Fig. 3, and Plate I., Vol. II., represent some of the most decided patterns. Fasciculated markings extend on either side of the median line from about the middle of the dorsum towards the apex, and these are usually brown. The color is yellowish brown at the margin of the sides, somewhat mottled with black interrupted stripes. The venter has a broad brownish band, extending from the pedicle to the spinnerets, sometimes with a marginal border of yellow. The epigynum (Plate I., Figs. 3, 3a) has a large and well arched porch, and a stout scapus, bent downward and forward, spoon tipped, furrowed, and of about equal width throughout, though slightly narrowed at the tip and widened at the base. The basal parts of the portulæ widen at either side, and are curved like a scallop shell.

MALE: The male (Vol. II., Plate I., Fig. 10) is much smaller than the female, specimens in hand being 4.8 mm. long. The cephalothorax is longer than the abdomen in some examples, of yellowish color, and the legs have the same hue, with yellowish brown annuli at the joints. The abdomen is white or whitish, with a slight tendency to be broken up into irregular markings. The palps are a whitish yellow, with dark yellow markings upon the digital joint, which is represented at Plate I., Fig. 36.

There is an undoubted resemblance between this species and the European Epeira quadrata; but after having compared the typical Epeira trifolium with an example of E. quadrata, female, sent me from Moscow, Russia, by Professor Waldemar Wagner, I regard them as distinct. Not to speak of other differences, the scapus of the epigynum is quite different. In Quadrata it is wide at the base, where it is slightly notched, and rapidly narrows to a rounded point; the tip is spooned, and the edges rimmed slightly throughout the entire length. The organ has somewhat the shape of a mason's trowel. On the contrary, the scapus of Trifolium is not so wide at the base, is not set upon the atriolum like the blade of a trowel upon its handle, but is continuous with the same. Moreover, it preserves a nearly equal width throughout the entire length of the scapus, being slightly wider at the base. Dr. Thorell, to whom I sent a specimen of Epeira trifolium, agrees in thinking the two to be distinct species. The spines in Quadrata are dark or blackish, but in Trifolium they are yellow, with brown bases. The midfront eyes of Quadrata also seem to be relatively nearer than in Trifolium.

DISTRIBUTION: Throughout the United States, along the Atlantic Coast from Canada through New England, New York, Pennsylvania, Maryland, Virginia. It has been found in Alabama, and probably inhabits the central Southern States; has been collected in Maryland and the District of Columbia, in Wisconsin, Colorado, Wyoming, and Montana, and as far south as New Mexico. I have the male from Utah (Professor O. Howard), the Big Horn Country, Tacoma, Wash. (Mr. M. S. Hill), and San Diego, Cal. (Mrs. Smith). Hentz's original description is from a specimen collected in Maine, and he makes no note of having found it in the Southern States, where most of his studies were pursued.

Epeira trifolium, variety candicans.   Plate I., Fig. 4, 4a.

I find a number of specimens of this species differing little from the typical form in detailed structure, but which in general appearance are strikingly different, approximating the form of the male. (Plate I., Fig. 4.) In these the abdomen is ovate, considerably narrower than long, instead of the globose form most prevailing. The color is yellowish white, with lighter shade upon the dorsum, which is rather flattened than arched. The skin is abundantly covered with long whitish yellow bristles. One specimen from California (Mrs. Smith) measures 10 mm. long; abdomen, 7 mm. long by 4.5 mm. wide; cephalothorax, 5 mm. long by 4 mm. wide; first leg, 15 mm. It has the spines dark, as in E. quadrata, instead of yellow with brown bases, as in E. trifolium.

No. 7. **Epeira Benjamina** WALCKENAER.   Plate I., Fig. 7; Pl. II., Figs. 4, 5.

1837. *Epeira Benjamina*, WALCKENAER . Ins. Apt., ii., p. 42; ABBOT, G. S., No. 126, 351.
1837. *Epeira mutabilis*, WALCKENAER . . Ibid., p. 73; ABBOT, G. S., No. 351.
1847. *Epeira domiciliorum*, HENTZ . . . J. B. S., v., p. 469; Id., Sp. U. S., p. 108; xii., 7.
1864. *Epeira Hentzii*, KEYSERLING . . . Beschr. n. Orbitel., p. 97; v., 10, 11.
1884. *Epeira domiciliorum*, EMERTON . N. E. Ep., p. 312; xxxiii., 17; xxxvi., 1-4.
1888. *Epeira Benjamina*, McCOOK . . . Necessity for Revising Nomenclature of Am. Spid.,
                                          Proceed. Acad. Nat. Sci., Phila., p. 5.
1889. *Epeira domiciliorum*, McCOOK . . Amer. Spid. and their Spinningwork, Vols. I., II.
1889. *Epeira benjamina*, MARX . . . . Catalogue.

FEMALE: Total length, 15 mm.; cephalothorax, 6 mm. long, 5 mm. wide; abdomen, 11 mm. long; 9 mm. wide at the base, diminishing to 3 mm. or less at the apex. The species differs much in size and color according to distribution and age, and in the typical specimens the fore part of the body is bright orange on the cephalothorax, thighs, and patella, lightening into yellow on the remaining joints of the legs, with decided darker annuli at the tips. The abdomen is yellow or yellowish brown, which in old age deepens, becoming reddish brown, a shade which the cephalothorax and legs also assume. See Plate II., Fig. 4, drawn from an old specimen.

CEPHALOTHORAX: Rounded, truncated at the indented base; the posterior slope smooth, the fosse deep; the cephalic suture marked, grooves sufficiently distinct; margins and sides of corselet and sides of the head covered with yellowish hairs; skin glossy, varying from uniform brown to yellow or orange yellow streaked with orange marks. Caput quadrate in front, and slightly depressed from the summit of the corselet. Sternum shield shaped, yellowish brown, lightened in the middle; in some specimens much brightened with a broad yellow band attenuated at the point; skin glossy, covered with yellowish gray hairs, with sternal cones raised in the middle. Lip wider than high, subtriangular; maxillæ somewhat longer than wide, tips subtriangular, and directed towards each other; color brown, or like the sternum, with lighter shade upon the tips.

EYES: Ocular quad on a prominence more decided in front, where it is slightly wider, the sides somewhat longer. MF separated by at least 1.5 diameter, somewhat larger than MR, which are separated by about one diameter; the ocular eminence is surrounded at the base by yellowish hairs, which are also on the face and eyebrows. Side eyes on tubercles, contingent; SF larger than SR. MF removed from SF by about 2.5 times their intervening space or 1.3 their alignment; front row somewhat recurved, rear row procurved; clypeus about two diameters of MF high, and the margin has rows of strong yellowish bristles.

LEGS: 1, 2, 4, 3; stout upon the femora, though somewhat attenuated at the terminal joints; the ground color yellow, with bands of bright orange at tips of joints and in the middle of the femora; slightly darker median annuli also mark the tibia. The surface is heavily clothed with yellow hairs and bristles, and numerous yellow spines with dark bases, which, however, are blackish on the terminal joints. Palps heavily armed throughout and colored as legs, but of lighter hue; mandibles conical, slightly arched at the base, articulating in the plane of the face; brown or yellowish brown.

ABDOMEN: When the female is not gravid the abdomen is subtriangular in shape, being much wider at the base. It is sparsely covered with long white bristles with dark bases, which are gathered in a tuft at the front where it overhangs the cephalothorax; also freely covered with short yellowish hairs. The dorsal folium (Plate I., Fig. 7) is an anchorlike figure, with strong dentations of light color branching from each side and diminishing towards the apex; this is flanked on either side by well shaded margins of dark color. This pattern shows much more distinctly on some specimens than others and tends to be obliterated with advancing age. (See Plate II., Fig. 4.) The venter has a wide, black subtriangular patch surrounded by a broad yellow margin, presenting the appearance of a military chapeau, sometimes interrupted, leaving but two large circular spots on either side. The spinnerets are bright orange or brown with an interrupted girdle of yellow, one large patch on each side forward of the base. The epigynum (Plate I., Figs. 7a, 7b) has a strong curved scapus almost as wide as the atriolum at its base, where the color is yellow or orange yellow; it is compressed about the middle and widened into a spooned bowl; it is furrowed on either side with a hard blackish brown rim almost to the base.

MALE: Plate I., 7c; Plate II., 5. The male resembles in color the female, though there is a tendency to lighter hues. The body is provided with the same long, strong, whitish bristles and hairs that mark the female. These are often strong on the eyebrows and the margin of the cephalothorax, and they form decided brushes upon the palpal joints. The tibia of the second leg is curved outward and provided on the inner and under side with a double row of short toothlike spines extending the entire length, flanked on one side by a single row, the long spines making a formidable clasping apparatus.

DISTRIBUTION: This is one of the most common of our American species and is widely and probably generally distributed throughout the United States. I have specimens ranging from New England to Florida along the Atlantic Coast, and as far west as California. I have taken it in Portland, Oregon, and Dr. Marx reports it from Nebraska, Texas, Utah, in Colorado at a height of twelve hundred feet, in Minnesota, and at various points along the Atlantic Coast.

No. 8. Epeira arabesca WALCKENAER.   Plate I., Figs. 8, 8a.; Pl. II., Figs. 6, 7.

1805. *Epeira arabesca*, WALCKENAER . . Tableau des Araignées, p. 63, No. 44.
1837. *Epeira arabesca*, WALCKENAER . . Ins. Apt., ii., p. 74; ABBOT, G. S., Nos. 331, 346.
1837. *Epeira mutabilis*, WALCKENAER . Ins. Apt., ii., p. 73, No. 58, in part; ABBOT, G. S., No. 355.
1864. *Epeira trivittata*, KEYSERLING . . Beschr. n. Orbit., Isis, p. 95, 6-9.
1884. *Epeira trivittata*, EMERTON . . . N. E. Ep., p. 311, xxxiii., 16; xxxvi., 2, 3, 5, 8.
1888. *Epeira arabesca*, McCook . . . . Proceed. Acad. Nat. Sc., Phila., p. 3.
1889. *Epeira trivittata*, McCook . . . . Amer. Spid. and their Spinningwork, Vol. I.
1889. *Epeira arabesca*, MARX . . . . . Catalogue, in loc.
1892. *Epeira trivittata*, KEYSERLING . . Spinn. Amerik., p. 172, pl. viii., 127.

FEMALE: Body length varies from 6 to 8 mm.; cephalothorax, 2.5 mm. long, 2 mm. wide; abdomen, 5 mm. long, 4 mm. wide at the base, narrowing much at the apex. Like Epeira Benjamina, which this species resembles in many respects, the general color varies from reddish brown to yellow.

CEPHALOTHORAX: Cordate, elevated at the centre, sloping abruptly to the indented base; the fosse a longitudinal slit; corselet grooves and cephalic suture distinct; color yellow to orange yellow or brown; freely covered with yellowish white hairs, especially on the sides of the caput, which form strong eyebrows at the side; the head depressed, sloping to the face; sternum shield shaped, somewhat longer than wide, sternal cones distinct; pubescent edges brown, with broad median yellow band upon the middle; lip low, obtusely triangular; maxillæ about as wide as long, obtusely triangular at the tips; labium and maxillæ yellow.

EYES: Ocular quad upon an eminence projecting in front, leaving MR scarcely elevated above the facial surface; the front somewhat longer than the rear, and the sides longer than either; the eyes about equal in size, though MR appear slightly larger; MF separated by about 1.3 diameter; MR by about a radius; eyes upon tubercles, barely contingent; SF larger than MF, and separated by about a little more than the area of the former, or 1.5 times their intervening space; the margin of the clypeus is obliterated by the ocular prominence, which comes to the very edge, the height being scarcely more than one diameter of MF; all the eyes are upon black bases, thus appearing dark; the front row recurved, the longer rear row procurved; a few whitish yellow bristles mark the eye space and margin of clypeus.

LEGS: 1, 2, 4, 3; yellow, with strong brown annuli; stout, heavily armored with strong brown spines and yellowish bristles and pubescence; palps colored and armed as legs; mandibles rather long, conical, with yellowish pubescence at edges; the basal cog well marked.

ABDOMEN: Ovate, widest at base, narrowing to the distal spinnerets; dorsum well arched; color yellow to yellowish white, with a brownish yellow folium, whose margins are marked from middle to apex by a V-shaped figure of semicircular brownish patches, with the concavity towards the base; the centre of the dorsum is lighter yellowish; the pattern a broad arrow shaped marking, with side offshoots (Plate I., 8); from the median line curved branchlets proceed on either side to the spinnerets; the surface is reticulated and covered thinly with soft, yellowish hairs. The lunette markings on the dorsum enable one to easily identify the species, though something similar may be found in other species, especially in the male of Epeira Benjamina. In some specimens the colors are reddish brown (Plate II., 6), particularly on the abdomen; in others they are yellow and gray; the venter has a truncated triangle of brown, flanked on either side by a broken yellow band, consisting usually of three spots on either side; the epigynum (Plate I., 8a) has a decided scapus, broadest at the base, not tapering, flattened and spooned at the tip, resembling that of E. Benjamina; the atriolum is narrow, and the portulæ hidden by the projecting scapus.

MALE: Length, 6 mm.; abdomen, 3.5 mm. long, 2 mm. wide; cephalothorax, 3 mm. long, 2 mm. wide; colors and markings closely resembling those of the female, except that the dorsal folium is often white, or whitish, instead of yellow, the colors generally inclined to be a little paler. Tibia-II is somewhat curved, and armed along the entire inner side by a double row of strong black clasping spines; these at the articulation with the patella are flanked on the inside by two very long, strong, straight spines, and on the outer side by three of a similar character, but shorter; the palp is represented at Plate I., Fig. 86. This spider has sometimes been confused with small specimens of Epeira Benjamina. The ocular quad of the latter species is relatively much wider in front; also, the midfront eyes are removed from the margin of the clypeus at least three times their diameter, while in E. arabesca they are about one diameter therefrom. Moreover, the abdomen of Benjamina is rather flattened and triangular ovate, while that of Arabesca is rather ovate. E. Benjamina is a much larger species.

DISTRIBUTION: The United States. I have collected this species along the entire Atlantic Coast, from New England to Florida, and as far west as Ohio. I have specimens from Georgia, North Carolina, Florida, and Texas. It has been taken in Alabama, New Mexico, Utah, and California. I have specimens from as far north as Wisconsin, and have captured it in Canada in the neighborhood of Montreal. It also inhabits South America, where it presents great variation in color and size. As far as my observation goes, it is not subject in the United States to great variation in size, being rarely found longer than 8 mm., or shorter than 6 mm.

No. 9. Epeira cucurbitina (CLERCK).   Plate III., Figs. 1, 2, 3; Pl. IV., Fig. 6.

1757. *Araneus cucurbitinus*, CLERCK . . Svenska Spindlar, p. 44, pl. 2, tab. 4.
1761. *Aranea cucurbitina*, LINNEUS . . Syst. Nat. Ed., 10, i., p. 620.
1775. *Aranea senoculata*, FABRICIUS . . Entom. Syst., t. 2, p. 426, No. 71.
1778. *Epeira viridis-punctata*, DE GEER, Mem., vii., p. 233, pl. 14.
1793. *The Gourd Spider*, MARTYN . . . Natural History of Spiders, p. 19, Spec. 12, pl. 2, Fig. 6.
1805. *Epeira cucurbitina*, WALCKENAER, Tabl. des Aran., p. 63, No. 46.

1806.   *Epeira cucurbitina*, LATREILLE, . . Genera Crust. et Insect., tab. 1, p. 107, No. 11.
1832.   *Epeira cucurbitina*, SUNDEVALL . Svenska Spindlarness, p. 245, No. 8.
1850.   *Miranda cucurbitina*, KOCH . . . Die Arachniden, v., 53, pl. 159, Figs. 371, 372.
1861.   *Epeira cucurbitina*, WESTRING . . Araneæ Svecicæ, p. 50, p. 53, pl. 159, Fig. 371,
               male; 373, female, 342.
1864.   *Epeira cucurbitina*, BLACKWALL . Spid. Gt. B. & I., 342, xxxv., f. 247.
1870.   *Epeira cucurbitina*, THORELL . . Syn. Ent. Spid., p. 23.

VARIETY: E. displicata HENTZ.

1842.   *E. cucurbitina-americana*, WALCK.. Ins. Apt., ii., p. 76; ABBOT, G. S., No. 178.
1847.   *Epeira displicata*, HENTZ . . . . J. B. S., v., 117.
1866.   *Miranda cucurbitina*, MENGE . . Preuss. Spinn., i., p. 68, pl. 10, tab. 14.
1871.   *Epeira cucurbitina*, BLACKWALL . Spiders from Montreal, Ann. & Mag. Nat. Hist.,
               4th ser., viii., 429.
1875.   *Eperia displicata*, HENTZ . . . . Sp. U. S., p. 117, xiii., 17.
1884.   *Eperia displicata*, EMERTON . . . N. Eng. Ep., p. 313, pl. 34, Fig. 4.
1889.   *Eperia displicata*, McCOOK . . . Amer. Spid. and their Spinningwork, Vol. I., 121.
1889.   *Eperia displicata*, MARX . . . . Catalogue, p. 544.
1889.   *Epeira cucurbitina*, MARX . . . . In part, Catalogue, p. 544.

FEMALE: Body length, 8 mm.; abdomen, 6 mm. long, 5 mm. wide; cephalothorax, 3 mm. long, 2 mm. wide. The adult specimens of this spider vary from 9 to 6 mm. in length. The typical color appears to be, for the legs and cephalothorax a yellowish brown; for the abdomen a light green, with a tinge sometimes of blue, for the background of the dorsum, upon which are drawn, on either side of the median line, two irregular scalloped bands of white or yellowish white coming to a point at the apex. On the sides a band of similar color and width passes entirely around the abdomen. The color, however, varies much upon the abdomen; I have specimens in which it is almost white, and others again (as Plate III., 3, 3a) from California where the ground color is a bright red. The species is briefly distinguished by four round black spots arranged in V-shape, beginning at the apex and widely open at the last spot, which is placed almost at the middle of the dorsum. The six pits which mark the muscular attachments, on either side of the median line, are also quite distinct, especially the two forward pairs. On most specimens longitudinal lines, curved or straight, mark the lower part of the dorsum.

CEPHALOTHORAX: Rounded, smooth, glossy; caput erected above the surface though sloping at the ocular area; sternum yellow, slightly covered with hairs, narrowed and pointed at the base, with slight sternal cones.

LEGS: 1, 2, 4, 3; uniform yellow brown, except that the feet are black. They are well clothed, but not excessively, with spines and bristles. Mandibles colored as the cephalothorax, project at the clypeus, where they are well rounded, and taper slightly towards the fangs.

EYES: Ocular quad rectangular, MF smaller than MR, about 1.5 diameter apart; MR separated by less than one diameter. The front row is slightly recurved and is shorter than the rear row, which is procurved; MF distant from SF about 1.5 their alignment, and MR from SR about twice their alignment; clypeus, 3 diameters MF high.

ABDOMEN: Well arched, especially when gravid, but a number of specimens are flattened upon the dorsum; shape oval, diminishing towards the base as well as towards the spinnerets, which it slightly overhangs. The venter is marked by a broad brownish band, with two round white or whitish yellow spots on either side of the median line. These spots in some specimens become interblended. The epigynum (Plate III., 3b, 3c) has a broad but decided scapus quite wide at the base and slightly spooned at the tip; it is well elevated above the venter.

MALE: Plate III., 2, 2a. In general color and markings the male corresponds with the female; it is 5 mm. in length. The tibia of the second pair of legs is without any special armature.

DISTRIBUTION: I have specimens from New England southward to Florida, and also from Utah (Professor O. Howard), from Southern California, and as far north as Wisconsin (Professor Peckham). Dr. Marx notes it as found in Alabama and Texas, indicating a distribution throughout the entire Gulf States. We may thus conclude that it is distributed over the entire United States. As it may be regarded as identical with the European E. cucurbitina, it has a wide European distribution, and is one of the best and earliest known species.

No. 10. Epeira vertebrata McCook. Plate III., Figs. 6, 7; Pl. IV., Fig. 1; Pl. V., Fig. 4.

1888. *Epeira vertebrata*, McCook . . . . Acad. Nat. Sci., Philada., p. 196.
1889. *Epeira vertebrata*, McCook . . . . Amer. Spid. and their Spinningwork.

FEMALE: Total length, 14 mm.; cephalothorax, 6 mm. long, 4.5 mm. wide, at the face 2 mm. wide; abdomen, 11 mm. long, 7.5 wide, narrowing to 2 mm. at the apex and 3 mm. at the base.

CEPHALOTHORAX: A rather long oval, twice as wide at the base as at the face, flattened on top and gradually depressed to the base; fossa a longitudinal slit; caput very gradually sloping to the face, like the corselet somewhat flattened on top; color yellow, with a broad brown stripe on the sides which entirely surrounds the face, and a median stripe of brown which is broadest on the caput; it thus presents the appearance of having four longitudinal yellow stripes alternated with three of brown.[1] The corselet is sparsely provided upon the sides with whitish gray hairs, which form borders on the margin, and are much thicker on the sides of the caput; the sternum is shield shaped, somewhat longer than wide, with a broad median band of yellow; much rounded at the apex, and indented at the sides; well covered with hair; sternal cones prominent; lip large, subtriangular, one-half as high as the maxillæ, which are slightly longer than wide, subtriangular at the tips; both organs brown, with yellow tips, and covered with dark curved bristles.

EYES: Ocular quad wider in front than behind, length greater than width, the fore part on an eminence much more decided in front; eyes about equal in size; MF separated by about 1.3 diameter; MR by about one diameter; side eyes on tubercles, smaller than the middle eyes, but about equal in size, barely contingent; SF removed from MF by 1.5 the alignment of the latter, or at least two times or more their intervening space; the clypeus is narrow, the ocular eminence almost touching the margin, which is removed by about 1.5 diameter from MF; the front row is slightly recurved; the longer rear row procurved.

LEGS: 1, 2, 4, 3; stout at the thighs, but tapering well to the tarsi, abundantly armed with strong blackish brown spines and with gray bristles and hair; color yellow, with dark brown annuli, and the thighs, particularly at legs-I and II, in many specimens almost entirely glossy brown, flecked with gray pubescence; the palps are marked and armed as the legs, but rather lighter in color; the mandibles are dark glossy brown, with tufts of hair upon the inside; conical, much narrower at the tips than at the base; the basal cog prominent, as is the corresponding dewlap upon the face; the tips and fangs are black.

ABDOMEN: An elongated oval, somewhat narrowed at the rounded base, and much narrowed at the apex to the distal spinnerets; the dorsum well arched, the ground color brown, or yellowish brown, with a median herring bone or arrow shaped pattern with a double head, the point of the anterior one reaching to the middle of the basal front. Beyond the second arrow head the median pattern continues to the apex with indented edges; the centre of this vertebralike figure is marked along the entire dorsum with an interrupted ribbon of brown mottled with yellow; the folial margin consists of a band of yellow broken into spots, sometimes outlined with rosy brown, in the centre of which are black patches, forming a V-shaped figure receding toward the apex somewhat like Epeira arabesca; the sides are broad, irregular yellow bands. The surface is strongly reticulated and the whole appearance beautiful. The venter is a broad anchor shaped band of brown, flanked on

[1] This effect in some plates has been obliterated by the colorists.

either side by a yellow ribbon, which passes around the orange yellow spinnerets. The epigynum has a narrow atriolum (Plate IV., 1c, 1d, 1e), with a strong brownish black corneous scapus, channeled along the lower side and spooned at the tip, resembling that of Epeira Benjamina and E. arabesca.

MALE: Plate V., Figs. 4a–c. Fourteen mm. long, but varying much in length, one specimen being but little more than half this measurement. The cephalothorax is a long oval, 6 mm. long, 5 mm. wide; the fossa a lengthened longitudinal slit; the corselet flat upon the top; the corselet grooves and cephalic suture very indistinct; the juncture of the caput almost obliterated, the head narrowing to the face; very little pubescence; the skin yellowish brown, smooth, glossy, with yellow stripes around the margins and longitudinal stripes upon the summit faintly, after the fashion of the female. The legs are long, very stout at the femora and tibia-I and II, but much thinner at the tarsus and metatarsus; tibia-I is somewhat thickened at the apex, and provided with a number of short, black clasping teeth, flanked by strong rows of black spines. The femur of leg-II has a number of black shortened spines underneath the apical part; the tibia is thickened at the base, curved, thickened again at the apex, and provided the entire length with numerous black clasping spines, which are much more numerous towards the apex, where they are clustered together in quadruple rows; strong, long black spines arm the base, and extend along either side. All the legs are provided with yellowish brown spines and yellowish bristles; the leg armature of the male of this species is extremely formidable. The abdomen is a long oval, marked upon the dorsum somewhat as is the female, but with the herring bone pattern more interrupted; the palps are represented as to shape and color by Plate V., Figs. 4a–c. Strong, blunt, curved coxal spurs mark the first legs at their articulation with the trochanter, and a long conical spur marks the middle of coxa-IV in a similar position. Length, 1, 2, 4, 3, as follows: 24.5, 21.75, 20.5, 13.5 mm.

This species has points of resemblance to E. arabesca, but is decidedly different in many respects. The coloring and shape of the cephalothorax alone at once distinguish it; in Arabesca this is high, peaked in the centre, abruptly sloping to the base and to the face, whereas in E. vertebrata the corselet is rather flat on top, and very gradually depressed to the base and the face. The dorsal pattern upon the abdomen is quite different, although presenting points of resemblance. The characteristics of the male are different, and the size of Vertebrata is much greater than of Arabesca.

The dark forms of this species (Plate IV., Fig. 1) resemble in the markings of the abdomen the lighter variety. The abdomen, however, in the specimens possessed by me appears to be of a more uniform oval shape throughout. The colors of the abdomen are black, with yellow markings. On either side are two broad broken bands of circular and irregular waving figures, which meet in front and at the apex. The cephalothorax and the legs are of a dark reddish brown, or even blackish. The bands on the cephalothorax are also quite black, as are the tips of the female. This may be the normal color of the female after depositing the cocoon, but I have so many specimens that are marked in this way that it seems well to note the differences.

DISTRIBUTION: A number of specimens of both sexes and various ages have been received from California (Mrs. Eigenmann, Dr. Davidson, Dr. Blaisdell). From the "Albatros" expeditions (Mr. C. H. Townsend) I have examples from Clarion Island, Lower California, and Galapagos Islands. At San Diego the species is abundant and is distributed to some extent northward along the Pacific Coast. It may be a subtropical species. I have raised several from imported eggs to maturity in Philadelphia.

No. 11. Epeira Ithaca (new species).   Plate IV., Figs. 3, 3a–d.

FEMALE: Total length, 10 mm.; abdomen, 7 mm. long, 5 mm. wide at broadest part; cephalothorax, 5 mm. long, 3 mm. wide. The colors for the fore part of the body are orange yellow and yellow; of the abdomen, yellow and brown. The colors of the male are similar.

CEPHALOTHORAX: A rounded oval, sloping gradually backward from the deep circular fosse and forward to the eye space; head quadrate; cephalic suture distinct; slightly pubescent; color orange yellow. Sternum cordate, dark brown, sternal cones distinct; labium semicircular; maxillæ well rounded, as wide as long.

EYES: The ocular quad is somewhat longer than the greatest width; the front wider and more elevated than the rear. (Fig. 3a.) MF are on tubercles, are separated about 1.75 diameter, are slightly longer than MR, which are separated about or less than one diameter. MF distant from SF about 1.5 their intervening space. MR from SR about two alignments. The side eyes are on tubercles, are divided by more than a radius; SR slightly smaller than SF, and placed well behind but a little to the side. The front row is slightly recurved, the rear row slightly procurved; MF from margin of clypeus about two diameters.

LEGS: 1, 2, 4, 3; uniform light orange yellow, the tips of the terminal joints alone having narrow darker annuli; not heavily pubescent, rather sparsely provided with short dark spines. Palps colored and armed as the legs; mandibles conical, color orange yellow.

ABDOMEN: Well arched, oval, broadest at the base, tapering to the distal spinnerets; color yellow, folium widest anteriorly, a white median patch on the basal front, mottled white and rosy centre and brownish indentations; sloping bands of alternate white and orange mark the sides. The venter has a somewhat triangular patch of black or dark brown (3b), nearly surrounded by a band of yellow widest at the spinnerets. Two lunettes of yellow color are on either side of the base of the spinnerets. The epigynum (3c) has a well defined scapus, flattened, of almost equal width throughout, and slightly spooned at the tip.

MALE: 7 mm. long; in color and markings resembling the female. The legs are uniform dark orange yellow, more heavily armored than female with brown spines and gray bristles. Tibia-II is not swollen, has no special clasping spines, but numerous long brown spines symmetrically arranged in rows. Tibia-I is similarly armed. The palpal digit, 3d, somewhat resembles that of E. patagiata.

DISTRIBUTION: New York; I have two immature females from Ithaca, N. Y. (Mr. N. Banks), and a mature male, and female lacking one moult of maturity taken by me at Alexandria Bay on the St. Lawrence River.[1]

No. 12. Epeira placida HENTZ.   Plate IV., Figs. 4, 4a, 5, 5a.

1847. *Epeira placida*, HENTZ . . . . . J. B. S., v., 475 ; Sp. U. S., p. 115, xiii., 12.
1884. *Epeira placida*, EMERTON . . . . N. E. Ep., 316, xxxiv., 2; xxxvi., 10-13.

FEMALE: This beautiful spider is one of the smallest of our indigenous Epeira. Among numerous specimens I have only one that equals 4 mm. in length. It is strongly colored, especially upon the abdomen, where the colors are white or whitish yellow, with a prominent, blackish brown median band. Total length, 3 mm.; abdomen, 2.5 mm. long by 1.5 mm. wide; cephalothorax, 1.5 mm. long, 1 mm. wide.

CEPHALOTHORAX: Cordate, the corselet rounded at the edges, rather high, pitched sharply backward, the head gradually depressed to the face, the fosse a longitudinal slit; in these features resembling E. forata, and differing from typical Epeira. The color yellow or dark yellow, with a broad median band of brown or blackish brown, widest at the eyes; the skin glossy, slightly pubescent with grayish hairs upon the sides of the caput; sternum shield shape, acute at the apex, wide at the base, brown or blackish brown; at the margins a broad, yellow median band; the labium short, triangular; the maxillæ as wide or nearly as wide as long, sharply truncated at the tips; color of labium and maxillæ yellowish brown.

EYES: Ocular quad upon a squarish eminence projecting in front, somewhat longer than wide; the width in front and behind nearly the same, but slightly greater in front; the

---

[1] In reviewing this description since preparation of plates I am inclined to think the above a small varietal form of Epeira marmorea.

eyes not greatly differing in size; MF removed about 1.3 diameter; MR about one diameter; side eyes upon tubercles; SF removed from MF by a space not greater than, and scarcely as great as that which separates MF, thus *differing entirely from the eyes of the genus Epeira*; the space between SR and MR is at least twice the interval between MR, and about equals the distance between MF; all the eyes are upon black bases, giving them a blackish appearance; the clypeus is about the height of one diameter of MF; the front row is slightly recurved, the rear row procurved.

LEGS: 1, 2, 4, 3; stout, yellow, with brown annuli at the tips; the spines comparatively few and very long; the palps are colored and armored as the legs; mandibles long, conical, somewhat widened at the tips; yellowish brown in color.

ABDOMEN: Ovate, decidedly narrower in front than behind, differing therein from typical Epeira, and widest about two-thirds the distance from the front; the central pattern is a bottle shaped figure, with undulating edges, being blackish brown, the centre a lighter brown or yellowish hue; on either side are irregular borders of white, then follows a light band of yellowish brown, then another white, and so to the sides, which are dark brown; the whole surface is beautifully reticulated; the dorsum is highly arched, almost a semicircle; the spinnerets are distal; the venter a brown patch, flanked on either side by a white or yellowish white space, the spinnerets being black, with dark brown bases, and a white patch on either side; the epigynum (Fig. 4a) has a light colored scapus, somewhat convoluted, with a wide, rounded, spoon shaped tip; the portulæ on either side are strong, brown, bowl shaped objects.[1]

MALE: Fig. 5. Little more than 2 mm. long, colored and marked much like the female. One specimen in hand has the angular point of the dorsal folium much darker in color, giving the appearance of rows of black spots, symmetrically arranged on either side. The same markings may be seen in a number of females. The legs are a uniform yellow; tibia-II without any special clasping spines upon the legs, with a few very long ones, all of dark yellow color.

DISTRIBUTION: I have specimens from New England, New Jersey, North Carolina, Georgia, and from as far north as Wisconsin. It has been located in Florida, and probably inhabits the Western and Northern United States as far west as Mississippi, and may be found beyond.

No. 13. Epeira foliata HENTZ.   Plate IV., Figs. 7, 8, 8a.

1847. *Epeira foliata*, HENTZ . . . . . . J. B. S., v., 475; Sp. U. S., xiii., 14; pl. xviii., 50.
1884. *Epeira foliata*, EMERTON . . . . N. E. Ep., p. 318, xxxvii., 6-10.
1889. *Epeira folifera*, MARX . . . , . Catalogue, p. 545.

FEMALE: Total length, 6 mm.; abdomen, 4 mm. long, 2.75 mm. wide; cephalothorax, nearly 2.5 mm. long, 1.75 mm. wide. The general colors (in alcohol) are brownish yellow or pale brown for the fore part of the body. Mr. Emerton describes this part as grayish or greenish yellow, darker toward the end of the joints, the first and second legs being darker and with dark rings in the middle of the tibia and tarsus. The abdomen is grayish or olive yellow, relieved by brown or red longitudinal stripes. Dr. Marx, in his catalogue, names this species E. folifera, on the grounds (Note 17) that Hentz's name of "foliata" had been previously appropriated by Walckenaer for E. cornuta; but as Walckenaer's name is a synonym, and is not retained by araneologists for the species to which it is affixed, Hentz's name must be regarded as the true one for this species. The species differs from the typical Epeira in having *both eye rows procurved*, and further in the subconical shape of the cephalothorax.

CEPHALOTHORAX: Cordate, peaked and high in the middle, sloping sharply back to the indented base: the fosse a deep longitudinal indentation; cephalic suture distinct; caput depressed, somewhat arched at the base; color yellow, covered with pubescence, sloping to

---

[1] The base of the scapus in the figure is drawn too broad.

the face. Sternum cordate, inclined to rectangular, raised in the middle, with sternal cones, slightly pubescent, yellow; labium subtriangular, though rectangular at the base; maxillæ but little longer than wide, triangular at the tip, dark yellow, as is also the labium.

EYES: Ocular quad upon a rectangular eminence, most prominent behind; the front somewhat wider than the rear, and the side about equal to the front, MF separated by about two diameters; SR, which are somewhat larger and yellow in color, by about 1.5 diameter. Side eyes are on tubercles; SF somewhat larger; separated by less than a radius. MF separated from SF by 1.3 the area thereof, or 1.5 the intervening space. The space between MR and SR is much greater. The margin of the clypeus is removed from MF by two or more diameters, the central eminence coming close up to the margin. The front row is somewhat procurved; the rear row, which is the longer, is much procurved.

LEGS: 1, 2, 4, 3; yellow, with slight annuli at the tips of the tibiæ; provided with yellowish white bristles and hairs, and yellow spines, which are numerous underneath the femora, especially of legs-I and II. The palps are yellow, and armored as the legs. The mandibles are long, conical, somewhat separated at the tips, pubescent, and thickly hirsute on the inner edges.

ABDOMEN: A triangular ovate widest at the base, narrowing somewhat to the apex; the spinnerets distal; dorsum arched; the folium a simple triangular pattern, with undulating margins; colors light pea green, shaded with yellow to yellowish white and reddish marginal and longitudinal lines; the venter a broad, yellowish brown patch, cretaceous, extending upward on the sides; the whole abdomen covered with soft yellow hairs; the spinnerets yellow, surrounded by a blackish base. The epigynum presents a semicircular atriolum, brown, glossy, rugose, from which extends a short, chitinous scapus, like the bowl of a spoon.

MALE: Fig. 8, 8a; 5 mm. long. In color and markings quite similar to the female. The annuli of the legs appear to be darker; tibia-II is curved, and thickened at the middle and toward the base; provided from the middle to the apex with a series of strong dark brown spines about seven in number, of which one underneath is much longer, having nearly the length of half the joint, and placed upon an elevated base. Underneath the femora is a row of six or more acute, brown, erect spines. The humeral joint of the palps is as long as or longer than the three terminal joints, of which the cubital is globose, armed with long spines; the radial is bilabed; the digital is rounded, corneous, yellowish brown, the embolus being wide, bifid, or strongly notched at the tip, presenting at once a prominent characteristic. The legs are stout at the femora, but much diminished in size at the metatarsus and tarsus.

DISTRIBUTION: I have from Mr. Thomas Gentry one female in a collection from North Carolina and Georgia. Hentz described it from Alabama, Emerton from New England, and Dr. Marx has examples from Washington, D. C., Florida, Savannah, Ga., New York, Columbus, and Texas. It is thus distributed along the entire Atlantic Coast, and probably throughout all the Gulf States as far as Texas. Its western distribution has not been determined, though it will probably be found as far west as the American Plains.

No. 14. Epeira balaustina McCook.　　　　　　Plate IV., Figs. 2, 2a, 2b, 2c.

1880. *Epeira balaustina,* McCook . . . Proceed. Acad. Nat. Sci., p. 198.
1889. *Epeira balaustina,* MARX . . . . Catalogue.

FEMALE: Total length, 16 mm.; abdomen, 9 mm. long, 7 mm. wide; cephalothorax, 7 mm. long, 5 mm. wide; width of face, 2 mm. The general colors are for the abdomen yellow, with black stripes and spots; cephalothorax, orange red; legs, orange and black.

CEPHALOTHORAX: Corselet rounded, high at the centre; head at the fosse elevated and sloping to the front, with little decrease of width; color brown, intermingled with orange; in one species this color is dark orange brown; skin glossy, and heavily clothed on the sides and head with long gray bristles. Sternum cordate, dark orange brown, covered along the margins with white bristles, higher in the centre, and about as wide as long;

labium subtriangular, and, like the maxillæ, dark brown, with yellow tips; mandibles colored as the cephalothorax, dark towards the fang, at which point also they narrow on the inside; clothed with white bristles.

EYES: Ocular quad, on a rounded prominence (Fig. 2b); MF decidedly larger than MR, separated by about 1.5 diameter; at the same distance from the margin of the clypeus; length of the quad about equal to the front, the rear narrowest; MR separated by about their diameter. Side eyes upon tubercles, but hardly so pronounced relatively as the central prominence. SF the larger; the two scarcely contingent. Clypeus 1.5 diameter; MF high; the front row is slightly recurved, almost aligned, and shorter than the rear row, which is slightly procurved.

LEGS: 1, 2, 4, 3; moderately stout, heavily clothed with white bristles, and at points with yellowish curved hairs, and provided with numerous strong white spines, set in dark and well elevated sockets; color, orange brown; the femora somewhat darker, or even blackish. Palps like the legs, heavily armed with spines and bristles. One specimen of the same species has bright orange legs, without decided annuli, but the femora of first, second, and fourth legs marked with black bands, which cover three-fourths of the surface of the first two and one-half of the fourth pair; the femur of third leg is without the dark bands, but has a slight median annulus.

ABDOMEN: Subtriangular in shape, longer than broad, arched upon the dorsum, but somewhat flattened upon the summit and rounding to the spinnerets, which are distal. It is heavily clothed with simple white bristles with brown pits, which are clustered more closely in a bushy tuft around the base; numerous smaller curved bristles of dark color are scattered over the entire dorsal surface; between these larger white ones and on the sides, mostly placed together, are golden yellow short curved bristles, which considerably modify the color. On one specimen, from San Domingo, the dorsum is bright yellow, with branching longitudinal lines from the middle to the apex; the sides are marked with yellow. On the specimen described the abdomen appears to have been a uniform yellow color, with a darker cordate band or folium occupying the greater part of the dorsum. The venter (2a, 2c) is a broad subtriangular patch, shaped like an old fashioned chapeau, of yellow or yellowish color, entirely girdled by an irregular ribbon of yellowish white; on the chapeau six dark or yellow spots are symmetrically arranged on either side of the long scapus. The venter, like the rest of the spider, is covered with numerous bristles and bristlelike hairs; along the edge of the gills these stand thickly and are white. Spinnerets dark orange; on a specimen from Florida, bright orange. The epigynum is most remarkable for the length of the scapus (2c), which reaches over the entire venter to the base of the spinnerets; it is narrow and tapers nearly to a point, is without groove or spoon, but exceedingly rugose and with a light line of hairs along the median of the lower surface. The vulval porch is scarcely wider than the base of the scapus, and the portulæ are not exposed to view. In only one specimen (Florida) is this organ preserved intact, in two others it is broken off. For this reason, in the description of this spider, as first given in the Proceedings of the Academy of Natural Sciences of Philadelphia, I erred concerning the form of the epigynum by describing the species from examples from which the long scapus had been broken off. The fourth specimen, which I subsequently found, was in perfect condition, and thus permits me to correct this error.

DISTRIBUTION: I have three specimens in my possession, all females; one from Florida (Fig. 2); one from San Domingo, collected by the late William H. Gabb; one from Swan Island, Carribean Sea, from Mr. C. H. Townsend; a fourth from the latter locality was sent by me to Dr. T. Thorell. From this showing the species would appear to be limited to the Gulf States and the tropical islands along the coast. It will probably be found widely distributed throughout the northern parts of South America. In general form it resembles closely E. ravilla and E. bivariolata, but lacks the circular blisterlike abdominal markings which characterize E. bivariolata. The female specimen marked in the Marx collections E. ravilla, and so recognized by Count Keyserling, lacks about one moult of maturity, and the epigynum (Plate V., Fig. 7a) is so different in length and structure from that of E. balaustina as drawn (Plate IV., Fig. 2c), that one doubts whether a final moult could overcome the difference. Should this resemblance prove to be identity, the distribution of the species would be extended, in the United States, to the Rio Grande of Texas, and also into Arizona, and its habitat become the semitropical and tropical parts of North America, including the West Indies; probably also the northernmost South American States.

No. 15. Epeira carbonaria Koch, L. Plate V., Figs. 1, 2.

1869. *Epeira carbonaria*, Koch, L. . . . Beitr. zur Kentn. den Arachn. Tyrols. Zeitschr. d. Ferdinandeums, p. 168.
1874. *Epeira carbonaria*, Simon, E. . . Arachn. de France, i., 92.
1875. *Epeira Packardii*, Thorell . . . Proc. Boston Soc. Nat. Hist., xvii., 490.
1884. *Epeira carbonaria*, Emerton . . . N. E. Ep., p. 315, pl. xxxiii., 18.
1892. *Epeira carbonaria*, Keyserling . Spinnen Amerikas, Epeiridæ, p. 204, tab. ix., 151.

FEMALE: Total length, 14 mm.; abdomen, 10.4 mm. long and 7 mm. wide; cephalothorax, 5 mm. long by 4 mm. wide; width at the face, 1.8 mm.

CEPHALOTHORAX: A rounded oval; fosse deep; cephalic suture distinct; head but little depressed at the base, rounding to the face; corselet brown, with yellow marginal stripe covered with rather long grayish white hairs, bristlelike at base and along sides of the caput. Head yellow, well covered with long, stiff gray hairs. Sternum shield shape, brown, heavily covered with long, gray bristlelike hairs; labium rather narrow, subtriangular, less than half the height of maxillæ, which are apparently somewhat wider than long, rounded, and yellow like the labium and surrounding coxæ.

LEGS: 1, 2, 4, 3, as follows: 18, 16.2, 15.4, 10.7 mm.; stout, strongly annulated, with dark brown rings at tips and middle of tibia and metatarsus; a brown hue diffused over upper surface of femora; strongly armored with long gray hairs, bristles, and numerous long black spines; palps colored and armed as legs; mandibles strong, curved at bases, yellow tipped with brown; slightly pubescent.

EYES: Ocular quad on a high rounded eminence, which is almost a square, but wider in front than behind; MR oval and decidedly larger than ME; the latter separated by about 1.5 to 2 diameters; MR about one diameter. Side eyes on tubercles, scarcely contingent; SR larger than SF, which are separated from MF by at least two and a half times the space that divides the latter. Clypeus high, its margin removed from MF by three and a half to four times a diameter of the latter, or considerably more than intervening space between MF and SF; front row recurved, rear row somewhat procurved. The face around the central eminence is thickly covered with stiff gray bristles which project over the margin.

ABDOMEN: A long oval, the dorsal field yellow interspersed with brown; the folium is outlined by thicker semicircular patches of grayish white bristlelike hairs, which heavily cover the entire organ. In the centre is a white herring bone marking with at least five angular projections and indentations symmetrically arranged on either side. This herring bone pattern is chiefly emphasized by the gray hairs, the field beneath being yellow margined with brown. From the middle of the folium downward are arranged symmetrically on either side rows of three circular spots, like buttons. The folium diminishes towards the apex; sides yellow, intermingled with brown amidst gray hairs; spinnerets at the foot of the apical wall which very slightly overhangs them; color brown, with lighter hue at the bases. The venter is covered with gray hairs over a brownish field, with yellow spots on either side heavily covered with gray hairs. The epigynum (Fig. 1b, 1c) is long and triangular, wide at the base and diminishing sharply to a point, upon which is a rounded dot. It is hollowed upon the lower anterior surface and is covered with rows of stiff white hairs.

MALE (Fig. 2): In color and marking closely resembles the female; length (two specimens), 9 mm., 8 mm. Cephalothorax a longer oval than the female. The legs differ in lacking the decided brown annuli of the female; are heavily armed with spines and bristles, double rows of spines are on the under side of the femora, longer and stronger on the sides. Tibia-II is curved, and its apical joint (Fig. 2a) provided with strong, thick, black clasping spines, with several strong ordinary spines; also a decided conical projection or tooth with a corneous tip, which, from the indentation thereon, would appear to be the seat of a strong spine. A strong curved spur marks the apex of coxa-I, next the trochanter; coxa-II has a toothed cone on the middle of the base; coxa-IV a toothed cone on the base underneath. The abdomen is very heavily covered with bristlelike hairs, which are arranged in a heavy tuft upon the base. The ocular quad appears to be relatively a little longer than in the female, presenting thus more the figure of a rectangle than a square. The palpal digit (Fig. 2b) is dark brown, and distinguished by a palm shaped claw projecting from the side outward.

DISTRIBUTION: I have a female of this beautiful species from Salt Lake, Utah (Professor O. Howard), and a male from Wisconsin (Professor Peckham), and from California (Mr. Curtis). Dr. Marx has a specimen from the District of Columbia. Professor Packard found a specimen as far north as Labrador. Emerton collected it on Mt. Washington, N. H., where it is common along the large slopes of Bear Rock; and it has been found on Mt. Lincoln, in the Rocky Mountains. It is a European species which prevails among the Alps.

No. 16. Epeira carbonarioides KEYSERLING.   Plate V., Fig 9.

1892. *Epeira carbonarioides*, KEYSERLING. Spinnen Amerikas, Epeiridæ, p. 206, tab. x., 152.

FEMALE: Total length (two specimens), 10 (8) mm.; abdomen, 5.5 (5.2) mm. long by 4.5 (3.6) mm. wide; cephalothorax, 4.5 (3.2) mm. long by 3 (2.7) mm. wide. Under the above name Count Keyserling has described a species from Western North America which strongly resembles Epeira carbonaria. His description does not distinctly indicate the differences between the two, and at this writing Plate X. of his work upon which this species is figured is not in hand. I am somewhat doubtful whether the integrity of the species can be maintained, having but a single pair, but it is a good variety at least, as indicated by the following differences: In Carbonarioides the ocular quad is slightly wider behind than in front, while in Carbonaria the relative lengths are reversed, the front being, if anything, the wider. The female is not so heavily covered with strong gray bristles as Carbonaria, particularly upon the cephalothorax and around the face, though this may be an individual difference. The brown annuli upon the legs of Carbonarioides are much wider relatively than upon specimens of Carbonaria observed by me; so much so that the joints appear dark brown with narrow yellow annuli, rather than yellow with brown annuli. The epigynum of Carbonarioides, though resembling that of Carbonaria in the single specimen in hand, is much narrower at the base, and, indeed, is relatively smaller throughout. The dorsal folium also differs somewhat, though this may well be an individual difference. In the male the palpal organs appear to show decided difference; tibia-II of Carbonarioides appears to be relatively longer, not so much thickened at the apex, and scarcely so numerously provided with heavy articulate spines. The coxæ of Carbonarioides, male, are distinguished by strong spurs and cones; the spurs upon the apical margin next the trochanter of coxæ-I, II, III, IV, strongest in I. The cones appear, at the bases of coxæ-I, where they are low down and directed towards coxæ-II; on II, upon the lower surface, directed downward; and in the same position on III; on IV it is low down and directed toward coxæ-III. The maxillæ are also distinguished by slight protuberances or swellings upon the inner margin.

CEPHALOTHORAX: Corselet a rounded oval, somewhat narrower toward the front; sufficiently high; caput not depressed; the fosse a semicircular pit; corselet grooves not prominent; cephalic suture distinct; the whole a glossy yellowish brown, the caput being yellow; surface not heavily pubescent, but some gray hairs on the caput (the pubescence in the alcoholic specimen has evidently been partly worn off). The sternum shield shaped, somewhat cordate; color dark brown, well covered with hairs, especially on the margin. Labium subtriangular, half as high as maxillæ, which are yellowish brown, subtriangular at the base, as is the labium.

LEGS: Order 1, 2, 4, 3; dark yellow, strongly annulated both at tips and middle of the joints, with wide dark brown patches; the feet black; armed with black or blackish brown spines and dense white hairs, which are longer upon the bases of the femora and underneath and inside thereof; palps colored and armed as the legs; mandibles strong, dark brown, conical.

EYES: Ocular quad on a high, rounded, brown eminence, the rear manifestly wider than front and equal to sides; MF smaller than MR and separated by 1.5 to 2 diameters; MR separated by about 1.3 diameter. Side eyes placed on tubercles, contingent, SF somewhat larger than SR. MF separated from SF by more than their area; clypeus high, being as much as, or more than, the area of MF, or about three times the diameter of MF; front row slightly recurved, the longer hind row procurved; the space between SR and MR equal to about 1.5 the area of MR. The ocular quad is free from hairs, and the long gray hairs which so profusely cover the interspaces of the ocular area in Epeira carbonaria are but slightly represented in the specimen under description.

ABDOMEN: A long oval, narrowed at the base as well as apex; dorsum arched to the spinnerets, which are distal, overhanging the cephalothorax at the base. The ground color is yellow, heavily clothed with white hairs, thickest on the base in front and upon the sides; the folium is a broad, brown, scalloped band passing from the base, slightly narrowing to the apex; in the centre is a yellow herring bone marking (Plate V., Fig. 9), with a broad arrow at the base and tapering towards the apex; color yellow, modified by the heavy pubescence. The sides are yellowish brown, clothed heavily with white pubescence, which passes down to the venter, where the white hairs are mingled with brown. The ventral band is a rectangular brown patch, with yellow margin and a broad yellow median band; spinnerets brown. The scapus of the epigynum is tolerably long, of almost equal width for about two-thirds of its length, when it tapers to a point. It is narrower relatively than that of Epeira carbonaria, grooved upon the lower side, and covered with strong white hairs.

MALE: The male differs little in general characteristics from the female; the head is more attenuated at the face; the color of the cephalic corselet suture is dark grayish brown; the bristlelike hairs upon the abdomen are long and intermingled with numerous black and yellow bristles, the herring bone median pattern is well defined and resembles that of the female. The legs are yellow or yellowish brown, strongly annulated, but not quite so widely as the female. The metatarsi are well provided with spines, which appear also on the under sides, especially of first and second legs. The tibiæ of second legs are curved, very little thickened toward the tip, and provided with additional clasping spines, which are strong but not numerous, arranged in two rows. The tip of the joint has about four of these black spines, and on the inside is a strong corneous tooth or notch, which has probably been the base of a long bristle. This part is similar to Epeira carbonaria, but not quite so strongly developed. The tibia about equals in length the metatarsus.

DISTRIBUTION: I have seen but one pair of this species apparently identical with that which Count Keyserling has described, and this was collected in Clear Creek County, Colorado. (Marx Collection.)

No. 17. Epeira bivariolata CAMBRIDGE. Plate V., Figs. 5, 6, 5a, 6a.

1889. *Epeira bivariolata*, KEYS. *in litt.* . Marx Catalogue, p. 543.
1890. *Epeira bivariolata*, CAMBRIDGE . . Biolog. Centrali-Amer., Aran., p. 27, pl. vi., 15.
1892. *Epeira bivariolata*, KEYSERLING . Spinn. Amerk., iv., Epei., p. 100, tab. v., 74.

FEMALE: Body length, 11 mm.; abdomen, 7 to 8 mm. long, 7 mm. broad; cephalothorax, 5 mm. long, 3 mm. wide. The general colors in alcoholic specimens are yellowish brown for fore part of body, and yellow with brown markings for the abdomen. In life, according to Cambridge,[1] the falces, legs, and lower part of the abdomen are orange, approaching light burnt sienna; on the abdomen, sharply outlined, bright green, metallic when first caught, bordered anteriorly by a narrow white line, posteriorly by ten black ocellæ with clear white margins.

CEPHALOTHORAX: Corselet well rounded; orange brown; the head depressed and covered with white hairs. Sternum orange yellow, slightly pubescent, with low sternal cones before coxæ-III; slightly elevated in the centre; about as wide as long. The labium and maxillæ as in Epeira.

LEGS: 1, 2, 4, 3, as follows: 23.1, 19.8, 17.1, 11.6 mm.; they range in color from light yellow to light brown. The joints are annulated, with reddish brown color; armed with bristles, and rather short spines. Palps as the legs. Mandibles long, conical, widely separated at the tips.

EYES: Ocular quad on a rounded prominence, the sides but little longer than the front (if any), and the latter wider than the rear; the four eyes about equal in size; MF separated about 1.5 to 2 diameters; MR separated one diameter. MF removed from SF about 1.3 alignment; SF from MR about twice their alignment; side eyes propinquate, about equal in size. The front row of eyes is aligned, or but slightly recurved; the longer rear row procurved; height of clypeus 2 to 2.5 diameters of MF.

ABDOMEN: Almost as wide at the base as long, bluntly pointed in front, forming thus a basal triangle. The dorsum is a rounded yellow triangle (in life green with metallic lustre), in the middle of which is a brown folium scalloped or triangulated at the margin and diminishing to the apex; a row of brown bristles marks the base, and the dorsum is covered with similar but shorter hairs with brown pits, which modify the color. On the dorsal median, well towards the apex, are two circular prominences, corneous, shining, each about one millimetre in diameter, and separated from each other about three diameters; a light yellowish line girdles their point of union with the abdomen; they appear like blisters upon the surface; are destitute of hairs. The venter is a broad patch whose margin is yellow, enclosing a subtriangular patch of brown, the whole reticulated; four brown dots arranged in a square mark the anterior part near the gills; spinnerets brown, surrounded at their base by a yellow brown band. The epigynum has a long, straight, rugose scapus, narrowing from the base towards the point (Fig. 5a); the base rather flattened, the apex rounded.

MALE: Fig. 6, 6a. About equal in length to the female, or a little longer. Cephalothorax, from 5 to 6 mm. long and 4 to 5 mm. broad. The MF eyes are relatively larger than MR and more widely separated than in the female. The palp is marked by a strong boot shaped projection from the digital joint. (Fig. 6a.) The cephalothorax is yellowish brown, strongly marked with gray hairs. The patella of the first and second pairs of legs is long and thin; the tibia of the second pair short, curved, and thickened at the end, and strongly armed with two rows of about four each clasping spines, while the metatarsus is also somewhat curved and armed at the base with one very long spine, and a shorter one at the apex. The metatarsus-I is long, thin, and also slightly curved; a strong bent spur is on coxa-I next the trochanter. The abdomen is longer than broad, oval, of a greenish or olive color (in alcohol), and marked at the base, along the sides towards the apex, with long grayish brown spinous bristles. The specimen in hand appears to want the corneous blisterlike plates in the middle of the apical part, as above described in the female.

DISTRIBUTION: Cambridge describes this species from Gautemala; Dr. Marx has an undeveloped female from Utah, one from Texas, a mature male from Florida like the above (which is from Summit Canyon, Utah), but lacking the dorsal corneous plates. The female above described is from Lake Klamath, Oregon. This would indicate a wide distribution in the subtropical parts of the United States, and indeed of North America; and that the species has found its way northward to Oregon, along the Pacific Coast, and eastward to the American Plains. (The Marx Collection.)

---

[1] Biologia Centrali-Americana, Araneidea, p. 27, pl. vi., 15.

No. 18. Epeira ravilla KOCH, C.   Plate V, Figs. 7, 8.

1845. *Epeira ravilla*, KOCH, C. . . . . . Die Arachnid., xi., p. 73, Fig. 890.
1889. *Epeira ravilla*, MARX . . . . . . Catalogue, p. 547.

FEMALE: Total length, 11 mm.; abdomen, 8 mm. long, 7 wide; cephalothorax, 5 mm. long, 4 wide; face, 2 mm. wide. General colors, orange yellow and brown for the fore part of the body, and whitish yellow (in alcohol) for abdomen. The specimens here described from Dr. Marx's collection were identified by the late Count Keyserling as Koch's E. ravilla, and after casual examination were drawn and lithographed on the strength of this authority along with examples of E. bivariolata Cambridge. Subsequently more careful study has satisfied me that there is no specific difference between these and the specimens recognized by Keyserling as E. bivariolata. I regard it as probable that Koch's E. ravilla and Cambridge's E. bivariolata are identical, the peculiar blisterlike spots on the lower dorsal field having been overlooked by Koch, or being wanting in his type, as is the case in some of Dr. Marx's specimens. However, I retain here the names and figures as originally engraved, and append description.

CEPHALOTHORAX: Corselet well rounded on the sides, truncate and indented at the base; fosse large and circular; head quadrate, depressed from the gently arched base, covered with gray bristles; color dull orange yellow with brown marginal bands, the eye space somewhat lighter hue and mandibles dark brown. Sternum orange, slightly longer than wide, sternal cones distinct, elevated in middle, marginal gray bristles; labium semicircular, maxillæ gibbous, both dark brown with yellow tips.

EYES: Ocular quad elevated, front width greater than rear and equal to length; MF somewhat larger than MR; separated 1.5 diameter, MR by less than one diameter. Side eyes propinquate, on decided tubercles, SF a little larger than SR; MF distant from SF about 1.5 their alignment, MR from SR by a greater space; front row recurved, the longer rear row procurved; height of clypeus about 2.5 diameter MF.

LEGS: 1, 2, 4, 3; stout, especially I, II; orange yellow; femora-I, II, almost entirely dark brown to blackish, and about half of femur-IV similarly marked; joints annulated; well armed with white and yellowish spines and gray bristles and hairs; palps of lighter hue, with bright yellow wings and strong gray bristles.

ABDOMEN: Oval, contracted both in front and behind; dorsum lightly arched on top, rounding thence to the spinnerets directly beneath the high apical wall. The dorsal pattern, Fig. 7, and on the anterior base, is a rhomboidal patch of whitish yellow color, the original hue of which may have been green or bright yellow; surface with stout white bristles, with raised brown sockets. In the median line on the lower part of the dorsal saddle are two circular blisterlike markings like those in E. bivariolata, separated about two mm., the lower one smaller than the upper. The sides are yellow, with lateral streakings of dark color. The venter has a chapeau shaped spot of brown surrounded with yellow; at the orange colored spinnerets and along the sides bordered by brown. The epignynum, Fig. 7a, in the specimen described, probably lacks one moult of maturity; scapus moderately long and pointed.

MALE: Fig. 8; resembles the female in color and pattern; length, 8 mm.; cephalothorax longer and wider relatively than in the female; fosse a longitudinal depression. The legs have less prominent bands of black on the femora and are strongly annulated throughout; tibia-II much swollen and curved, and armed with stout, brown clasping spines arranged in two rows on the side, and one underneath; ordinary spines whitish, with brown bases; metatarsus curved, thin, one stout brown spine on the inside near the base, two near the apex; patellar joints long and rather thin; a long brown spur on apex of coxa-I; the palpal digit is remarkably developed (Fig. 8a), the corneous processes resembling those of E. bivariolata. The abdomen clothed at the base with a cluster of long gray bristles; circular blisters on the dorsal apex as in female.

DISTRIBUTION: Texas; a male from El Paso, Isleta, Rio Grande. (Marx Collection.) Dr. Marx also locates it in Arizona. It is probably an inhabitant of the entire Pacific Coast, or at least the tropical or semitropical sections. Koch's E. ravilla was from Mexico.

No. 19. Epeira volucripes KEYSERLING. Plate VI, 1, 2.

1884. *Epeira volucripes*, KEYSERLING . . Verh. d. z. b. Ges. Wien., p. 528, pl. 13, Fig. 27.
1892. *Epeira volucripes*, KEYSERLING . . Spin. Amer. Epeir., p. 199, tab. ix., 147.

FEMALE: General colors dark yellow and brown for the fore part; on the abdomen yellow, with blackish to brown markings. Total length (two specimens), 11 (9.5) mm.; cephalothorax, 4 (3.5) mm. long, 3.5 (3) mm. wide; head, 2 (1.3) mm. wide; abdomen, 8 (6.5) mm. long, 7 (6.2) mm. wide.

CEPHALOTHORAX: Corselet a rounded oval; brown, with lighter tints on caput and summit of corselet; well clothed with white hairs. Sternum cordate, as wide as long; color brownish yellow, with a bright yellow median band; lip and maxillæ brown, tipped with pale yellow.

EYES: Ocular quad on an eminence projecting forward; front as wide as long, and wider than rear; MF separated about 1.5 diameter; MR smaller than MF, and separated less than one diameter; SF from MF about 1.3 their alignment; side eyes on tubercles; SF larger than SR, the latter placed well to the sides, and marked with strong gray eyebrows; front row very little recurved, the longer rear row procurved. Clypeus about two diameters MF high; grayish bristles on the margin.

LEGS: 1, 2, 4, 3; yellow, with brown apical and median annuli; thickly covered with gray bristles and hairs, interspersed with strong brown and yellow spines; palps colored and armed as the legs. Mandibles brown, conical, rounded at the base, divergent at tips.

ABDOMEN: Subtriangular; widest at the base; somewhat arched to the distal spinnerets. The dorsal base overhangs the cephalothorax, is pale yellow, heavily clothed with gray hairs; the folium has a yellow median herring bone pattern, flanked by lunettes of black or blackish brown in two rows approximated towards the apex. The folium is not unlike that of E. arabesca in its general form. The dorsal color extends over the sides in irregular loops, margined by a scalloped black band that extends underneath. The venter is a broad squarish band of yellowish brown, with two bright spots near the spinnerets and blackish color within. The epigynum is rather short for such a large species, subtriangular, wide at the base, spooned at the somewhat rounded top. (Plate VI., Figs. 1a, 1b.)

MALE: Plate VI., Figs. 2, 2a. 7.5 mm. long; in general color and markings resembles the female. The abdomen is somewhat lighter in color; tibia–II is not especially developed in size or in clasping armature, being simply marked by several long strong spines, one of which, on the inside, is longer and stronger than the others, and placed upon a slight process; the joint is also somewhat bent.

DISTRIBUTION: Savannah, Ga. (Marx Collection.) Keyserling reports the species from New Hampshire and Tennessee; and from Haiti and Panama, Central America. This indicates an elastic temperament, at once adapted to the rigors of New England and the fervors of the tropics. I believe, however, that the species is not common. in northern latitudes, but belongs more especially to the southern fauna.

No. 20. Epeira tranquilla, new species.[1] Plate VI, Figs. 3, 3a.

FEMALE: Total length, 4 mm.; abdomen, 2.5 mm. long, 2.5 mm. wide; cephalothorax, 2 mm. long, 1.5 mm. wide, and 1 mm. wide at the face.

CEPHALOTHORAX: Cordate; corselet rounded at sides, high in the middle, shelving behind; head much depressed, sparingly clothed with gray pubescence; corselet grooves rather distinct; cephalic suture distinct; color yellow, with brownish flecks along the margins and in the sutures; pubescence gray; the hairs thick on the caput, particularly at the sides, where they form strong gray eyebrows. Sternum shield shaped, wide at the base, where it is rather squarely truncate; obtusely triangular at the apex; color yellow, with darker marks on the margin and an interrupted median band of brownish yellow; labium

[1] In his catalogue Dr. Marx refers to this spider as E. Heidmannii, *in litt.* (page 544); and again as E. tranquilla Keyserling, *in litt.* (page 548). I have hence adopted the name Tranquilla.

triangular, wide, more than half as long as maxilla; base brown, with a touch of bright yellow in the centre; tips yellow; maxillæ gibbous, as wide, or nearly as wide, as long.

EYES: Ocular quad on an eminence higher before; the front slightly wider than the rear, and somewhat narrower than sides; eyes about equal in size; MF separated by about 1.5 diameter; MR by about one diameter; side eyes on tubercles, barely contingent, subequal; the space between SF and MF about equals the area of the latter, or 1.5 their intervening space; the distance between SR and MR is much greater; the clypeus has the height of about 1.3 diameter MF, and is marked by rows of long whitish hairs, which are also found in the eye space; front row is slightly procurved, almost aligned; rear row longer and much procurved.

LEGS: 1, 2, 4, 3; color yellow, with dark brown annuli, armed, but not numerously, with blackish spines and yellowish bristles and hairs; palps similarly colored and armed, but rather lighter yellow; mandibles dark brown, conical, retreating toward the face, rather long.

ABDOMEN: Triangular ovate, widest across the shoulders, narrowing somewhat toward the front and apex; the folium is a broad blanket of yellowish white, which covers the entire dorsal field, extending up in a triangular form toward the front, the apex almost touching the cephalothorax on either side; the base is marked with dark brown; the surface is reticulated, and rather sparsely covered with short gray hairs; the ventral pattern is a mottled dark brown figure, with an interrupted band of yellow on either side; the epigynum (Fig. 3a) has a short wide scapus, rounded at the tip, where it is spooned, the channel extending along the lower surface of the scapus; the atriolum is rather high, the portulæ on either side prominent.

DISTRIBUTION: District of Columbia, one specimen only. (Marx Collection.)

No. 21. Epeira punctigera DOLESCHALL. Plate VI, Figs. 4, 4a-b.

1857. *Epeira punctigera*, DOLESCHALL . Tweede Bijdrage t. d. Kenn. d. Arac. Ind. Arch.
1864. *Epeira triangula*, KEYSERLING . . Isis, Dresden, Beschreib. Orbit., p. 98, tab. v., 12-14.
1871. *Epeira indigatrix*, KOCH, L. . . . Arach. Austral., p. 66, pl. v., 8-9.
1877. *Epeira vatia*, THORELL . . . . Studi I., Ragni di Selebes, p. 382-384.
1878. *Epeira punctigera*, THORELL . . . Studi II., Ragni di Amboina, p. 59.
1881. *Epeira punctigera*, THORELL . . . Studi sui Rag. Malesi e Papuani, p. 104, var. vatia.
1889. *Epeira punctigera*, MARX . . . . Catalogue, 547.
1892. *Epeira punctigera*, KEYSERLING . . Spin. Amer. Epeir., p. 136, pl. vii., 100.

FEMALE: Rather uniform in color, the dorsum a brighter yellow or whitish, the colors evidently faded in alcohol. Total length, 9 mm.; cephalothorax, 4 mm. long, 3.5 mm. wide; abdomen, 5 mm. long, 4 mm. wide. Keyserling describes a female 11.2 mm. long, with an abdomen 8 mm. long by 7.2 mm. wide.

CEPHALOTHORAX: The corselet well rounded at margin, rather high, and peaked at the crest; depressed and wide at the face; fossa not deep, being rather a slit upon the sloping posterior side; cephalic suture distinct; corselet grooves rather indistinct; color dark uniform yellow, with a yellowish patch at the caput, the base covered with whitish yellow hairs; sternum yellow, with a broad lighter yellow median band; shield shaped, triangular at the apex, compressed in the middle; sternal cones distinct before coxæ-I, III; strong yellow hairs; labium large, subtriangular; maxillæ wide as long, subtriangular at tips, which are inclined towards each other; sparsely covered with curved dark bristles.

EYES: Ocular quad on high rounded eminence, front as wide as sides, narrowest behind; the four eyes about equal, and on black rings; MF separated by nearly two diameters; MR by about one; side eyes on not very prominent tubercles, barely contingent; SF larger than SR; clypeus height about 2.5 diameter MF; front row slightly recurved, almost aligned, the rear row longer and procurved; space between SF and MF equals at least 1.5 area of MF; space between SR and MR more than twice the area of MR.

LEGS: 1, 2, 4, 3; stout, well clothed with pubescence; strong yellow spines, with brownish bases; underneath each femur is a single row of long bristlelike spines, and tufts of spinous bristles on the coxæ; color yellow, slightly tipped at joints with brown; palps marked and armed as legs, but apparently without annuli; mandibles articulated in facial plane, conical; separated at the tips, margins provided with long spinous bristles of yellow color.

ABDOMEN: Triangular ovate on top, widest at base; dorsum but slightly arched in the specimen in hand; the dorsal folium consists of an anchorlike figure with double flukes; the muscular · pits strongly marked; at the margins touched with brown or blackish color; quadruple lines branch from the middle towards the apex, and darkish lateral lines encompass the sides, which are yellow; pubescence on the dorsum rather scant, but on the base a tuft of whitish gray bristles, around which are clustered whitish or yellow hairs. The ventral pattern a bluntly triangular patch, narrowing towards the yellow spinnerets, which it encompasses; at either corner are broad circular or irregular patches of yellow or yellowish white; the base of spinnerets surrounded by an interrupted ring of yellow patches, especially marked with golden yellow hairs, deepening at places into brown; the epigynum (Fig. 4a) has a prominent scapus, which is widest at the base, having a sort of pedestal of yellow color, and enlarged at the middle point, where the channel begins, and again widening towards the tip, which is rounded and spooned; it is bent venterward, the color deep brown and glossy.

MALE: I have no male from which to describe, and have drawn the figure of the male palp (4b) from a manuscript drawing of Count Keyserling.

DISTRIBUTION: Fort Bridger, Wyoming; Washington State. (Marx Collection.) If we accept this species as entirely identical with E. punctigera of Doleschall, it is of wide distribution throughout the world, including the Mauritius (Keyserling); Australia, where Koch describes it, and the Malasian Islands, where it is described by Thorell. It is found only on our western coast, which may indicate that its distribution from the east to the west has been by commerce across the Pacific Ocean, in which respect it resembles E. Theisii.

No. 22. Epeira mormon KEYSERLING.   Plate VI, Figs. 5, 5a.

1882. *Epeira mormon,* MARX *in litt.* . . . Catalogue, p. 546.
1892. *Epeira mormon,* KEYSERLING . . . Spinn. Amerik., iv. Epeir., p. 182, pl. ix., 134.

FEMALE: Total length, 5.5 mm.; cephalothorax, 2.5 mm. long, 1.7 mm. wide, narrowing to 1 mm. at the face; abdomen, 3.5 mm. long, 2.5 mm. wide. The general colors of this species are uniform yellow in the fore part, with brown and brownish yellow and cretaceous longitudinal stripes upon the abdomen.

CEPHALOTHORAX: Cordate, high in the middle, which is peaked, sloping backward sharply to the truncate base; corselet grooves distinct; cephalic suture deeply marked; color yellowish brown, with brown median band from the fosse to the face; skin glossy, slightly pubescent; sternum black or blackish brown, heart shaped, deeply indented at the edges; sternal cones marked; high in the middle, pubescent; labium subtriangular, large; maxillæ as wide as or wider than long; obtusely triangular at the tips, which are yellow, while the base is brown in color, thus resembling the labium.

EYES: Ocular quad on a low eminence, most prominent before; decidedly wider in front than behind, and the sides rather longer than the front. MF somewhat larger than MR, separated by about 1.5 diameter; MR propinquate, separated by less than a radius. Side eyes on very slight tubercles, propinquate, not greatly differing in size. MF separated from SF by a space about equal to their area, or 1.3 times the intervening space. The face is yellow, the vertex evenly rounded. The clypeus has a height about 1.5 or more diameter MF. The front row is slightly recurved, the longer rear row slightly procurved.

LEGS: 1, 2, 4, 3; uniform yellow, with a slight darkening at some of the tips; pubescence and bristles yellow; well provided with acutely yellowish spines, with rows of especial strength underneath the femora.

ABDOMEN: A long oval, rounded both at base and apex; color yellow; the arched dorsum traversed by a folium of blackish brown, with indented margins that inclose a field traversed by three longitudinal stripes of lighter brown. Short gray hairs cover the skin, and are tufted at the base, which well overhangs the cephalothorax. The ventral pattern is a broad blackish band, with lighter hues in the middle, which encompasses the spinnerets, which are of lighter color, and decidedly overhung by the rounded apical wall. . The epigynum (Fig. 5a) has a short subtriangular scapus, which extends barely beyond the genital cleft, and is flanked on either side by a slight bowl shaped projection apparently marking the outer walls of the portulæ.

DISTRIBUTION: District of Columbia; Utah. I have seen but two specimens, which are in Dr. Marx's collection, but the habitat of these indicates a wide distribution from the Atlantic Coast to the Rocky Mountains.

No. 23. Epeira reptilis, new species.   Plate VI, Figs. 6, 6a-c.

1889. *Epeira reptilis*, MARX *in litt.* . . . Catalogue, p. 547.

MALE: Total length, 4 mm.; cephalothorax, about 2 mm. long, 1.5 mm. broad; abdomen, about 2.3 mm. long, 1.4 mm. broad. The color is a bright yellow to orange, with a lateral longitudinal band of yellow on the caput, and lateral bands along the cephalic grooves; fosse a deep longitudinal cleft; cephalic suture sufficiently distinct; slightly pubescent. The sternum (Fig. 6b) cordate, longer than broad, yellowish brown color. Labium, tip triangular, compressed at the base; maxillæ somewhat longer than wide, obtusely triangular at tip.

EYES: Ocular quad on a rounded elevation, dark brown in color, the eyes themselves black. The quad wider in front than rear, the sides longer than either; eyes subequal; MF separated by 1.5 diameter, MR by about 1. Side eyes on slight tubercles, contingent, about equal in size. MF removed from SF by about 1.3 their area, or nearly twice their intervening space; front row recurved, the longer rear row slightly procurved; clypeus about two diameters MF in height.

LEGS: 1, 2, 4, 3; yellow or reddish brown, well armored with long brownish spines; tibia-II without any special thickening or clasping armature, except a few additional ordinary spines towards the tip, which are also found upon tibia-I. The palpal digit is represented at Fig. 6c.

ABDOMEN: A long oval, rounded at the base and apex, of about equal width throughout; colors yellow, with a silvery gloss. A light folium, almost obliterated in alcohol, passes along near the distinctly marked muscular indentations, and on the apical part of the dorsum are two black longitudinal patches, passing to the apex. The surface is pubescent. The ventral pattern is a black or blackish brown band, which encompasses the spinnerets, of like color, and which are somewhat overhung by the abdomen. The entire skin of this spider is glossy.

DISTRIBUTION: Crescent City, Fla. The female of this species has not been taken. (Marx Collection.)

No. 24. Epeira forata, new species.   Plate VI, Figs. 7, 12.

1889. *Epeira forata*, MARX *in litt.* . . . Catalogue, p. 545 (KEYSERLING *in litt.*).
1889. *Larinia nigrifoliata*, MARX . . . Catalogue, p. 550.

FEMALE: Total length, 5 mm.; cephalothorax, 2.5 mm. long, 1.3 mm. wide; abdomen, 4 mm. long, 3 mm. wide. General colors of fore part of body uniform yellow; abdomen cretaceous white, with a black folium.

CEPHALOTHORAX: Cordate, high in the middle, sloping backward to the base of the corselet; the fosse a deep conical pit; corselet grooves tolerably distinct; cephalic suture distinct; caput narrow at base, rounded, smooth, but much depressed to the wide face. Sternum shield shape, yellow, slightly pubescent, flat in the middle; skin glossy; only a few whitish-yellow bristlelike hairs; labium subtriangular, about half the height of maxillæ, which are gibbous, somewhat longer than wide, yellow, with a few bristlelike hairs thereon.

EYES: Ocular quad on a rounded eminence, and nearly a square in form, but widest in front (Fig. 7b); eyes not greatly different in size, but MR appear somewhat the larger; MF separated by about 1.5 to 2 diameters, and MR about 1.5 or a little less; the side eyes on slight tubercles; separated by about a radius; SF somewhat larger than SR; MF separated from SF by about the area of the former, or 1.5 the dividing interval thereof; SR separated from MR by at least 1.3 the space between SF and MF; clypeus height about 1.5 diameter MF; the front row is almost aligned, the rear row longer and procurved.

LEGS: Uniform yellow; sufficiently stout, especially at the femora; armed tolerably freely with bristles and sparingly with long, thin, yellow spines; palps colored and armed as legs; mandibles conical, rather long, yellow, and flecked with yellowish brown.

ABDOMEN: Triangular ovate, the base rounded, high, and overhanging the corselet; the dorsum arched to the distal spinnerets; there are no shoulder humps, but across the base the dorsum is widened, forming from that point toward the apex an isosceles triangle, of which the field is white, and the folium a similar triangle of black, mottled with white spots; a semicircular patch of black marks the upper part of the base, on which are several transverse curved rows of white spots, from the centre of which issue bristlelike hairs; similar hairs are distributed over the dorsal field; the whole surface of the abdomen is beautifully reticulated; the sides are white, as is also the venter; the spinnerets are amber yellow, surrounded at the base with an irregular border of black. The epigynum (Fig. 7c) has the general characteristics of that of E. miniata (scutulata), the scapus being long, though greatly convoluted, sometimes wrinkled, of uniform width throughout, terminating in a deep wide bowl; the whole light yellow; the portulæ on either side are rounded and hollow bowls of glossy dark brown, open toward the genital cleft, toward which also they narrow; at the part next the scapus and on either side thereof issues a stout, curved horn of similar color and character.

MALE: Plate VI., Fig. 12. Resembles the female in color and substantially in markings; the cephalothorax is cordate, indented at the base, where it is relatively wider than the female, and narrower at the face; the cephalothorax is smooth and glossy; the fossa a longitudinal indented slit; the legs are uniform yellow in color, without any special clasping apparatus, and with only a few long, strong yellowish brown spines upon the tip of legs-I and II, which are arranged in about four rows, both above and beneath; the maxillæ are relatively longer and feebler than in the female, being much narrower at the tips and concave on the front. The ocular quad is quite prominent, much projecting in front, where it is decidedly wider than behind, the front eyes being upon black bases, the side eyes upon tubercles; SF somewhat larger than SR. The humeral joint of the palps is long, the radial and cubital joints short; the digital rounded; the cymbium yellow, covered with bristles, with two long, strong, curved spines thereon; the embolus much curved. (Fig. 12a.) Length of abdomen, 3 mm.; cephalothorax about the same. The abdomen is a long oval, narrow at the apex, the folium being black, broadest at the base, where there are two circular spots like eyes in the midst of the black.

DISTRIBUTION: Santa Rosa, Cal. (The Marx Collection.)

No. 25. Epeira Theisii WALCKENAER.   Plate VI, Figs. 8, 9.

1841. *Epeira Theis*, WALCKENAER . . . Ins. Apt., ii., p. 53, pl. xviii., 4.
1847. *Epeira mangareva*, WALCKENAER . Ins. Apt., iv., p. 469.
1863. *Epeira Oaxacensis*, KEYSERLING . Neuer Orbit Sitz. d. Isis, p. 121, pl. v., 15-18.
1871. *Epeira mangareva*, KOCH, L. . . Arach. Austral., p. 85, pl. vii., 4, 5.
1877. *Epeira mangareva*, THORELL . . . Studi, I., Ragni di Selebes, p. 395.
1878. *Epeira triangulifera*, THORELL . . Studi, II., Ragni di Amboina, p. 65.
1878. *Epeira mangareva*, THORELL . . . Ibid., p. 65.
1881. *Epeira Theisii*, THORELL . . . . . Studi, III., Rag. Melesi e Papuani. Principal form E. Theisii, and varieties E. mangareva and E. triangulifera.
1889. *Epeira Theis*, MARX . . . . . . Catalogue, p. 548.

FEMALE: Two specimens, total length, 8.5 (14.2) mm.; abdomen, 5 (10) mm. long, 2.5 (6.9) mm. wide; cephalothorax, 3.5 (5.5) mm. long, 2.5 (4.2) mm. wide. General colors on the fore part a yellowish brown; abdomen, yellow. (Fig. 8.) The species is notable for the wide difference in size of its individuals, as may be seen by the two measurements given above.

CEPHALOTHORAX: Corselet rounded at margin, high in centre, fosse long and deep; cephalic suture well marked; corselet grooves indistinct; caput somewhat raised from the corselet, but slightly depressed at the face, which is squarish and wide; color uniform yellow, with a dark median longitudinal ribbon from the face along the caput. Sternum slightly longer than wide, with feeble sternal cones; yellow, and lighter cretaceous interrupted median band; lip triangular, about half the length of maxillæ, which are somewhat longer than wide, and subtriangular at tips, which are directed towards one another; color yellowish brown.

EYES: Ocular quad somewhat longer than wide, and of about equal width in front and behind. MF upon slight projecting tubercles, separated by about 1 to 1.3 diameter; about equal in size to MR, which are separated by about the same distance. Side eyes on tubercles, with a cluster of wide bristles behind them as eyebrows; SF somewhat larger than SR, barely separated; SF removed from MF about 1.3 their area, or about 2.5 the dividing interval of MF; clypeus low, less than one diameter MF in height. Front row recurved; rear row longer and procurved.

LEGS: 1, 4, 2, 3, as follows: 21.1, 19.7, 18.5, 12 mm. Uniform yellow color, abundantly armed with bristles and hairs, and well provided with spines of dark brown color; palps armed and colored as legs; mandibles long, conical, and not divergent; color of legs.

ABDOMEN: A long oval; the dorsum arched, and somewhat narrower at apex than base; nearly twice as long as wide; color of field yellow, with a broad median longitudinal cretaceous band, which is narrowed at the apex, having a ragged marginal border of black, and a median lanceolated band of yellow. Venter with a median yellow band, with yellow cretaceous border; epigynum (Fig. 8a) has a rather narrow atriolum, out of which rises a stout scapus wide at the base, somewhat constricted at the middle, terminating in a long spoon, widest at the base and rounded at tip, which is dark brown, glossy, chitinous.

MALE: Fig. 9, 9a. In color and markings generally resembles the female, but lighter on the abdomen; the corselet differs in its decidedly rounded character, which, however, does not appear upon the sternum; corselet uniform glossy yellowish brown, and somewhat elevated; cephalic suture less distinct than in female; clypeus a little higher than in female, and ocular quad more evidently slightly narrowed behind; SF not relatively so widely separated from MF. Legs yellow and darker at joint; tibia-II curved, thickened, with a double row of black clasping spines extending the entire length of the joints on the inside, longer and stouter at the base, and flanked by several ordinary black spines. There is also a double row of spines underneath femur-II, and a small but feebler armature underneath femur-I; spines are abundant upon all joints, of a dark or blackish brown color. Coxa-I has next the trochanter a long, strong, brownish spur, curved at the point; coxa-IV has a conical spur nearer the middle of the joint.

DISTRIBUTION: Male and female, San Diego, Cal. (Marx Collection.) There is probably no American spider with wider exotic distribution than this species. It is reported from New Guinea, where Walckenaer first described it; Cape York, Singapore, Amboia, and is found in New Holland. It is thus dispersed throughout the Australian archipelago and the islands of Malasia. It may have secured its lodgment in California by importation from these quarters, but will probably be found in Central and South America.

No. 26. Epeira juniperi EMERTON.   Plate VI, Figs. 13, 13a–c.

1884. *Epeira juniperi*, EMERTON . . . . N. E. Ep., p. 313, xxxiv., 6; xxxvi., 14, 15, 16.

FEMALE: Total length, 5.5 mm.; abdomen, 4 mm. long, 3 mm. wide. Specimens obtained from Florida through Dr. George Marx retain their color well in alcohol, the fore part being a light yellow, the abdomen green, with bright yellow spots.

CEPHALOTHORAX: Corselet rounded, sloping sharply behind, rather high in the middle; base of caput on a level with corselet, but depressed in front; fosse deep; color in some specimens green, or greenish yellow, in others yellow; sternum shield shape, almost as wide as long, with slight sternal cones, colored, as also labium and maxillæ, pale yellow; labium triangular; maxillæ gibbous, somewhat longer than wide.

EYES: Fig. 13b. Ocular quad on a rounded eminence, the front wider than rear and about the length of sides; MF larger than MR and black; separated by about 1.75 their diameter; MR separated by at least or more than two diameters. Side eyes on separate tubercles whose bases are contingent, and separated by about one diameter, about equal in size; MF from MR about their area, and from margin of clypeus by about 1.5 diameter; front row recurved, rear row slightly procurved, and somewhat longer.

ABDOMEN: A well rounded oval, a little longer than wide, skin reticulated, without shoulder humps, covered rather scantily with long yellowish bristles, color green, with base bright or grayish yellow. In some specimens the green is more decided, and the yellow shades into green on the dorsal field, base, and sides; no distinct folium marks the dorsum, but a median line of darker green traverses the middle part of the field, from the side and end of which issue brownish lines to the apex. The venter is not distinctly marked, but is covered with green reticulations, as are the sides; spinnerets distal, yellowish brown; the epigynum (Fig. 13c) has a long convoluted scapus, brown below the tip, which is an oval bowl; atriolum wide at base; semicircular portulæ show on each side, from which issue rounded tongues curved like a ram's horn.

DISTRIBUTION: New England. Specimens received also from Florida, through Dr. Marx. The species is no doubt distributed along the entire Atlantic Coast, and will probably be found much more widely dispersed.

No. 27. Epeira Wittfeldæ, new species.[1]   Plate VII, Figs. 6, 6a–d; 7, 7a–d.

FEMALE: Total length, 9 mm.; cephalothorax, 4 mm. long, 3 mm. wide; 1.7 at the face; abdomen, 6 mm. long, 5 mm. wide. This spider in its general characteristics reminds one of E. strix. In its somewhat stout legs, strongly armed with spines, in the head and eyes, the two species are much alike. The dorsum of the abdomen, however, is more roundly arched in E. Wittfeldæ, and the epigyna differ. The colors and general forms of the two animals are also different.

CEPHALOTHORAX: Cordate, rounded at the sides, indented behind; smooth and glossy; sparsely provided with hairs; corselet grooves indistinct; cephalic suture distinct; fosse a longitudinal slit, within a circular pit; color yellow; the caput reddish brown, glossy, not depressed, slightly arched to the whitish face, whose margins just below the lateral eyes drop into a decided dewlap. Sternum shield shape, flat in the middle, with faint sternal cones, slightly pubescent, yellow. Labium wide, rounded at the sides; tip triangular; maxillæ oval, obtusely triangular at tips, inclined toward each other; wider than long; both of these organs yellowish brown.

EYES: Ocular quad on a dark brown eminence, most prominent in front, where it projects over the clypeus; rear eyes but little elevated above the facial surface; the quad is decidedly wider in front than behind (Fig. 6b), and the side about equals front width; MF larger than MR, separated by about 1.5 diameter, MR but little more than a radius. SF are on blackish tubercles, not greatly differing in size, separated by about a diameter; SR placed low down upon the face, so that with the four front eyes they appear to constitute an aligned group; MF separated from SF by 2 to 2.5 their intervening space, or 1.3 their area, and separated by a much greater relative distance from SR; the forehead is low, almost obliterated; the clypeus shows but a narrowed space between the margin and the base of the central eminence, the margin being removed about one diameter of MF from the latter; front row slightly recurved, rear row procurved.

---

[1] Wittfeld; proper name; after the late Miss Anna Wittfeld, of Merrit Island, Florida.

LEGS: 1, 2, 4, 3; stout, comparatively short; uniform orange yellow or yellow, without annuli; armed copiously with long yellowish white bristles, and rather short brownish spines; palps colored and armored as legs; mandibles dark glossy brown, corresponding with color of face, and strongly curved at the base; conical, somewhat separated at the tips; but little pubescent, except on inner edges.

ABDOMEN: Ovate; dorsum well arched to spinnerets, which are placed directly underneath the high apical wall; apex but little if any narrower than base; color yellow or yellowish green, which fades in alcohol to yellow; dorsal folium limited by an indistinct undulating ribbon; with a median longitudinal line branching from about the middle; skin beautifully reticulated and covered quite thickly with short, soft, yellowish hairs; at the sides the greenish yellow color of the dorsum is broken into elongated lobes of dark yellowish brown extending to the venter. The epigynum (Fig. 6c), viewed from the spinnerets, shows a yellow scapus curved over from the base of the atriolum, like the clasp of a padlock; viewed from the side (6d) the scapus is seen to be free at the apex; the portulæ on either side are glossy, blackish brown.

MALE: Fig. 7. Somewhat smaller than female, being 7 mm. long; otherwise similarly marked, but in the specimens in hand the abdominal folium more decidedly marked. Legs uniform orange yellow; tibia-II not thickened, without any special clasping armature, and not more numerously provided with dark brown spines than leg-I; coxa-I has a marginal spur on the apex; in the palps (Figs. 7a, 7b) the digit is globular, the cymbium yellow and covered thickly with yellowish bristles; the embolus a simple curved hook; the radial joint boat shaped, covered with a number of long yellowish bristles; the cubital joint also short and rounded; these, like the humeral joint, are uniform yellow in color.

DISTRIBUTION: I have collected this specimen in Florida, the only point from which I have specimens.

No. 28. Epeira Thaddeus HENTZ. Plate VII, Figs. 3, 4, 5.

1847. *Epeira Thaddeus*, HENTZ . . . . . J. B. S., Vol. v., p. 473; Id., Sp. U. S., p. 113, pl. 13, Fig. 6.

1879. *Epeira baltimoriensis*, KEYSERLING. N. Spinn. Amer. i, Verh. d. z. b. Ges. Wien, p. 305, pl. 4, Fig. 8.

1884. *Epeira thaddeus*, EMERTON . . . N. E. Ep., p. 309, pl. 34, Fig. 9.

1888. *Epeira thaddeus*, McCOOK . . . . Amer. Spiders and their Spinningwork.

FEMALE: Total length, 7 mm.; abdomen, 6 mm. long, 5 mm. wide; cephalothorax, 3.5 mm. long, 2.3 mm. wide; measurement taken from one of the largest specimens. The color varies from yellow and cretaceous to the most brilliant purple and pink, marking the species as one of the most beautiful of our indigenous spider fauna. The legs vary in color from yellow to orange yellow and to yellowish brown, a variation which also marks the cephalothorax.

CEPHALOTHORAX: Corselet rounded, truncated, indented at base; dorsum high, arched; fosse distinct; corselet grooves feeble; cephalic suture distinct; color yellow, flecked with brown, slightly pubescent; head colored as corselet, with dark median band in some specimens; caput slopes from the crest, is slightly pubescent, a band of long hairs down the middle; sternum shield shape, as broad as long, indented at edges, with sternal cones, one in the middle, which is raised; color uniform brown, rather glossy; labium, large, triangular, and maxillæ broad as long, subtriangular at tip, colored like sternum.

EYES: Ocular quad on a prominence; front decidedly wider than rear, and about equal to the side in length. Eyes about equal in size; MF oval, on a black base; MR round and yellow; MF separated by about 1.5 to 2 diameters; MR by about their radius; side eyes on tubercles, contingent, but equal in size; MF separated from SF by a little more than their area, and from clypeus margin by about one diameter; front row recurved, longer rear row procurved.

LEGS: 1, 2, 4, 3; stout, yellow to orange yellow, with brown annuli at tips of joints; well provided with bristles and hairs, rather sparsely with stout spines; palps yellowish brown, armed as legs; mandibles conical, little divergent at tips, and, like the face, somewhat glossy.

ABDOMEN: Globular ovate thickest and widest at the base, arched on dorsum, tapering to distal spinnerets; in gravid females the width is as great as the length; surface reticulated; dorsal field without a folium, but with three or four circular longitudinal patches in the median line; an irregular band of color traverses the margin; a deeper color marks the side; the apical half of the dorsum is sometimes marked by median lines of yellow, and has several black spots on either side tapering V-shape to the spinnerets. Venter pattern a triangular patch of yellow, marked with six or eight circular spots symmetrically arranged on either side of the median line. Spinnerets brown; epigynum (Figs. 3a, 3b, 3c) has a tolerably long, yellow, curved scapus somewhat wrinkled, of nearly equal length throughout, slightly tapering and rounded at the tip. The parmula is a thin, elevated plate, scrolled at the top, placed between it and the spinnerets; this is yellow at the base, brown and glossy at the apex and edges, not quite as long as the scapus.

MALE: Figs. 5, 5a. The males of this species appear to be rather scarce; at least it has proved so in my collecting. They are colored and marked substantially like the female. Underneath femur-I is a row of four spines, which appear to be smaller in femur-II; tibia-II not thickened at the apex, but slightly curved, with a row of four strong rounded spines on the apical half; digital bulb globular; radial joint with a strong curved brown spur; cubital joint about the same length as radial; mandibles relatively longer and slighter than in female; the first two femora rather darker than corresponding joints of female.

DISTRIBUTION: Throughout the Eastern and Middle United States, having been collected in New England, New York, Pennsylvania, Maryland, District of Columbia, Virginia, Georgia, North and South Carolina, Alabama, Florida, Tennessee, and as far to the north and west as Wisconsin. I have no specimens from the Pacific Coast.

No. 29. Epeira Pegnia WALCKENAER.   Plate VII, Figs. 8, 9.

1837. *Epeira Pegnia*, WALCKENAER . . Ins. Apt., ii., p. 80, No. 69;[1] ABBOT, G. S., No. 375.
1837. *Epeira tytera*, WALCKENAER . . . Ins. Apt., ii., p. 81, No. 70.[2]
1865. *Epeira globosa*, KEYSERLING . . . Beitr. z. K. d. Orbit. Berh. d. z. b. Ges. Wien, xv.,
           p. 820, pl. 18, Figs. 19–21.
1876. *Epeira triaranea*, McCOOK . . . . Proceed. Acad. Nat. Sci., Phila., p. 201, and Ameri-
           can Spiders throughout.
1878. *Epeira globosa*, McCOOK . . . . Proceed. Acad. Nat. Sci., Phila., p. 127.
1884. *Epeira triaranea*, EMERTON . . . N. E. Ep., p. 315, pl. 34, Fig. 9; pl. 36, Figs. 6, 7.
1889. *Epeira globosa*, MARX . . . . . . Catalogue.
1892. *Epeira globosa*, KEYSERLING . . . Spinn. Amer. Ep., p. 159, tab. 8, Fig. 117.

FEMALE: Total length, 5 mm.; abdomen, 4 mm. long, 3 mm. wide; cephalothorax, 2 mm. long, 1.5 mm. wide. The general colors are yellow and brown for the fore part, and yellow and gray for the abdomen. The abdomen is placed so nearly at right angles to the plane of the cephalothorax as much to shorten the apparent total length.

CEPHALOTHORAX: Corselet rounded at sides, truncated behind, elevated at centre, the grooves indistinct, cephalic suture well marked; color yellow, with sometimes streaks of

---

[1] No. 375 of Abbot's MSS. drawings is undoubtedly Epeira globosa Keyserling, and the drawing corresponds well with Walckenaer's description of Epeira Pegnia. Walckenaer makes two varieties of the species. Abbot's No. 464 seems not a good variety, but rather to be a Theridioid spider. His variety B is Abbot's No. 375, a female. Abbot's No. 389 (Var, B, Walckenaer) is an immature male. Abbot's No. 555 is also a remarkably good drawing of this species, but I have not been able to find it referred to in Walckenaer's descriptions.

[2] This, according to Abbot's drawings, appears to be the same species as the above—a female.

brown on the sides, and a brighter yellow band along the margin; dorsum slightly pubescent, except at the caput, which is well covered with yellowish gray hairs. Sternum shield shape, about as wide as long, covered with yellowish hairs, brown, with a broad yellow median band. Labium and maxillæ light yellow, and as in Epeira.

EYES: Ocular quad on a central prominence, wider in front than behind, the side about as long as front; eyes about equal in size; MR amber yellow, MF blackish; MF separated by about 1.5 diameter; MR by less than 1 diameter; side eyes on slight tubercles, barely contingent; SF somewhat larger than SR; MF separated from SF by about or less than their area, and set close to clypeus margin, less than one diameter MF therefrom. Front row recurved, the longer rear row slightly procurved.

LEGS: 1, 2, 4, 3; stout, well provided with bristles and hairs, and rather sparingly with blackish spines; the color is yellow, with brown annuli at tips of joints and middle of tibia and metatarsus. Palps colored and armed as legs; mandibles conical, tapering to widely separated tips; the base rounded and with a brown chitinous cog.

ABDOMEN: Triangular ovate, almost as wide as long, widest at the base; carried nearly perpendicularly; dorsum arched to distal spinnerets. Color yellow, with a cretaceous folial pattern, resembling rudely a butterfly, with outspread wings, on the base, and the body extending backward along the median line. This figure is margined with black, and irregular lines of black pass from the dorsum along the sides toward the venter, leaving the sides marked by a broken band of yellow. Four brownish lines pass from the middle of the dorsum longitudinally to the apex, and spots of black are symmetrically arranged on either side receding to the spinnerets. In front the abdomen is marked by two or three rows of black circular spots; the whole surface is strongly reticulated. The venter has a broad patch of white or whitish yellow, marked in the middle by a lateral band of blackish brown; the epigynum (Fig. 8a, 8b) is marked by a decided scapus.

MALE: Closely resembles female in color and markings; the inner apical half of tibia-II provided with a double row of strong clasping spines, and somewhat thicker than tibia-I. The color, in some examples at least, tends to be lighter than in the female, and the markings upon the abdomen are more cretaceous.

DISTRIBUTION: My collection places this species from New England southward along the Atlantic seaboard, and westward through Pennsylvania and Ohio, as far northwest as Wisconsin (Professor Peckham); along the Pacific Coast at Santa Cruz, Cal. (Mr. Harford), and San Diego (Mrs. Eigenmann and Mrs. Smith). It will probably be found distributed throughout the entire United States.

No. 30. Epeira labyrinthea HENTZ.   Plate VII, Figs. 10, 11, 12.

1847. *Epeira labyrinthea*, HENTZ . . . . J. B. S., v., 471, pl. xxxi., 3.
1875. *Epeira labyrinthea*, HENTZ . . . . Sp. U. S., p. 111, pl. xiii., 3.
1884. *Epeira labyrinthea*, EMERTON . . N. E. Ep., p. 314, pl. xxxiv., 8.
1889. *Epeira labyrinthea*, McCook . . . Amer. Spiders and their Spinningwork.

FEMALE: The specimens vary much in size, from the large examples on the Pacific Coast to those which inhabit the Atlantic Coast and interior; I describe from the latter. Total length, 5.5 (8) mm.; abdomen, 3.5 (5.5) mm. long, 2.5 (6) mm. wide; cephalothorax, 2.3 (3) mm. long, 2 mm. (2.5) mm. wide. A large female from the Pacific Coast measures 7 mm. long, of which the abdomen is 4.5 mm. long by 4 mm. wide; the cephalothorax, 3 mm. long, by 2.3 mm. wide in the middle, narrowing at the face to one mm.

CEPHALOTHORAX: Blackish brown, with lighter yellow patches on corselet and face; the margin of lighter color; the caput, especially around the eye space, covered with long. white, bristlelike hairs; sternum shield shape, pointed at the apex, but little longer than broad, elevated in the middle, traversed by a wide yellow band, the margins of which are brown; the surface covered with hairs and bristles, and broken by slight sternal cones; the lip more than half the length of the maxillæ, which are rounded and as wide as or wider than long; colored as the sternum, except lighter tips.

LEGS: 1, 2, 4, 3, as follows: 12.2, 10.2, 9, 6.3 mm.; color brown at tips of joints, stout, strongly provided with brown spines, and more freely with yellowish white bristles and pubescence; palps as the legs, but of lighter color; mandibles conical; color, brown, flecked with yellow.

EYES: Ocular quad on an eminence; somewhat narrower behind, sides not longer than front; MF smaller than MR, and separated by about 1.5 diameter or more; MR by about their radius; side eyes on tubercles, barely contingent; SR somewhat, but little, larger than SF, with strong gray eyebrows; MF separated from SF by about their area, and from the margin of the clypeus by about 1.5 their diameter; front row recurved, rear row longer and procurved; all the eyes on dark bases.

ABDOMEN: A rounded ovate, little longer than wide; arched upon the dorsum, narrower in front than behind, though in some species the difference is small; gravid females are rather widest at the middle and sloping at either extremity; color brown or brownish yellow, thickly covered with white and yellowish long hairs; central folium wide, the margins consisting of brown longitudinal bands, which inclose a herring bone pattern yellow in color, as shown in the figures, Plate VII. The ventral pattern is blackish brown, with a median band of bright yellow and marginal bands darker yellow; spinnerets brown, the base darker and encircled by yellow spots. The epigynum (10a) has a wide scapus, somewhat compressed at the base and depressed abruptly at the apex, forming a small oval tip. The portulæ are well displayed on either side, and the atriolum strongly pubescent.

MALE: Fig. 11. Length, 3.5 mm.; in color and pattern differs little from female, but apical parts of femora dull brown; femur-I provided with two rows of formidable spines, particularly long on the outside; tibia-II not thickened, but a few strong spines clustered around the apex. The cephalothorax is about one-fifth longer than broad; in front not quite half as broad as in the middle; dorsum flat, with slight lateral grooves, but a deep median longitudinal fosse; clypeus low; the arrangement of the eyes similar to that of the female, except that the space between SF and MF is relatively less, being hardly greater than the distance which separates MF. Total length, 3.5 mm., but some specimens measure as much as 4.1 mm.

DISTRIBUTION: This species appears to be widely distributed throughout the United States, and probably inhabits every part thereof. My specimens are from New England, along the Atlantic Coast southward to Florida, and through the entire Middle and Central States. I have numerous specimens from California, and along the Pacific Coast (Messrs. Harford, Orcutt, Dr. Davidson, Mrs. Smith, Mrs. Eigenmann), and specimens from the Barbadoes and West India Islands, and from Venezuela (Professor Peckham) and other parts of South America. The tropical and California specimens (Figs. 12, 12a) are usually larger, the annuli upon the legs darker and wider, particularly upon the femora. The colors generally upon the head, cephalothorax, and abdomen are more pronounced, the browns being darker and the blackish colors deeper. The abdomen is more pointed at the apex, wider and more rounded at the base, forming an inverted cone, at the apex of which well underneath are the spinnerets. The spinningwork of these spiders is like that of their Eastern congeners, but the cocoons are longer.

No. 31. Epeira anastera WALCKENAER.   Plate VIII, Figs. 1–4.

1837. *Epeira anastera*, WALCKENAER . . Ins. Apt., ii., No. 4, p. 33; ABBOT, G. S., No. 381.

1837. *Epeira eustala*, WALCKENAER . . Ins. Apt., ii., No. 12, p. 37; ABBOT MSS., Nos. 119, 120, 361.

1847. *Epeira bombycinaria*, HENTZ . . . J. B. S., 476; Sp. U. S., p. 117, xiii., 16.

1863. *Epeira parvula*, KEYSERLING . . . Beschr. n. Orb. Isis, p. 131.

1884. *Epeira parvula*, EMERTON . . . . N. E. Ep., p. 317, pl. xxxiv., 12; pl. xxxvii., 1, 2.

1888. *Epeira eustala*, McCOOK . . . . . Proceed. Acad. Nat. Sci., Phila., p. 199.

1889. *Epeira eustala*, MARX . . . . . . Catalogue, p. 544.

1892. *Epeira bombycinaria*, KEYSERLING, Spinn. Amerik., iv., Ep., p. 145 pl. 7, Fig. 107, fem.

It is not strange that the remarkable diversity of markings and color upon the dorsum of the abdomen of this spider should have led Baron Walckenaer to erect a number of species thereupon from the manuscript drawings of Abbot in his possession. In my paper heretofore referred to,[1] I mentioned this fact, and therein gave Epeira eustala as the proper title of this species. The more thorough studies which I gave Abbot's MSS. in the summer of 1892 showed me that the first description in order is given under the name of E. anastera, on page 33, and is No. 4 of Walckenaer's descriptions of the Orbweavers. This corresponds with Abbot's No. 381, which is sufficiently accurate to be recognized as the species under consideration. I have only given in the synonyma these two first occurring names. For the benefit of future students I place in the foot notes the various titles of the species which Walckenaer has given, with their corresponding numbers in Abbot's drawings.[2]

FEMALE: Total length, 7.3 mm.; abdomen, 5.3 mm. long, 5 mm. wide; cephalothorax, 2.9 mm. long, 2.4 mm. wide in the middle, narrowing in front to about a millimetre. The general colors of the fore part of the body vary from yellow and yellowish brown to orange brown; the abdomen from yellow to yellow and dark brown. This organ, however, is greatly varied in markings and shades of color, as illustrated in Plate VIII., Figs. 1a–1f. Some of the specimens are entirely without color upon the dorsal field, which is white, with a blackish central spot, or with blackish interrupted triangular lines. These variations are not wholly dependent upon the moulting, as they exist in mature females, numbers of which have been compared with a view to determining this fact. Hentz describes the species as rather inclined to be nocturnal in its habits, being motionless during the day, but active after sunset. It runs with great speed, leaping like an Attus (E. prompta). Its snare is usually pitched upon shrubs and bushes and among grasses and weeds, and resembles that of E. strix. I have seen it sitting upon its hub in a position which is sometimes assumed by its congeners, the abdomen partly resting upon the broken lines of the hub. The spider matures in June and July, and the young are found later in the season.

CEPHALOTHORAX: High in the middle, sharply sloping to the truncated base; sides rounded; dorsal fosse a rather deep slit placed on the sloping base below the crest; cephalic suture distinct; head sloping toward the front; skin smooth, provided with yellow hairs; corselet brown, with flecks of yellow, the head yellow or yellowish brown, as is also the face, except at the posterior part of the ocular quad; partly covered with whitish hairs. Mandibles colored as face and head. The sternum is shield shape, scarcely longer than wide, with marked sternal cones; the centre flattened, the color yellow, with patches of yellow around the margins. Labium subtriangular, wide at the base, yellowish brown, as are the maxillæ, which are wide as long.

EYES: Ocular quad on a marked prominence, this portion of the face, indeed, seeming to be contracted; the front slightly wider than rear, and about the width of the sides, the quad forming nearly a square. MF somewhat smaller than MR, separated by twice their diameter or more, and by an even less space from MR; MR separated by about 1.5 to 1.7 their diameter. Side eyes propinquate, separated by about or less than a radius, nearly equal in size, but less than those of the central group; MF separated from SF by about 1.3 their area, or more than twice the distance between them; height of clypeus about 1.5 diameter MF or more; front row is slightly procurved, and decidedly so viewed from behind and above; the hind row is also procurved, and decidedly longer than the front row, being set well to the side of SR instead of behind it, so much so that the two side eyes appear to form the extremities of the front row. (Fig. 1q.)

---

[1] Notes on the Nomenclature of Orbweavers, Acad. Nat. Sci., Phila., page 199.

[2] Epeira cepina W. No. 13, p. 37; Abbot, Nos. 173, 157. E. apatroga W. No. 23, p. 43; Abbot, Nos. 371, 373, 376. E. spatulata W. No. 24, p. 44; variety C; Abbot, Nos. 171, 366. E. illustrata W. No. 25, p. 45; Abbot, Nos. 186, 187, 188. E. decolorata W. No. 28, p. 49; Abbot, Nos. 345, 399. E. vividia W. No. 38, p. 54; Abbot, No. 474. E. triflex W. No. 48, p. 60; Abbot, No. 112. E. trinotata W. No. 62, p. 75; Abbot, No. 272. E. subfusca W. No. 63, p. 67; Abbot, No. 273. Thus this one most variable species has been described by Walckenaer as eleven distinct species, and a number of varieties in addition is included under these descriptions. A like confusion marks the descriptions of Professor Hentz, his Epeira prompta, E. hebes, E. foliata, and E. bombycinaria being probably variously marked specimens of the same species.

Legs: The legs 1, 2, 4, 3, as follows: 10.6, 9.3, 9, 5.7 mm. They are yellow or yellowish brown, varying to orange brown, with dark annuli not only upon the tips of the joints but between them. They are well covered with yellowish hairs, long upon the femora, and not numerously, with light colored spines with dark bases, which are rather long and thin. The palps are colored as the legs, but not so strongly annulated.

Abdomen: The abdomen is subtriangular, widest at the base, which is somewhat contracted at the middle front, where it greatly overhangs the cephalothorax. The posterior part is truncated, and the spinnerets set immediately beneath the apical wall. The apex in many specimens is smooth (Fig. 1h), but others, particularly those received from the Pacific Coast, have a blunt caudal tubercle. (Fig. 1n.) The color is yellow, mottled with black spots upon the margin. The folium consists of a dark brown triangular figure, open at the base, where it is widest, and with a toothed or zigzag margin, which narrows towards the apex. A median line of dark color extends from the front to the apex, with dentations corresponding with the margin, and flanked on either side by lighter color. The sides are marked by a wavy yellow band; the venter is usually a broad reticulated patch of yellow or yellowish brown between the gills and the spinnerets, marked on either side of the median by a row of three or four black circular spots. The epigynum (Figs. 1k and 1m) has a long conical scapus, very wide at the base and diminishing to a sharp point.

Male: Fig. 3. Length, about 4 mm.; abdomen, 2.3 mm. long, 2.1 mm. broad. On the abdomen the male shows the same variety of markings as in the female, with perhaps, judging from the specimens in hand, a tendency to an excess of cretaceous upon the dorsal field. The abdomen appears to lack the caudal tubercle which marks the apex of the female. The cephalic fossa is deeper and longer than, and the cephalic suture not quite as distinct as in the female, and the head apparently more depressed and more contracted at the face. The eyes are as in the female, except that the central prominence is more marked, side rear eyes if anything set even a little lower than in the female; moreover, the ocular quad is a trifle wider in front than behind, instead of being a trifle narrower, as with the female. Digital joint of the palp is distinct, as at Fig. 3a.

Distribution: I have collected this spider along the Atlantic Coast from New England southward to Florida, and westward through Pennsylvania and Ohio and in Texas. It is found to the north in Wisconsin (Professor Peckham), and I have numerous specimens from the Pacific Coast (Dr. Blaisdell, Mr. Curtis, and others), and from Utah. It is found throughout our Southern States, in Mexico, and Central America. It may therefore be regarded as a continental species, and is probably found with some variations in the northern belt of the South American States.

E. anastera, variety conchlea McCook.   Plate VIII, Fig. 1n.

In the proceedings of the Academy of Natural Sciences, 1888, page 199, I distinguish under the above name those examples of E. anastera which have a decided caudal tubercle upon the dorsum of the abdomen. The specimens bearing this characteristic seem as a rule to be larger and to have more sharply outlined and darker markings, but in other respects do not substantially differ from the typical form as above described. The variety is abundant, and indeed prevalent, in California and Florida.

No. 32. E. eustalina Marx, variety E. anastera.   Plate VIII, Fig. 1p.

1889. *Epeira eustalina*, Marx . . . . . Catalogue, p. 545 (Keyserling *in litt.*).

Under the above name Count Keyserling in his manuscript notes described as a separate species what I take to be simply one of the numerous variations of E. anastera. After examining examples named by Keyserling I see no reason for establishing thereon a new species.

THE ANGULATA GROUP OF EPEIRA.

One meets a series of small Epeïroids, mostly of the Angulata group, which are characterized by several common features, and some of which are difficult to distinguish one from another. They all possess a peculiarly shaped epigynum, which is generally characterized by having a long convoluted scapus of cretaceous or whitish yellow color, of about equal breadth throughout the stalk, and widening at the tips into a broad spoon or ladlelike oval. This peculiarity at once strikes the observer, and compels him to place the species together in one group. Moreover, he observes that they are all small, being about 5 mm. in length, a little more or less. These species are furthermore found to resemble one another in the general shape of the face and arrangement of the eyes; an agreement which extends to the form of the cephalothorax, which is somewhat oval, rounded at the margins of the corselet, pitched high in the middle, and sharply slopes before and behind. The resemblance is further seen in the strong, well arched shape of the caput, rather squarish in its general contour, and wide at the face.

Looking at the abdomen, the series is at once seen to be divided into two sections, of which one, like Epeira juniperi and E. linteata, has an ovate abdomen, smooth upon the surface, that is, without shoulder humps. Comparing the above two species, one remarks a difference in the shape of the atriolum, which in E. juniperi is divided in the posterior part, leaving the portulæ rather distinct, and having curved, pointed, or ram's horn processes issuing from the inner side of the bases. E. linteata has an atriolum that is more or less continuous, being somewhat bowl shaped, from the middle of which the scapus arises, and this is shorter than E. juniperi's. The abdomen of Linteata is also more triangular in shape than that of Juniperi.

Passing to the other, or Angulata section, the differences are not so marked, and the species are often difficult to determine. The species which I take to be typical of the well known Hentzian Epeira scutulata, which must now yield to the prior name of Walckenaer, E. miniata, is distinguished by two leaflike appendages (Plate VIII., Fig. 8c), which arise from the base of the atriolum near the issue of the scapus, and are held aloft upon a short stalk, which, like the leaflike process, is black. E. Mayo has the same characteristic scapus (Plate VIII., Fig. 11a), though perhaps a little more rounded or ladlelike; but the atriolum is without the leaf shaped appendages, and sends out two broad curved sides, which in some specimens unite underneath the tip of the scapus, seeming to form a continuous bowl. In other specimens these are seen really to be separate, and to form simply flanking walls of the portulæ. In E. Bonsallæ (10b), on the contrary, the epigynum strongly resembles that of E. juniperi, having the ram's horn appendages to the inner bases of the portulæ. The tip of the scapus is not quite so circular as that of Mayo, but this may be an individual characteristic. The abdomen of E. Bonsallæ is at once distinct from that of Juniperi, by being subtriangular, having short shoulder humps, and possessing V-shaped rows of brownish spots approximating at the apex. E. Pacificæ differs from the other species of the section to which it is most closely allied, by the strong character of the dorsal folium; by the deeper brown bands upon the legs, and the median annuli; by the generally stronger and darker colors of the whole animal; and, moreover, by the form of the epigynum, the scapus of which is much convoluted, issuing from the base of a bowl shaped atriolum, which is continuous both in front and behind, and not dividing underneath the tip of the scapus, as is the case in Mayo, from which species it also differs by the general markings and color.

It is possible that future students, who may be favored with more numerous specimens from which to judge, may find that these characteristics, some of them at least, are more individual than specific; or, that instead of giving good grounds for specific distinction, may simply establish varietal forms of one common species, of which E. miniata may be held as the principal form. I have presented these differences as they appear to me, in the hope to simplify and economize the labors of naturalists, who are sure to find, as I have found, much labor and perplexity in discovering good characteristics by which to distinguish this perplexing group.

No. 33. Epeira linteata, new species.   Plate VIII, Figs. 5, 6.

FEMALE: Total length, 4 mm.; abdomen, 2.6 mm. long, 2.6 wide, narrowing to 1 mm.; cephalothorax, 2 mm. long, 1.6 wide; about the face, 1 mm. wide.

CEPHALOTHORAX: Corselet a rounded oval, peaked in the centre, sloping sharply down behind, with a smooth flat surface, and with almost an equal slope toward the front; fosse almost obliterated; cephalic suture not prominent; corselet grooves indistinct; color yellow, with streaks of brown; caput and face covered with long yellowish-white hairs. Sternum shield shape, truncated in front; sternal cones; yellowish brown; slightly pubescent; labium sub-rectangular at base and triangular at tip; maxillæ rounded at margins; as broad as long; color as the sternum.

EYES: Ocular quad without a decided prominence; its eyes about equal in size; the front about equal to the length and a little wider than the rear. MF dark colored, separated by about 1.5 diameter; MR amber, and separated by about one diameter. Side eyes about equal in size, placed on a slight tubercle; SF separated from MF by about the area of MF; the two eyes contingent, and SR so nearly on a line with SF that they seem to be grouped rather with the front than the rear row. The space between SR and MR equals at least three, perhaps four, times the space between MR. The front row of eyes is almost aligned, being only slightly recurved; the rear row is much procurved; clypeus high, its margin separated from MF eyes by about twice their diameter.

LEGS: 1, 2, 4, 3; stout for so small a species; armored with yellow spines and long bristlelike hairs; the joints appear to lack annuli, but my specimens have been much injured in the alcohol. Mandibles subconical, diverging slightly at the tips, a little receding beneath.

ABDOMEN: Subtriangular, and without dorsal tubercles. At the widest part it is about as broad as long; is highly arched, rounding from the cephalothorax to the apex, and marked by a broad, white, capelike folium, with denticulate edges, extending from the spinnerets to the base, and overlapping the basal front in a scalloped point. (Plate VIII., Fig. 5b.) This white patch is broken in the centre by a dark interrupted band, dividing on either side of the median line. The sides, like the centre, are yellow brown, marked with reticulations at the venter. The epigynum has a short scapus broadened at the tip (Figs. 5c, 5d), and appears to proceed from a bowl shaped atriolum, that in one specimen at least (5d) seems to pass out beneath.

MALE: The male (Fig. 6) differs little in its markings from the female. It is somewhat smaller in size, and is distinguished by a palpus whose digital bulb is given at Fig. 6b.

This species resembles in many particulars E. miniata, but differs in the form of the eyes (5f), especially in the location of the rear lateral eyes. It also lacks the well developed dorsal tubercle upon the base of the abdomen which characterizes E. miniata, and the epigynum is distinctly different. The species is probably a little more closely related to Emerton's Epeira alboventris.

DISTRIBUTION: The only specimens which I have, four females and one male, were collected in North Carolina.

No. 34. Epeira corticaria EMERTON.   Plate VIII, Figs. 7, 7a–d.

1884. Epeira corticaria, EMERTON . . . N. E. Ep., p. 300, pl. 33, Fig. 14; pl. 35, Fig. 9.

FEMALE: Total length, 7 mm.; abdomen, 5 mm. long, 5 mm. wide; cephalothorax, 2.5 mm. long, 1.5 mm. wide.

CEPHALOTHORAX: Longer than broad; rounded at the edges; cephalic suture deep; somewhat flattened on the top; the fosse well marked; corselet grooves not prominent; color yellow, with a patch of brown around the fosse and on the base of caput; head but little depressed, wide in front. Sternum dark or blackish brown, with decided sternal cones, slightly covered with yellow pubescence; labium subtriangular; maxillæ slightly longer than broad, and subtriangular at the tip.

EYES: Ocular quad on a decided prominence, about as wide as high; MF about equal to MR in size; MF separated by about 1.5 diameters, and by about an equal space from MR. The latter separated by about one-half diameter; side eyes separated by a slight space. SF larger than SR, the latter being placed well to the side of the former, and in their curvature. The front row recurved, rear almost aligned; SF separated from MF about 1.3 the space which divides the latter, or about the area of the latter; SR from MR by about twice the area of the latter; clypeus about 1.5 to 2 diameters MF.

LEGS: 1, 2, 4, 3; stout, well armed with yellow spines and bristles with dark bases; joints with brown apical and median annuli; palps colored and marked as legs; mandibles conical, yellow.

ABDOMEN: Triangular ovate; the length about equals the breadth across the base, at which point are two well-developed tubercles. The color varies from dark brown to yellowish brown. The tubercles are beautifully reticulated, as is also the subtriangular basal front, which slopes rather sharply to the cephalothorax; in the middle of the base is a cruciform marking of yellow color. The folium is somewhat undulated laterally, but individuals vary in this respect. The dorsum is well arched, the spinnerets distal. A squarish patch of brown marks the venter, bordered with a broad reticulated yellow ribbon; the spinnerets are brown, except a slight ring of yellow. The epigynum (Fig. 7d) has a well developed scapus, broad at the base and narrowing toward the top; it is much wrinkled and curved, as shown in the side view (7b), and extends well over the portulæ.

DISTRIBUTION: Massachusetts and New York; probably all of New England and the northern Middle States. Specimens received from Mr. Nathan Banks, Ithaca, N. Y.

No. 35. Epeira miniata WALCKENAER.   Plate VIII, Figs. 8, 9; Pl. X, Figs. 7, 8.

1837. Epeira miniata WALCKENAER . . Ins. Apt., ii., No. 17, p. 39; ABBOT, G. S., NOS. 228,
                                          229, 230.
1837. Epeira cingulata WALCKENAER . Ins. Apt., ii., No. 18, p. 40; ABBOT's "Belted
                                          Spider," Nos. 232, 365.
1837. Epeira guttulata, WALCKENAER . . Ins. Apt., ii., No. 65, p. 78; ABBOT, G. S., No. 233.
1837. Epeira bivittata, WALCKENAER . . Ins. Apt., ii., p. 78, No. 66; ABBOT, G. S., No. 234.
1850. Epeira scutulata, HENTZ . . . . . B. J. S., p. 19, iii., 3; Sp. U. S., p. 121, xiv., 3.
1879. Epeira punctillata, KEYSERLING . Neue Spinn. aus Amer., i., Verh. Zool. Bot. Ges.
                                          Wien, p. 304, pl. iv., 7, male palp.
1889. Epeira scutulata MARX . . . . . Catalogue, p. 547.
1892. Epeira scutulata KEYSERLING . . Spinn. Amerik. Ep., p. 129, tab. vi., 96, fem.

In giving the synonyma of this beautiful species I have been guided by my studies of Abbot's MSS., and my notes thereon show that I regard Nos. 228, 229, and 230 as the same species, and consider that 230, an immature male, also belongs to the same species. No. 228 resembles those forms in my collection in which two oval white patches stand out prominently upon the dorsal base of the abdomen between the tubercles. The colors are yellow, the round whitish patches red. The dorsal tubercles are strongly suggested in No. 228, but are represented in No. 229, which is distinctly marked by the V-shaped rows of black spots on the margin of the folium. No. 230 of Abbot is a yellow specimen with brownish folium, and a red bow along the abdominal front. The V-shaped spots show distinctly in this example, which, like No. 227, is an immature male. Abbot's Nos. 232 and 233 appear to be the same species, although the abdominal tubercles are indistinctly indicated, or so nearly wanting that one hesitates to decide. They are beautifully colored. No. 233 has the cephalothorax and legs of greenish yellow; the abdomen green and yellow, with a median lateral stripe of brown with yellow centre; the eight V-shaped spots are brown within yellow circles. No. 233 (Epeira guttulata, Walck.) has the cephalothorax, legs, and abdominal front yellow, the abdomen orange brown, with eight black V-shaped spots within yellow circles. I place this number in the synonyma with less confidence than the others, but on the whole think there is not much doubt that it is correctly placed as here. Abbot's No. 234, Epeira bivittata, Walck. (II., page 78), is probably the same; the central folium is vermilion or lake color, as are also the eight V-shaped spots. The remainder of the abdomen is green, with yellowish-white bands. The legs and cephalothorax are also green, with orange yellow annuli.

Abbot's No. 365, which Walckenaer regards as identical with his Epeira cingulata (Abbot's No. 233), is without doubt a beautiful example of Hentz's E. scutulata. Abbot has designated it in his brief notes as "none so pretty;" he beat it off of a sumac bush. The general color of the abdomen is green; but a T-shaped figure of delicate ash color defines the dorsal folium. The V-shaped spots are included within contiguous circles of dark pinkish color, and the circular spots upon the dorsal front are similarly marked. The cephalothorax and legs are light yellow. The abdominal tubercles on this specimen are plainly indicated in Abbot's drawing—so much more distinctly indeed than in No. 228 (E. miniata) that I have hesitated whether it would not be right to accept this as the type of the species, although the description occurs much later in order in Walckenaer's publication. However, I have little doubt that No. 228 is identical with Hentz's E. scutulata, and therefore conclude to give it the priority.

FEMALE: Total length, 4.5 mm.; abdomen, 3 mm. long, nearly 4 mm. broad; cephalothorax, 2 + mm. long, somewhat less in width; head about half as wide as cephalothorax.

CEPHALOTHORAX: Corselet rounded ovate; cephalic suture, fosse and grooves distinct; caput slightly depressed, pubescent, with a few bristles; sternum slightly longer than wide, with sternal cones, somewhat arched, pubescent, yellowish brown. Labium triangular, base wider than length; maxillæ as wide as long.

EYES: Ocular quad about as wide in front as long and wider than rear; MF black, separated about 1.75 diameter; MR amber, about equal to MF, and separated one diameter; side eyes barely contingent, equal, SF removed from MF less than alignment of the latter, or about 1.3 their intervening space; front row slightly recurved, rear row longer and slightly procurved; clypeus about 1.5 diameter in height.

LEGS: 1, 2, 4, 3; stout for so small a species, clothed freely with strong hairs and dark spines; color varying from yellow to light brown, and without distinct annuli; the palps resemble the legs; the mandibles are conical, not divergent.

ABDOMEN: Triangular ovate, much broader at base than apex; the base, which overhangs the cephalothorax, slopes downward thereto, forming a triangular front, thus leaving the abdomen divided by a ridge into two well defined slopes. The anterior part is in some species darker in color, has a lateral row of circular spots, and on the crest two bright white oval spots; beyond this sometimes rosy tints. Shoulder humps well defined, darker in front, in the rear tipped at times with white; dorsum abruptly arched to the spinnerets. The color in many specimens is grayish yellow, in others pale yellow, and in some quite white, with reticulated markings. An indistinct triangulated folium marks some specimens; on each dorsal margin is a row of four brownish yellow spots within white circles, which converge to the apex in V-shape. In some specimens these are quite distinct, in others apparently wanting. The ventral pattern is a dark brown median band, with light yellow and gray margin; spinnerets dark; epigynum with a short somewhat sinuated scapus, well rounded at the tip.

MALE: Total length, 3 mm.; cephalothorax slightly longer than broad, and ridged in the centre as in female; abdomen slightly longer than broad; shoulder tubercles less prominent than in female. Adult specimens in hand (California) have a triangular dorsal folium with scalloped edges sharply marked by dark brown interrupted lines, punctuated at points with black spots. In several immature specimens from Florida and elsewhere the markings closely resemble that from which Hentz described the species, an immature male; V-shaped black dots mark the margin of the dorsum, narrowing to the apex, and four somewhat similar spots are on the base in front. In some immature specimens from California, when freshly taken, the abdomen was prettily tinted with pink hues. (Plate VIII., Fig. 9a.) Length of legs (1, 2, 4, 3), 7, 5.2, 4.7, 3.1 mm.; tibia-II is not thickened or otherwise modified, and the coxæ are without spurs.

DISTRIBUTION: I have specimens from New Jersey, Georgia, North Carolina, Florida, and California; Hentz described it from Alabama; Dr. Marx has it from the District of Columbia, Illinois, and Texas. This indicates that it inhabits the entire southern portion of the United States, as far north as the District of Columbia, and westward to California. I have no specimens from any of the Northern or Middle States, except one from Wisconsin, marked doubtful. The coloring and dorsal markings vary greatly in specimens under observation, the difference not dependent upon moulting changes, as it shows in mature examples. Colors range from dark yellow, with blackish or brown spots, to white; pale yellows, pinks, and even greenish tints appear on fresh specimens. The V-shaped circular spots on the dorsum are sometimes distinct, and again disappear, giving place to a triangular folium with interrupted margins.

No. 36. Epeira Bonsallæ, new species.   Plate VIII, Fig. 10.

FEMALE: Total length, 5 mm.; abdomen, 3.5 mm. long, 3 mm. wide; cephalothorax, 2 mm. long, 1.75 mm. wide.

CEPHALOTHORAX: Corselet well rounded, elevated in centre, deep rounded fosse; corselet grooves sufficiently distinct; cephalic suture deeply marked, separating the head decidedly from the corselet; the head slightly depressed toward the face, where it is not narrowed; color, yellow to yellowish brown, with a lighter stripe on the margin. Sternum shield shaped, with slight sternal cones, color yellow, slightly clothed with long yellow hairs. Maxillæ subglobose, decidedly broader than long; cut square at the tip; labium triangular, colored as maxillæ, and is about half their height.

EYES: Ocular quad almost a square; but very slightly narrower in front than behind; MF black, MR amber color; MF separated by about 1.5 their diameter; MR slightly larger and separated by about 1.3 diameter; front row slightly recurved; rear procurved. (Fig. 10a.) Side eyes on slight tubercles, propinquate, SF the larger. SF separated from MF by a little more than the area of the latter; SR from MR by at least 1.3 the area of the latter; MF from the clypeus margin by about 1.5 their diameter.

LEGS: 1, 2, 4, 3; stout, yellow, without decided annuli, armed with strong bristlelike hairs and strong long spines. The palps are similarly armed.

ABDOMEN: Subtriangular, with slight basal tubercles, at which point the width about equals the length; dorsum rounded, and well arched to the distal spinnerets; color green, with a folium yellow at the margin, green in the centre, except at the median line, which again is yellow. Branching longitudinal lines mark the apical part of the folium, and on either side within the yellow irregular folial margin is a row of four brownish spots, approximated toward the apex in V-shape. The venter is greenish yellow, except the epigynum, which is brown, of which the scapus is rounded, wrinkled, of nearly equal thickness throughout, except at the tip, where it broadens out into a heart shaped spoon of at least twice the width of the base. A minute tonguelike appendage extends from each of the portulæ.

DISTRIBUTION: This species, of which I have but one specimen, was received from California. It strongly resembles E. miniata, of which it may possibly be a variety, or variant form.

No. 37. Epeira Mayo, new species.   Plate VIII, Fig. 11.

1889. *Epeira Mayo*, MARX *in litt.* . . . Catalogue, p. 546 (KEYSERLING *in litt.*).

FEMALE: Two specimens, one 5 mm. long, the other 4.5. In general form and character this species resembles closely E. miniata. It appears to me to be a variety thereof after studying the type upon which Count Keyserling in his MSS. notes established the above. The abdomen is more ovate than E. miniata, not so wide relatively across the base, nor so sharply ridged in the dorsal crest. The V-shaped folial spots are wanting, and no distinct folial pattern appears, only irregular, waving, pale lateral lines, which give an indistinct suggestion of a folium. The epigynum differs in the greater length and decided convolution of the scapus. The cochlear in E. miniata is more prolonged and slightly compressed in the middle, while that of E. Mayo is almost a circular bowl. The leaflike appendages of the atriolum in E. miniata are wanting, or at least are folded down, in E. Mayo.

DISTRIBUTION: Two females, one in my collection from Wisconsin and one in the Marx Collection from Minnesota. The known patria is thus the Northwestern United States.

No. 38. Epeira bispinosa KEYSERLING.   Plate IX, Figs. 3, 3a-b.

1884. *Epeira bispinosa*, KEYSERLING . . Neue Spinn. Amerik., vi., p. 531, xiii., 30.
1892. *Epeira bispinosa*, KEYSERLING . . Spinnen Amerik. Epeir., p. 124, vi., 92.

FEMALE: Total length, 5 mm.; cephalothorax, 2.5 mm. long by 2.1 mm. broad; abdomen, 3.5 to 4 mm. long by 4.5 to 5 mm. wide at the base. Colors, fore part of body brown, abdomen yellow. This spider is distinguished from E. miniata by the arrangement and relative size of the eyes, especially of the middle group, the midfront eyes of Bispinosa being almost twice as widely separated as those of Miniata and decidedly smaller; the ocular quad is relatively wider in front than rear, and the clypeus much higher. The legs also are annulated, and the epigynum has a longer and more convoluted scapus, like E. Mayo.

CEPHALOTHORAX: A long oval; the crest high; corselet brown, with a lighter marginal band; caput yellow, flecked with brown; the whole sparsely covered with hair. Sternum almost as broad as long; raised in the middle; with decided cones; yellowish brown color; labium long, subtriangular.

LEGS: 1, 2, 4, 3, as follows: 9.4, 8.10, 5.2, 7.3 mm. Color, yellowish brown, with annular markings; well provided with hairs, bristles, and spines.

EYES: Ocular quad on a well rounded prominence, the sides about equal to front, the latter wider than rear; MF separated from 2 to 2.5 times their diameter, and are much smaller, from one-third to one-half, than MR, which are separated from one another about a diameter and a half. The side eyes on prominent tubercles, well separated; SF somewhat larger than SR, and about or a little more than the size of MF, from which they are divided by about the area of MF. Front row recurved, rear row longer and slightly procurved; clypeus is high, about the space between MF.

ABDOMEN: Triangular ovate; with prominent shoulder humps, basal front subtriangular; dorsum arched from crest to distal spinnerets; color yellow, surface reticulated, lines of brownish color extending longitudinally in the middle of the dorsum, and laterally along the sides and underneath to the venter; ventral pattern squarish, yellow, reticulated, and bordered with brown; spinnerets dark brown, mottled at the edges with whitish and yellow spots. The sides are yellow, with alternate black and white, running vertically from the venter to the dorsum. The epigynum (Plate IX., 3b) has prominent bowl shaped portulæ; the scapus long, much sinuated, and terminates in a wide spooned bowl.

DISTRIBUTION: Southern California; Fort Yuma, Ariz. (Marx Collection); Keyserling locates it in Central America, Panama, and Hayti. The specimen is probably widely distributed throughout the semitropical parts of North America.

No. 39. Epeira Pacificæ, new species.   Plate XI, Figs. 15, 16.

FEMALE: Total length, 5 mm.; abdomen, 4 mm. long, 4 mm. wide; cephalothorax, 2 mm. long, 1.5 mm. wide.

CEPHALOTHORAX: Oval; corselet well rounded at the edges, rather high in the middle, sharply sloping backward, caput inclined forward to the face; corselet grooves distinct; cephalic suture decidedly marked; color, dark brown, with flecks of yellow on the median base; grayish white hair rather sparsely distributed over the surface. The caput is strong, squarish, wide at the face, well rounded, colored as the cephalothorax, with a yellow patch at the eye space; sprinkled with gray hairs, more abundant upon the face and eyebrows. Sternum shield shape, not greatly pointed at the apex; longer than broad; sternal cones distinct, covered slightly with gray hairs; a uniform brown; labium triangular; maxillæ as wide or wider than long, subtriangular at the tips; both colored as the sternum, but a rather lighter hue.

EYES: Ocular quad on a rather squarish rounded eminence, of which the front is more pronounced; MF upon separate tubercles, black, about the size of MR, which are bright amber color; the quad barely longer than wide, slightly wider in front than behind; MF separated by about 1.75 diameter; MR by about 1.5; side eyes upon slight tubercles, about equal in size, contingent; SR well to the side and separated from MR by 1.3 the area of the latter. MF separated from SF by little more than their area, or 1.3 their interval; the height of the clypeus equal to about 1.5 diameter MF; front row but little recurved, rear row longer, and from the same aspect procurved, but viewed from above but a little, is aligned.

LEGS: 1, 2, 4, 3; yellow, with darker annuli at the tips, and slight annuli at the middle of the metatarsus and tibia; freely provided with grayish yellow hairs, bristles, and dark long spines; the palps are colored and armed as the legs; mandibles subcylindrical, long, strong, brownish yellow, pubescent.

ABDOMEN: Subtriangular, thickened at the base, which rises almost perpendicularly from the corselet, forming an angle with the dorsal field; the latter arched to the apex, which slightly overhangs the spinnerets. .The abdomen is thus divided into two subtriangular fields; the front is mottled cretaceous and blackish, punctuated with numerous black spots; reticulated, with a lanceolated cretaceous band between the shoulder humps; the dorsal field is cretaceous or yellow, which color extends along the inside and posterior of the rounded, strongly marked shoulder humps; between the humps, in one specimen, are two oblong bright white patches, which take the place of the cretaceous extension above referred to. The dorsal field is marked by a decided folium, which occupies half thereof; it is outlined by a cretaceous undulating margin, within which extends a parallel line of black to the apex; beyond this are yellowish shades extending to the median pattern, which is an interrupted cretaceous or yellow ribbon. The venter is a brownish yellow ribbon, compressed in the middle, with wide semicircular patches of reticulated white and yellow on each side, which merge into similarly colored stripes along the sides of the abdomen. The epigynum has a long, yellowish white, convoluted scapus (Fig. 15a), whose stalk is of about equal width throughout, and terminates in a widened and rounded spoon. The atrialum rises from the margin of the genital cleft as a hollow bowl or shell, from the base of which the scapus originates.

MALE: Fig. 16. Resembles in general form and markings the female. The cephalothorax is more oval, the head relatively narrower at the face, the legs have wider and darker annuli at the tips; the abdomen is relatively narrower at the base than the female, though with the same general conformation, and with smaller shoulder humps; the folium is outlined at the margin by an interrupted waving band of black, margined by white. A brownish yellow median line shoots between the shoulder humps, sending forth four longitudinal branches of like color along the middle of the folium to the apex. Tibia-II is not thickened, and has no distinctive clasping spines; is simply provided with three rows of long brownish spines. Tibia-I is greatly longer than tibia-II, and has more spines thereon. Femora-I and II are covered, except near the bases, with wide dark brown annuli; the palpal digit is globular, glossy, and brown; the cymbium blackish, covered well with grayish hairs.

DISTRIBUTION: I have three females and one male of this species, received from California; and two other females and an immature male from San Diego, Cal.

THE LARGER ANGULATA.

The smaller Angulata above described might, perhaps, be properly assigned to a subgenus. Its principal characteristics, apart from the small size of its species, would be a high and subconical corselet; an epigynum with a scapus much convoluted or wrinkled, terminating in a ladlelike tip; an abdomen of marked triangular form; a diminished interval between the midfront and sidefront eyes as compared with the typical Epeira.[1] To this group might also be assigned E. forata, E. linteata, and ·E. juniperi, although their abdomens are not distinctly triangular, and lack shoulder humps.

The larger Angulata, whose descriptions follow,[2] may be regarded as among the more typical Epeira. They all make the typical Epëiroid wheel shaped web, which is often very large, and occupy, especially during the day, a leafy nest above and at one side of the snare, which they command by a taut trapline attached to the hub thereof. In some parts of the United States the most common Orbweavers belong to this group. One at least, E. angulata, is distributed throughout the entire country; and two, E. diademata and E. angulata, are common in Europe, the former, indeed, being the best known garden Orbweaver.

No. 40. Epeira gemma McCook.   Plate X, Fig. 6; Pl. IX, Figs. 1, 2.

1888. *Epeira gemma*, McCook . . . . . Proceed. Acad. Nat. Sci., Phila., p. 193.
1889. *Epeira gemma*, MARX . . . . . . Catalogue, p. 545.
1892. *Epeira gemma*, KEYSERLING . . . Spinn. Amerik., Epeir., p. 115, vi., 85.

FEMALE: Total length, 19 mm.; abdomen, 16.7 mm. long, 16 mm. wide; cephalothorax, 6.9 mm. long, 6 mm. in the middle, and 2.8 mm. wide. One large specimen in my possession measures over 20 mm. in length. General colors yellow, with brown or darkish markings, and for the fore part yellow and brown.

CEPHALOTHORAX: Cordate, the base indented; corselet rounded at the edges, rather low, the fosse a deep rounded pit; corselet grooves tolerably distinct; cephalic suture sufficiently marked; color yellowish brown, with a gray median band, which is chiefly marked out by long gray bristles; gray hairs abundant over the surface, which is glossy. The caput is depressed, flattish upon the top, subtriangular at the base, wide, squarish at the face, colored as the cephalothorax, with yellowish longitudinal bands upon the sides. Sternum heart shaped, pointed at the apex, about one-fourth longer than broad, raised in the middle; sternal cones before coxa-III, and more rounded ones before coxae-I, II; a decided cone opposite the lip; color ruddy brown, covered with golden yellow and long bristlelike hairs. Labium subtriangular; maxillae gibbous, somewhat longer than wide; both these organs brown, with yellow tips, and the maxillae with a few brownish spines and yellow bristles.

EYES: Ocular quad on a well rounded prominence, length not greater than width in front, the front decidedly wider than rear; MF on separate tubercles, larger than MR, separated by at least 1.5 diameter; MR separated by not more than one diameter. Rear eyes on tubercles, not contingent; SF somewhat larger than SR; SF removed from MF by about 1.3 their area, or at least twice or more their intervening space; SR from MR by 2.5 to three times the area of the latter. The height of clypeus about 1.5 to 2 diameters MF, with a row of strong yellowish bristles along the margin; the space between the eyes is also sparingly marked with shorter bristles; front row slightly recurved, rear row slightly procurved.

LEGS; 1, 2, 4, 3, as follows: 23 (24.2), 22.5 (23.2), 21.5 (21.6), 14.75 (15.5) mm. Stout, thickly covered on all sides with long yellowish bristles and yellow spines, which are numerous along the metatarsus and tibia; these are particularly abundant underneath femora-II; color yellow, strongly annulated with brown at tips of joints and along femora underneath. Palps colored and armed as legs; mandibles strong, conical, brown, glossy, with yellowish tips and yellowish white bristles upon the inner sides.

ABDOMEN: Triangular ovate, slightly longer than wide, highly arched on the dorsum from the cephalothorax, which it overhangs, to the distal spinnerets; basal front subtriangular, high, marked at the summit by two large conical humps; color blackish brown,

---

[1] I venture to propose for this group the subgeneric name "Burgessia," in honor of the late Mr. Edward Burgess, of Boston, the editor of Professor Hentz's "Spiders of the United States," who was favorably known as an entomologist, and later was an eminent designer of sailing yachts.

[2] Epeira gemma, angulata, diademata, Nordmanni, cavatica, Silvatica.

interspersed at irregular intervals with yellow spots. Along the dorsal median extends a narrow band of yellow, upon which are placed two angular or lance-head markings, pointed forward, the first of which is placed about the middle of the basal part and the second near the crest. This color band continues more or less regularly along the dorsum to the apex, and in some examples reminds one of Epeira diademata's pattern. About the middle of the dorsum is a shield shaped figure with scalloped edges, blackish brown in color for the most part, though interrupted by yellow lines of a herring bone pattern; a narrow yellow border encompasses the folial shield; color of dorsum and posterior half of shoulder humps yellow. Dark brown waving and interrupted lines extend along the sides, between which are small round spots, distributed laterally along the sides with more or less regularity. A broad brownish band extends along the venter from the spinnerets to the epigynum, bordered along either side by a yellow band more or less interrupted in various specimens, with a median band which is sometimes divided longitudinally. The epigynum (Plate IX., 1d) is provided with a rather short scapus, wide at the base, and terminating in a well defined spoon. It has a wide subtriangular atriolum, but rather small for such a large species; the scapus is short, but little differing in width throughout, and terminates in a rounded, heart shaped, brown, chitinous tip, which is spooned.

MALE: Plate IX., Fig. 2. Although I have received a large number of females of this species I have but one mature male thereof. It is small, compared with his mate, having a total length of 8 mm. The cephalothorax is a longer oval than in the female, having at the base a width of 3.5 mm. The color is yellowish brown; it is less pubescent than the female, the fossa a longitudinal slit, with a rectangular depression around it. The eyes are arranged about as in the female; the mandibles are much feebler, comparatively longer, and semiconcave upon the front surface, with a rounded cog at the base. The legs are yellow, with decided median annuli upon the tibia and metatarsus; spines yellow, with brown bases, and are particularly long underneath femur-I, where they are grouped and clustered about the middle. Tibia-II is not swollen, and has no special clasping apparatus. The palpal digit is globular, with a strong hooked process at the base of the cymbium. (Fig. 2a.) The abdomen is shaped as in the female, though much smaller, with decided shoulder humps, and strong spinous bristles of yellowish bright color sparsely scattered over the surface.

DISTRIBUTION: This is one of the largest orbweavers of the Pacific Coast, and is found from San Diego northward as far as British Columbia. Numerous specimens have been received from Mrs. Eigenmann, Mr. Orcutt, Dr. Davidson, Dr. Blaisdell, Mr. Curtis, and others; from Utah (Professor Orson Howard); and the Marx Collection notes it from Dakota, Montana, and Louisiana. The spider makes a large circular web characteristic of the Angulata group to which it belongs, and rests in a nest of rolled leaves or dome shaped rubbish placed on the upper side of its snare.

No. 41. Epeira bucardia, new species.[1]   Plate IX, Figs. 4, 4a-d.

FEMALE: Among collections sent from Southern California I have a single specimen, a mature female, which on the whole appears best classified with the genus Epeira. Its total length is 5 mm. The color is bright yellowish brown, mottled with black and darker brown.

CEPHALOTHORAX: A rather long oval, corselet well peaked in the centre; head slightly depressed; caput strongly marked with dark longitudinal lines; sternum smooth, glossy, dark brown (Fig. 4a); almost as wide as long, with slight sternal cones.

EYES: In their grouping the eyes approximate more closely Zilla than those of the typical Epeira. (Fig. 4c.) They do not greatly differ in size, the side eyes being smaller and the midrear pair the largest of all. The drawing of these eyes is unfortunately very imperfect.

---

[1] "Bucardia," an ox heart. The single specimen of this species was unfortunately lost during the drawing of plates, and I am not able to revise my original notes of description. The plate having been printed before this loss was discovered, I cannot omit the species, as I otherwise would have done, but submit description as above without revision.

LEGS: 1, 2, 4, 3; yellowish brown in color; strongly annulated at the joints, and between the joints of femora and tibiæ. The palps are yellow; the mandibles conical, parallel, but divergent at the tips; dark brown.

ABDOMEN: Subtriangular, or cordate; as wide across the base as the length; strongly marked shoulder humps (Fig. 4b, side view) upon the dorsum, which is yellow, mottled with undulating and transverse black bands, which constitute an irregular folium, occupying most of the dorsum. The apical part of the abdomen is marked by a black foliated figure, in the centre of which may be seen the dark branching lines so common in spiders. The base of the abdomen appears to be divided into two parts by a natural constriction, which, although it may have been caused by the shrinking of the skin in the abdomen, has been represented in the figure. Underneath the color is yellow, mottled with black. (Fig. 4a.) The ventral pattern is a broad, black, brown median band, bordered with yellowish white, in the lower portion of which, near the spinnerets, are two whitish spots. The epigynum has a horseshoe shaped atriolum (Fig. 4d), the ends and ridge of which are black and corneous. The scapus is short, narrow, but widening at the tip like a spoon.

DISTRIBUTION: Southern California, one female.

No. 42. Epeira Nordmanni THORELL.   Plate IX, Figs. 5, 6, 7; Pl. XI, Fig. 5.

1870. *Epeira Nordmanni*, THORELL . . . Synonyms European Spiders, p. 4
1884. *Epeira Nordmanni*, EMERTON . . N. E. Ep., p. 301, pl. 33, Fig. 6.

FEMALE: Total length, 15 mm. for largest specimen, smallest adult specimen, 9 mm.; abdomen, 11 mm. long, 9 mm. wide; cephalothorax, 5.5 mm. long, 4 mm. wide; width of the face, 2 mm.

CEPHALOTHORAX: A rounded oval, somewhat flattened on top; fosse a deep semicircular pit; corselet grooves distinct, but interrupted; cephalic suture distinct; caput slightly depressed; surface smooth, without pubescence; color, brownish yellow, with dark shades of brown. Sternum shield shaped, bluntly pointed at the apex; raised in the middle; sternal cones prominent, especially opposite coxæ-I, III, most prominent of all in front of labium; sparsely covered with white hairs. Labium wide, obtusely triangular at tip; at least half as high as maxillæ, which are slightly longer than broad, tips subtriangular; color of sternum, lip, and maxillæ dark brown, tipped with yellow.

EYES: Ocular quad elevated; length slightly greater than frontal width, narrowest behind; MF largest, separated by about one diameter; MR separated by 0.7 diameter. Side eyes on high tubercles; barely contingent; SF larger than SR; MF separated from SF by about 1.3 their area; SR from MR by from 2 to 2.5 area of latter. Clypeus about 1.5 diameters MF high; front row slightly recurved, the longer rear row slightly procurved.

LEGS: 1, 2, 4, 3; stout; armed with numerous yellow spines with brown bases, and stout bristles; color, yellow or cretaceous, with apical brown annuli; femora-I, II, in some specimens are bright brown, in others yellowish brown. The palps stout, heavily armored and yellow, except at tips of digital joints; mandibles conical; strong, slightly divergent at tips, colored as cephalothorax.

ABDOMEN: Triangular ovate; two prominent shoulder humps; the basal front slopes toward cephalothorax, forming an equilateral triangle; dorsum arched in a long triangle to the spinnerets; the apical wall about two-thirds the thickness of the base. Color yellow to yellowish brown; marked on the dorsal front by a light Y-shaped figure; the arms of the Y extend between the shoulder humps, thence following the median line. This bright yellow band passes onward with interruptions nearly to the apex. On either side of the lower half is a row of five semilunar dashes of brown color, converging to the spinnerets; in some specimens on the lower part of the abdomen these marks are almost entirely united by a band of brown, as in Plate IX., Fig. 5. The yellow color of the sides is crossed by four or five series of brown lines, drawn from common points on the dorsum at the margin of the folium, and widening as they pass around the sides toward the venter. Ventral pattern a dark brown oval band, marked by two bright yellow roundish spots near the middle, and two triangular yellow marks at corner of the gills, and four patches at base of the dark brown spinnerets. The epigynum (6a) has a prominent scapus, wide at the base, narrowing toward the tip, which is spooned; and in some specimens the seminal chambers are prominently displayed, as in Plate XI., Figs. 5a, 5b.

MALE: Plate IX., Fig. 7; Plate XI., Figs. 5a, 5b. Differs little in markings from female; cephalothorax a warm yellow or yellowish brown; legs bright yellow, with brown annuli; abdomen with cretaceous field and shield shaped folium, with denticulate edges; shoulder humps not pronounced, and appear in some specimens almost wanting. Tibia-II curved, thickened, and provided on the inner and under side, from about midway, with a double and partly triple row of black, toothlike clasping spines. Coxa-II has at its base a long rounded spur, and coxa-I at the articulation with the trochanter a brown, short, chitinous, curved spur. The palpal bulb is rounded, the base provided with a blunt, curved spur; the radial joint much widened at the base, the apex a truncated cone; the cubital joint is short, rounded, and provided with two long spines.

DISTRIBUTION: This is a European species, and may have been introduced by commercial intercommunication. My specimens locate it along the Atlantic Coast from New England to North Carolina, in the Adirondack Mountains, and in Pennsylvania. Dr. Marx notes it as far north as Maine, and through Massachusetts to Pennsylvania. It will probably be found to affect more closely mountain regions or high elevations.

No. 43. Epeira cavatica KEYSERLING. Plate X, Figs. 1, 2; Pl. XI, Fig. 6.

1881. *Epeira cavatica*, KEYSERLING . . Verh. Zool. Bot. Ges. Wien, p. 269, xi., 1.
1884. *Epeira cinerea*, EMERTON . . . . N. E. Ep., p. 302, pl. 33.
1889. *Epeira cavatica*, McCOOK . . . . Amer. Spiders and their Spinningwork. (Ibid., E. cinerea, ad part.)
1892. *Epeira cavatica*, KEYSERLING . . . Spinn. Amerik., Eper., 118, vi., 87.

FEMALE: Total length (two specimens), 18 mm., 13 mm; abdomen, 12.8 mm., 10 mm. long, 11.3 mm. and 7 mm. broad; cephalothorax, 7 mm. long, 5.8 (6) mm. broad; width of head, 2.5 mm.

CEPHALOTHORAX: Corselet yellow, with brownish hue at sides and on grooves and median fosse; moderately high; sparsely covered with long, fine pubescence; fosse and cephalic suture deep; head depressed, flat on top, rather narrow at the face. Sternum codiform; longer than broad, with sternal cones; covered with bristles; color of sternum, labium, and maxillæ, brown, the latter lighter and yellow at the tips.

EYES: Ocular quad elevated, the width in front greater than length, and narrowest behind; MF separated by 1.5 diameter; MR somewhat smaller than MF, separated by two-thirds diameter; lateral eyes smaller, nearly equal in size, separated by about their radius; MF from SF about 1.5 their area. Clypeus 1.5 to 2 times diameter MF; front eye row slightly recurved, longer rear row procurved.

LEGS: 1, 2, 4, 3, as follows: Two specimens, 29 (32) mm., 27.6 (31) mm, 25 (28.5) mm., 17.2 (19.5) mm. Color yellow, with median distal brown annuli. In some specimens the median rings are extended almost to cover the metatarsi and patellæ; long and not robust, thickly covered with bristles and gray pubescence except at the tarsi, and with yellow spines, brown at the base. Palps yellow, armed as legs; mandibles yellow or brown.

ABDOMEN: Yellowish or brownish color, and in many specimens, on account of the numerous gray hairs and white bristles which densely cover the dorsum, assuming in life a quite gray appearance; dorsum arched in gravid specimens, subtriangular, with shoulder humps; on the basal half a rather indistinct folium, broadest in front, narrowing toward the apex, the edges bordered by a yellow scalloped band; the sides with dark black stripes extending to the venter, which has a broad black band in the centre passing from the gills and surrounding the spinnerets; this at each side bordered by a narrow, yellow, curved stripe; at base of spinnerets on either side two small yellow spots; epigynum rather short, wide at the base, channeled throughout, somewhat narrowing toward the tip; a line of hairs marks the middle, and the seminal chambers (Plate XI., 6) are well displayed. The scapus is longer than that of E. gemma, which it resembles.

MALE: Total length, 12.5 mm. (in one specimen 15 mm.); abdomen, 6.3 mm. long, 4.9 mm. wide; cephalothorax, 7 mm. long, 5.4 mm. wide; face, 2 mm. wide; resembles in form and color the female. (Plate X., 3; Plate XI., 9b.) The legs are much longer relatively than those of the female, as follows: Two specimens, 38 (41) mm., 34.5 (40) mm., 19.5 (23) mm., 29.5 (32) mm. These members are indeed enormously long; in one specimen the first leg measures 43 mm., the second 43 mm. The thighs and femur of legs-I, II, are formidably armed with spines; a slight hook marks coxæ-I. The tip of tibia-II is not thickened, nor armed with special clasping spines; but legs-I and II are both provided with numerous long spines, bristles, and hairs, especially on the femora beneath.

DISTRIBUTION: I have not collected this spider elsewhere than in the Adirondacks of New York, in New England, and in New Jersey, where it was colonized. It has been taken in Maine, New Hampshire·(Mrs. Mary Treat), and is found in New England generally. Dr. Marx records it as collected in Kentucky and Tennessee. Its distribution is probably limited to the more northern belt of States, particularly along the seaboard, and it will probably be found to prefer mountainous and elevated locations.

No. 44. Epeira angulata (CLERCK).   Plate X, Figs. 3, 4, 5; Pl. XI, Figs. 2, 3, 4.

1757. *Araneus angulatus*, CLERCK . . . Aran. Svec., p. 22, i., tab. 1, Figs. 1, 2, 3.
1757. *Araneus virgatus*, CLERCK . . . . Ibid., p. 41, ii., tab. 2.
1761. *Aranea angulata*, LINNÆUS . . . Faun. Suecica, Ed. ii., p. 487, 1909.
1775. *Aranea angulata*, FABRICIUS . . . Systema Entom., ii., p. 414, 29.
1778. *Aranea angulata*, DE GEER . . . Mem. des Ins., vii., p. 221, pl. vii., 1.
1789. *Aranea reticulata*, ROEMER . . . . Genera Insectorum, Linn.
1805. *Epeira angulata*, WALCKENAER . . Tableau des Aranéides, p. 57, ad part.
1832. *Epeira angulata*, SUNDEVALL . . . Svenska Spindlarness, p. 232, No. 1.
1837. *Epeira angulata*, KOCH . . . . . Uebers. des Arach. Syst., heft 1., p. 2.
1837. *Epeira quercetorum*, KOCH, C. . . Ibid., i., p. 2.
1837. *Epeira pinetorum*, KOCH, C. . . . Ibid., i., p. 3.
1850. *Epeira angulata*, KOCH, C. . . . . Die Arach., xi., 77, Figs. 892, 893.
1857. *Epeira angulata*, BLACKWALL . . Ann. & Mag. Nat. Hist., 2d ser., xx., 502.
1861. *Epeira angulata*, WESTRING . . . Aranese Svecicæ, p. 23.
1864. *Epeira angulata*, BLACKWALL . . Spid. Gt. Brit. & I., ii., 360, pl. 27, Fig. 259.
1865. *Epeira eremita*, KOCH, C. L. . . . Herr-Schaeff., Deutsche Insekten, 131, 23, 24.

VARIETY bicentenaria McCOOK.

1884. *Epeira angulata*, EMERTON . . . N. E. Ep., pl. xxxiii., 12; pl. xxxv., 2.
1888. *Epeira bicentenaria*, McCOOK . . Proceed. Acad. Nat. Sci., Phila., p. 195.
1889. *Epeira bicentenaria*, MARX . . . Catalogue, p. 543.
1889. *Epeira angulata*, MARX . . . . . Catalogue, p. 542.
1892. *Epeira angulata*, KEYSERLING . . Spinn. Amerik., Ep., p. 114, vi., 84.

FEMALE: Total length (two specimens), 17 mm. (21); abdomen, 10 (14) mm. long, 9 (12) wide; cephalothorax, 8 (10) mm. long, 7 (8.2) wide; facial front, 4 (4.8) mm. wide. With some hesitation I have placed this spider as a variety of the well known European E. angulata, which it closely resembles. I note these differences: On the venter, in front of the female genital cleft, are two decided cones (Plate XI., 1d), between which the scapus passes.[1] In a large number, from all parts of the United States, which I have examined, there is no trace of this peculiarity. The American variety, especially in the Southwest, is much larger than the typical European form, and much more hirsute; the legs particularly are thickly covered with bristles and formidable spines, the latter stouter and stronger than

---

[1] I have examined only one specimen, from Moscow, Russia, sent me by Professor W. Wagner, but in this these cones are very decided.

those of the European examples. The clypeus of the American specimens appears to be somewhat higher.

CEPHALOTHORAX: A rounded oval; the fosse semicircular; skin glossy; color dark brown, with yellow patches on the caput base; corselet grooves rather indistinct; cephalic suture distinct; covered with yellowish-white hairs. Sternum cordate; sternal cones not very prominent, except one in front of the labium; skin glossy, covered freely with gray hairs; dark brown color, with yellow median band. Labium subtriangular at the tip, rounded at the sides; maxillæ broad as long.

LEGS: Stout; 1, 2, 4, 3, as follows (a large specimen): 35, 31, 26, 19 mm.; another example measures 31.0, 30.2, 27.9, 19.8 mm.; joints strongly annulated, both at tips and middle; heavily clothed with yellow spines with brown bases, and with dark bristles; palps stout, yellow, with brown annuli. Mandibles conical, parallel; dark brown, with yellow fronts.

EYES: Ocular quad elevated; length about equal to width; broader in front than rear; MF somewhat larger than MR, and separated by about or less than 1.5 their diameter; MR separated by about one diameter. Side eyes on tubercles; separated by about their radius; SF slightly larger than SR. Space between SF and MF about 1.5 area of the latter, or at least three times intervening space of MF; both eye rows slightly procurved; clypeus margin distant from MF 3 to 3.5 diameter of latter.

ABDOMEN: Subtriangular; wide at the base across the shoulder humps; front subtriangular, overhanging the cephalothorax; color grayish yellow or yellow with velvety brown markings; the surface dotted over with numerous short, whitish, thick hairs among the pubescence. On the high basal front is a yellow pattern, often assuming the shape of a lyre or the letter U. The folium is shield shaped, with scalloped edges, forming in the middle part a scalloped band of yellow, which unites with a broad band of like color across the dorsal base and the posterior face of the shoulder humps. Wide scalloped bands of yellow, mottled with brown, extend from the shoulder humps, narrowing toward the apex, from which brownish belts, mottled with yellow, extend to the venter. The ventral pattern is a broad trapezoid of brown, with yellowish margins, and three dark, rounded, yellowish spots along either side of the marginal line at the corners and middle; spinnerets distal, though slightly overhung by the high apical abdominal wall.

In the epigynum (Plate XI., Figs. 4b, 4c, and Figs. 4d, 4e) the atriolum is distinguished by a high tubular pedestal; the scapus is long, subcylindrical toward the basal part, which is about equal in length throughout, and beyond the middle part widens slightly into a long spoon shaped tip, whose bowl in some species appears to be more decidedly marked from the shaft than in others.

MALE: Resembles the female in color and markings (Plate XI., Figs. 2, 3); is 6 mm. in length. The tibia of the second leg enlarged, and armed with rows of black, short, clasping spines; underneath the femora, especially femur–I, are rows of long acute spines. The abdomen bears the shoulder tubercles, and has a folium resembling that of the female. A specimen from Russia (Professor W. Wagner) is somewhat longer, but otherwise resembles the American specimen (Plate XI., Fig. 3) collected in Connecticut.

DISTRIBUTION: This species is one of the largest and, in certain parts, the most common of our spider fauna. Along the Pacific Coast and in Texas it reaches enormous proportions, one specimen from Texas having an abdomen which measures 18 mm. in length and a cephalothorax 12 mm. long and 9 mm. wide. I have specimens from various parts of California (Mrs. C. R. Smith, Mrs. Eigenmann, Mr. Orcutt, Drs. Blaisdell and Davidson); have collected it in the Adirondack Mountains and the Alleghenies of Pennsylvania, as well as in New England. The collection of Dr. Marx has specimens as far to the Northwest as Portland, Ore.; I have specimens from Wisconsin (Professor and Mrs. Peckham); it may therefore be considered as inhabiting the entire United States. Its distribution throughout Continental Europe is quite general, and it probably inhabits the northern shores of Africa, Palestine, and other parts in Asia. It is thus one of the most cosmopolitan of our orb-weavers. It appears to have experienced little change in form and general characteristics, but is substantially the same in all climates and environments.

No. 45. Epeira silvatica Emerton.   Plate X, Fig. 9; Pl. XI, Fig. 9.

1884. *Epeira silvatica*, Emerton . . . . N. E. Ep., p. 30, xxxiii., 13.
1892. *Epeira silvatica*, Keyserling . . . Spinn. Amerik., p. 117, tab. vi., 86.

Female: Total length, 14.5 mm.; abdomen, 8.8 mm. long, 7.5 mm. wide; cephalothorax, 7 mm. long, 5.7 mm. wide; face, 2.9 mm. wide.

Cephalothorax: Corselet moderately arched; median fosse large; cephalic suture distinct; mouth parts, palps, legs, and sternum dark reddish brown, and well covered with white hairs.

Eyes: Ocular quad elevated; front about as wide as sides, and a little wider than rear; MF separated by about 1.75 diameter; MR somewhat smaller, and separated by about their diameter; side eyes not contiguous; on tubercles; SF a little smaller than SR, and removed from MF by about 1.75 area of latter. Front eye row slightly, rear row distinctly, procurved; clypeus about 2 MF high.

Legs: 1, 2, 4, 3, as follows: 24.4 mm., 22.3 mm., 21.6 mm., 14.5 mm.; with both median and distal brown annuli; stout; metatarsi-I, II, III, shorter than their tibiæ; well armored with hairs and bristles, and yellowish white spines, with brown bases.

Abdomen: Ovate; with conical shoulder humps; the dorsum arched to the distal spinnerets; ground color yellow, densely covered with small brown lines and spots; folium indented at the margin, wide in front, narrowing to the apex, color brown, with slashes of black or blackish. On the basal front is a median yellowish spot; the sides are streaked with brown undulating bands inclined towards the spinnerets. The venter is covered with whitish hairs; is brown, with two indistinct small light bands bent towards each other. The pubescence, which is very dense upon the whole abdomen, consists of whitish bristles with dark base, and shorter ones, mostly curved, brown at the base and yellowish at the tip.

Distribution: New England, New York in the Adirondacks and central parts, Pennsylvania, Eastern Ohio; Dr. Marx reports it at Washington, D. C., Colorado, and Fort Yukon, Alaska. It may be distinguished with some degree of accuracy from E. angulata by the form of the ocular quad, which is relatively higher and less narrowed behind; the difference between MF and MR in size is also less pronounced; moreover, the epigynum is relatively wider at the base, longer, narrower, and more attenuated, terminating in a slightly hollowed tip. Emerton's description is not satisfactory; he makes no mention of the eyes, and presents no figure thereof; his drawing of the epigynum does not appear to be accurate, at least I am not able to find any specimens marked as figured by him; in the absence of a typical specimen one is therefore unable to sharply indicate distinctive characteristics.

No. 46. Epeira diademata (Clerck).   Plate X, Fig. 10; Pl. XI, Figs. 10, 11.

1678. *Araneus rufus cruciger*, Lister . . De Aran. Angl., p. 28, tit. 2, Fig. 2.
1757. *Araneus diademaus*, Clerck . . . Aran. Suec., p. 25, No. 2, pl. 1, Fig. 4; Araneus Peleg. ib., pl. 1, Fig. 5, p. 27.
1778. *Aranea cruciger*, De Geer . . . . Mem. Hist. Ins., t. 7, p. 218, No. 1, pl. ii., f. 3, 6, 7.
1793. "*Crown spider*," Martyn . . . . Aranei, pl. 2, Fig. 5.
1802. *Aranea Myagria*, Walckenaer . . Faune Parisienne, t. 2, p. 193, No. 8; ibid., A. diadema, p. 193, 9.
1806. *Epeira diadema*, Walckenaer . . Aranéides de France, pl. 10, Fig. 3.
1830. *Epeira diadema*, Sundevall . . . Svenska Spindlarness, p. 235, No. 2.
1834. *Epeira diadema*, Hahn . . . . . Die Arach., ii., 22, pl. 45, f. 110.
1837. *Epeira diadema*, Walckenaer . . Ins. Apt., ii., p. 29.
1850. *Epeira diadema*, Koch . . . . . Die Arach., xi., 103, t. 384, f. 910.
1861. *Aranea diadema*, Linnæus . . . Fauna Suecica, ed. 2, 1993.
1861. *Epeira diademata*, Westring . . . Araneæ Svecicæ, p. 26.
1864. *Epeira diadema*, Blackwall . . Sp. Gt. B. & I., ii., p. 358, pl. 26, Fig. 258.
1869. *Epeira diademata*, Thorell . . . Europ. Spiders, p. 53.
1889. *Epeira diademata*, McCook . . . Amer. Spid. and their Spinningwork.

FEMALE: Total length, 14 mm.; abdomen, 9 mm. long, 7.5 mm. broad; cephalothorax, 6 mm. long, 5— wide; face, 2 mm. wide. The colors vary much. I have collected specimens in England, Scotland, and Wales having the bright yellows shown in the plates; but have taken others in Norway with dark or blackish folial colors.

CEPHALOTHORAX: Corselet oval; color orange brown, with a yellow marginal band, lateral stripes; fosse, a longitudinal slit, with deeper indentation in the centre; corselet grooves sufficiently distinct; cephalic suture deep; caput depressed at the face; slightly pubescent. Sternum cordate, hairy, with sternal cones, elevated in the centre; brownish color. Labium and maxillæ as in Epeira; color brown, with yellowish tips.

EYES: Ocular quad elevated; wider in front than behind; eyes not greatly different in size; MF separated by about two diameters; MR by about one. Side eyes on tubercles; not contingent; SF (round) larger than SR (oval), somewhat smaller than MR; separated from MF by about 1.3 area of the latter. Front row slightly recurved, rear row longer and procurved; height of clypeus about two diameters MF.

LEGS: 1, 2, 4, 3; orange yellow, with brown annuli at tips of joints; well armored with gray bristles and yellow and brown spines on metatarsus; spines with black bases; palps and mandibles orange yellow, the latter conical and parallel.

ABDOMEN: Triangular ovate; longer than broad; dorsum arched to the apex and somewhat flattened; apical wall high, perpendicular, and almost as thick as the base. The color varies from velvety brown to yellowish gray, with numerous yellowish white bristlelike hairs. Between the shoulder humps on the basal front is a yellow cruciform marking, interrupted in its members at their crossing. The folium has an interrupted scalloped margin of yellow receding towards the apex. On the sides are lateral lines of yellow, and longitudinal ribbons of yellow mottled with brown. Further down the sides are dark brown, covered thickly with yellowish hairs. The ventral pattern is a broad rectangular ribbon of brown, lighter in the centre, and flanked on either side by a broad yellow ribbon, which extends with an interruption around the brown spinnerets. The epigynum (Plate XI., 10, 10d) has a long wrinkled scapus, broadest at the base, but somewhat diminished towards the tip, which is rounded and spooned.

MALE: I have no male of this species from America, and the only specimen in my collection from Ireland (Mr. Thomas Workman) is much damaged. I have therefore given in Plate XI., Figs. 11, 11a, Blackwall's figure of this sex, correcting, however, the defective drawing of the second leg, whose tibia is more robust, and provided with two parallel rows of short, strong, black clasping spines on the anterior surface. The male bears a marked resemblance to the female, but is smaller, 8 mm.

DISTRIBUTION: I have specimens from Minnesota (Mr. Ainsley), Wisconsin (Professor Peckham); Dr. Marx has specimens from Vancouver's Island on the west and Newfoundland on the east. It is not improbable that this species has been introduced from Europe by immigrants, inasmuch as so few examples have been reported. It, however, is fixed upon our shores, and may be expected to occur along the northern tier of States and Territories from ocean to ocean, and in the future will doubtless be distributed southward at least to the semitropical States. E. diademata is one of the longest and best known species of Europe, where it is common, and is known as the "Cross Spider." It is probably found in the contiguous parts of Africa and Asia. It has been the subject of numerous studies by anatomists and histologists, having been generally accepted as the representative type of Orbweavers.

No. 47. Epeira Peckhamii, new species. Plate XVIII, Figs. 5, 6.

FEMALE: Total length, 6 mm.; abdomen, 4.5 mm. long, 3.5 mm. wide across the base; cephalothorax, 2.5 long, 2 wide. On my first casual study of this species it seemed to me a Zilla, and is so named in the plate, which was printed before closer study disclosed my error.

CEPHALOTHORAX: Corselet oval, almost as wide as long; rounded at the edges; elevated in the centre; fosse a deep circular pit; corselet grooves indistinct; cephalic suture well marked; color yellowish brown, darker at the fosse, and with four stripes of brown upon the caput. The caput is about as high as the corselet at its base, but is depressed at its face, which is yellow and slightly pubescent; the head, like the corselet, is glossy. The sternum is but little longer than broad, heart shaped, blackish brown, with sternal cones, with tufts of long bristlelike hairs. Labium short, semicircular, about one-third the height of the maxillæ, which are gibbous, rather longer than wide.

EYES: Ocular quad elevated only in front; the rear wider than the front, and the sides longer than either; eyes are about equal in size; MF separated by less than one diameter, while MR are separated by about one diameter. Side eyes are contingent; placed on a black patch, with a slight tubercle. The space between SF and MF is noticeably greater than that between MF, and is little greater than the area of SF, and about equals the space between SR and MR; front row recurved, rear row slightly longer and procurved; clypeus height about 1.5 diameter MF.

LEGS: 1, 2, 4, 3; relatively long and thin, narrowing much toward the feet; color yellow, with slight brownish annuli at the joints. The leg armature is rather sparse; long aculeate spines are scantily distributed on the femora and tibia. The antennæ are marked and colored as the legs. Mandibles conical, parallel, but little separated at the tips.

ABDOMEN: Oval, somewhat wider at the base than the apex; slight, rounded shoulder humps; the dorsum arched; color shining cretaceous or yellowish; the folial margin drawn in black or blackish brown, and irregularly undulating, with a median cretaceous herring-bone pattern, which blends with the cretaceous color on the front. Branching longitudinal lines pass from near the base to the spinnerets, which are distal. An irregular band of cretaceous yellow, which, however, is sometimes greenish, marks the sides, and beyond this is a mottled band of black and brown. The ventral pattern is a long blackish band, which includes the spinnerets, flanked on either side by a yellowish ribbon, which passes around the spinnerets in interrupted patches. The epigynum is without scapus, and might be described as a simple short scoop, whose convexity is directed toward the spinnerets.

MALE: Fig. 6. In general form and marking resembles the female. Length, 5 mm.; legs long; uniform brownish yellow, sparsely pubescent, a few spinous bristles underneath femora and tibia; about three rows of long, rather slight spines on femora and tibia, also on the inside of metatarsus-I a row of four spines; no special clasping spines or thickening on tibia-II. The palp (Fig. 6a) has a well rounded dark brown digital joint almost as long as the cubital and humeral; the femoral joint is as long as the three terminal ones. The basal processes of the digital joint are much specialized. The mandibles are rather stout and widely separated at the tips; the sternum shield shape, wide at the base, narrowing at the apex, somewhat longer than wide, with strong sternal cones before coxæ-II, III; color blackish brown, with a lighter shade or yellowish in the middle; labium subtriangular, about one-half the length of the maxillæ, which are decidedly longer than wide, and about the same width throughout.

DISTRIBUTION: Wisconsin, male and female, sent by Professor Peckham. Dr. Marx has sent me specimens from Biscayne Bay, Fla. These widely separated points would indicate a general distribution over the Middle United States.

No. 48. Epeira nephiloides CAMBRIDGE. Plate XIV, Figs. 5, 6; Pl. XXII, Figs. 3, 3a-c.

1889. *Epeira nephiloides*, CAMBRIDGE . . Biol. Cent. Amer., Aran., p. 32, vii., 1, 2.

FEMALE: Total length, 10.5 mm.; cephalothorax, 4.5 mm. long, 3.75 mm. wide, 2 mm. at the face; abdomen, 6.5 mm. long, 4 mm. wide across the base, narrowing to 2 mm. at the apex.

CEPHALOTHORAX: Cordate, truncated and indented at the base; corselet edges rounded, flat on the summit; the fosse semicircular; corselet grooves sufficiently distinct; cephalic suture distinct; color (in alcohol) dull yellow, with streakings of brown from the corselet along middle of caput; slightly pubescent; a few long gray hairs at the caput base, which is lowly arched upon a level with the corselet, and slightly depressed at the face. Sternum broadly shield shaped, the edges rounded, and apex obtusely triangular; sternal cones long and distinct; the centre raised, and opposite the labium well elevated; color brownish yellow, somewhat darker at margins; covered sparsely with dark bristlelike hairs; labium subtriangular, thick, and large, more than half the height of maxillæ, which are obtusely triangular at tip, rounded at sides, somewhat longer than wide; labium and maxillæ brown, with yellow tips.

EYES: Ocular quad on a squarish prominence, more decided in front; quad front slightly wider than rear, and almost equal to sides. The single specimen from which I describe has an abnormal formation of MF eyes, one being twice as large as the other; its normal form I cannot therefore determine; the small MF eye is much less than MR, and the larger one is somewhat larger. According to Cambridge "the fore central eyes are almost as large as the hind centrals." The side eyes are about equal in size, separated by about a radius; SF from MF by about 1.5 area of latter, or at least twice or more the intervening space. The distance between MR and SR is greater. Clypeus height about 1.3 diameter of large MF eye, or twice diameter SF; margin with a row of white bristles; front row slightly recurved, rear row procurved.

LEGS: 1, 2, 3, 4; color yellow, or orange yellow, with blackish apical annuli at ends of tibia and metatarsus; a broad black median annulus on metatarsus, and a lighter one on tibia; the femora are without annuli, but have a deeper hue of yellow; armored rather sparingly with short yellowish spines, and abundantly with bristles and hairs; the palps yellow, with a slight brownish tinge at the tip; provided with curved spines and bristles; the mandibles long, conical, arched at the base, with decided basal cog and a slight swelling about the middle; color yellowish brown at base, darker brown at tips, glossy and pubescent on inner sides.

ABDOMEN: Triangular ovate, much wider at base and narrowing toward apex, which is rounded; spinnerets distal; dorsum slightly arched at the base, but rather flat on top; dorsal field a yellow, gourd shaped pattern, without other marks, covered quite freely with long, white, bristlelike hairs. The sides (Plate XXII., Fig. 4) are marked by long tooth-like streaks of yellow; the venter (Plate XXII., Fig. 5) has a broad, black, central band irregularly oval, surrounded on all sides by a belt of yellow, which also encompasses the spinnerets in a broken ring; spinnerets orange brown, bases blackish brown. The epigynum shows an atriolum but little prominent, a scapus short, wide at the base, and diminishing at the rounded tip, covered beneath with short gray hairs. This organ, as figured by Cambridge (Plate XIV., Fig. 6b) is much longer and narrower than in the specimen before me.

DISTRIBUTION: Fort Canby, N. M., Santa Barbara, Cal. (Marx Collection.) Cambridge describes it from various points in Guatemala. In some of the specimens from Central America the legs are bright burnt sienna in color, and the abdomen is emerald green, closely dotted with minute chrome yellow spots; over all a very dark cellular network of dark lines. (See Fig. 6.)

No. 49. Epeira spinigera CAMBRIDGE.   Plate XIV, Figs. 3, 4.

1889. *Epeira spinigera*, CAMBRIDGE . . . Biolog. Cent. Amer., Aran., p. 43, v., 9, 10.

FEMALE: Total length, 9 mm.; cephalothorax, 3 mm. long, 2.3 mm. wide; abdomen, 6 mm. long, 4 mm. wide. General colors: the fore part of the body a uniform orange yellow; abdomen, a pale yellow and black. In the elevation of the head, the character of the mouth parts, the wide space between MF and SF, and the thornlike abdominal tubercles this species resembles Wagneria tauricornis; but the curvature of the eyes, the form and spinous armature of legs are different, and more closely approach Epeira.

CEPHALOTHORAX: Oval; corselet rounded at margins; fossa a lateral pit curved backward; corselet grooves indistinct; cephalic suture very distinct; caput highly arched above the level of the corselet; head quadrate; face wide and strong; color, orange yellow, with flecks of brown on the corselet and base of caput; covered with short, yellowish white hairs. Sternum cordate, one-half longer than broad, with sternal cones, color yellow, numerous bristlelike hairs; labium dark brown at the base; maxillæ gibbous, longer than broad, bluntly triangular at the tips, which are inclined inward.

Eyes: Ocular quad elevated; length somewhat greater than width, rear slightly wider than front; MF separated by about one diameter; MR about 1.5 their diameter; side eyes on black tubercles; barely contingent; SF somewhat larger than SR; SF removed from MF by about 2.5 the area of the latter, or at least five or six times the intervening space of same; SR removed from MR by a slightly greater distance than separates SF and MF; clypeus height about two diameters MF, the base of the rounded eminence coming close to the margin; the front row is scarcely curved, the eyes aligned; the rear row is a little longer and slightly procurved; the forehead is high and well rounded.

Legs: 1, 4–2, 3; rather short, but drawn too short and pointed in the figure; stout; color dull (greenish) yellow, except the feet, which are brown; well provided with pubescence and whitish yellow bristles, somewhat sparingly with rather short yellowish brown spines; palps armed and colored as the legs; mandibles strong, conical, yellow, with slight brown at the tips.

Abdomen: A long triangular ovate, widest at the summit of the base; the front raised high above the cephalothorax; the dorsum not arched, but rather flat or curved inward to the apex, which terminates in a rounded cone or caudal part, although Cambridge describes this part as "not in a caudal form." Strong conical shoulder humps, which terminate in sharp brown points, mark the base; the folium is not distinct; the general field is yellow, reticulated, with brownish margins, from which issue longitudinal lines to the apex, which is touched with blackish brown, and covered with a tuft of yellowish hairs; the spinnerets are placed far underneath the projecting apex of the abdomen, brown in color, surrounded by a base of yellowish spots; the venter is brown, with marginal ribbons of yellow merging into the sides, which are brown, with yellowish lateral patches; the epigynum presents a subtriangular cup shaped scapus, very wide at the base, rounded at the top, brown and chitinous; the basal part thereof is an irregular quadrilateral, somewhat wrinkled; smooth and hollowed on the lower part, like the scapus. Cambridge's drawings of this seem very defective, probably from an immature species.

Distribution: Biscayne Bay; two specimens, female. (Marx Collection.) Cambridge describes it from Panama. The species is thus probably distributed along the coasts of Central and subtropical North America.

Genus MARXIA, new.

I have thought it necessary to make a new genus to receive the species originally described by Walckenaer as Plectana 'stellata. Subsequent writers have relegated this species to Epeira, on the grounds of the strong likeness in the mouth parts, the general grouping of the eyes, and form of the legs. The peculiar tuberculated condition of the abdomen has not been regarded by these authors as of general value. To me it seems unreasonable that an organ of such prominence, which contains the vital organs, and especially the spinning apparatus, by whose functions the animal is most sharply differentiated from members of its class, and, indeed, all other animals, should count for nothing in classification. It has been urged that the abdomen, by its softer covering, is more plastic, and therefore presumably more liable to changes, through environment and other influences, than the harder cephalothorax. Yet I have not found that, in point of fact, the abdomen is less persistent in its peculiar forms, as characteristic of various species, than other parts.

I have therefore considered that a peculiarity so striking as that shown in the tuberculated margins of the dorsal field, if not sufficient ground in itself for separating these specimens from Epeira, at least should be considered as one distinctive feature. In addition, however, Marxia is well separated from Epeira by the form of the cephalothorax, which is strongly elevated at the caput above the corselet. In Epeira, on the contrary, the head is on a level with the crest of the corselet, or more frequently depressed therefrom.

Moreover, the eyes of Marxia are removed from the margin of the clypeus by a space equal to four or more times their diameter, making a high clypeus; whereas in Epeira the midfront eyes are close to the clypeus margin, not being separated usually more than 2 or 2.5 diameters, and often less, making thus a low clypeus. Further, both eye rows are decidedly procurved, which is the case with the front row of Epeira alone. Again, the typical species of Marxia is distinguished from the typical Epeira by the peculiar furry covering of the caput and margin of the cephalothorax, and especially of the abdomen. The latter organ is so closely covered that the skin is entirely concealed in parts thereof. Moreover, the skin of the abdomen is marked by symmetrically arranged rows of hard, smooth circular spots or dimples, resembling those which are found in Acrosoma. Upon these differences I found the genus Marxia; the diagnosis in other respects corresponding with that of Epeira.

No. 50. Marxia stellata (Walckenaer). Plate XII, Figs. 4, 5.

1805. *Plectana stellata*, WALCKENAER . . . Tabl. d'Aran., p. 65, Fig. 54.
1837. *Plectana stellata*, WALCKENAER . . . Ins. Apt., ii., 171. (Bosc, Carolina Spiders.)
1850. *Epeira stellata*, HENTZ . . . . . . J. B. S., p. 22; Id., Sp. U. S., p. 125, xiv., 12.
1864. *Epeira stellata*, KEYSERLING . . . Beschr. n. Orbitel., p. 140, vi., 24, 25.
1884. *Epeira stellata*, EMERTON . . . . N. E. Ep., 319, xxxiv., 17; xxxvii., 3, 4, 5.
1889. *Epeira stellata*, McCOOK . . . . . Amer. Spiders and their Spinningwork, Vol. I.
1890. *Epeira stellata*, MARX . . . . . . Catalogue, p. 548.

FEMALE: Length varies in adults, but described specimen measures 12 mm.; cephalothorax, 5 mm. long, 4 mm. wide; abdomen, 11 mm. long, 9 mm. wide, measuring from the tips of the cones in front and rear.

CEPHALOTHORAX: A rounded oval, truncated and indented at the base, high in the middle, sharply sloping from crest to base; caput squarely truncate at base, well arched to the head, which is much elevated and bulging at the sides, forming low humps; fosse a lateral pit; color varying from reddish brown to yellowish brown, and provided, particularly along the head and margin of the clypeus, with thick, golden yellow, plumelike hairs. The sternum is shield shaped, rather rounded at the edges, somewhat longer than wide; with sternal cones; elevated and flat in the middle; heavily clothed with strong, yellowish white, bristlelike hairs; color brown, with a broad median patch of yellow or yellowish brown. Labium subtriangular, half the height of the maxillæ, which are rounded on the outer margins and bluntly triangular on the tips; as wide as or wider than long; color brown, with yellowish tips.

EYES: Ocular quad on a high prominence; glossy yellowish brown, smooth within the eye space, but covered at the base with plumelike golden yellow hairs, which extend along the face, and are longer and stronger at the margin of the clypeus, a heavy cluster marking the middle point thereof. The quad is somewhat narrower behind than in front, where it is about equal to the sides in length; MF separated by about two diameters, MR by about 1.5; slightly smaller, and on black bases; side eyes on the outer side of strong tubercles, separated by their radius or more; of nearly equal size, though SF appears somewhat larger.[1] Both rows are procurved; the clypeus is high, the margin being separated from MF by 1.3 their area, or at least 2.5 times their intervening space. The distance between MF and SF is twice the area of MF.

LEGS: 1, 2, 4, 3; color yellow, with dark brown wide median and apical annuli; clothed with hairs and bristles, and rather sparsely with strong, rather short, yellowish spines, particularly on the basal joints. The palps are stout, armed and colored as the legs; the mandibles conical, wide, arched at the base, which project beyond the margin of the clypeus, and are covered with short white hairs similar to those on the face; color dark glossy brown.

---

[1] In the specimen the rear eye is lacking upon one side, but the normal number is upon the other.

ABDOMEN: Somewhat longer than wide, semiglobose, and provided with a number of conical tubercles disposed as follows: One on the middle front of the base, projecting forward over the corselet; one upon the apex immediately opposite the frontal cone; on the shoulders a double cone, which appears sometimes as simply a tubercle with a cleft top, but again as two distinct cones flattened against each other on their contiguous faces. On each side, symmetrically arranged on the margin of the dorsal field between the shoulder cones and the dorsal apical cone, are three tubercles, somewhat smaller than the above described. On the rear of the abdomen, and immediately beneath the dorsal posterior one, is a tubercle corresponding with those upon the sides. We have thus in all eleven tubercles, counting each of those upon the shoulders as one, or thirteen, counting each shoulder cone as two. These are colored yellow, as is the dorsal field, and are provided with small yellowish white plumose hairs, which are also profusely scattered over the surface of the abdomen, giving a furry appearance thereto. The folium varies in different individuals, being in some rather indistinct, consisting of a triangular patch of brown or brownish yellow, extending from between the shoulders to the apical cone, toward which it narrows. In other specimens, particularly in young examples, is a well marked triangular patch. With all but gravid specimens the frontal cone appears rather depressed toward the cephalothorax, making a triangular space, of which it is the apex, the sides drawn thence to the shoulder cones, thus dividing the frontal from the dorsal field. When the specimen is gravid this frontal triangle is rounded out and does not appear so conspicuous. Frequently two lines of yellow diverge from this frontal cone backward, surrounding a triangular patch of brown, the base of which terminates between the shoulders; both the dorsum and sides are covered with blackish brown dimples, arranged in rows more or less symmetrically, and extending around the front and sides. The venter has a broad brownish patch, marked with bands of yellow; the epigynum (Fig. 4b) has a rather wide atriolum, brown, glossy, chitinous, the scapus not prominent, but sufficiently distinct, of equal length throughout, but slightly narrower and rounded at the tip, where it curves toward the body; it is glossy, yellowish brown, with a few hairs at the base, and apparently not grooved.

MALE: Fig. 5. Resembles the female in general form, pattern, and colors, the dorsal folium of the abdomen more particularly approaching the forms of immature females; the number of tubercles is the same. The palpus is well rounded at the digital joint; a curved spur at the base of cymbium, with a small tooth near its base; the radial joint wide and bilobed, quite shoe shaped, but short, as is also the rounded cubital. The mandibles are relatively longer and narrower than in the female, and are not so prominently arched toward the base. Tibia-II is not thickened or provided with clasping spines, but has three rows of dark brown ordinary spines surrounding the joint, with several of the same character at the apex. Tibia-I is longer than II and is similarly armored; coxa-IV has a conical spur at the articulation of the trochanter.

DISTRIBUTION: This species has a wide range throughout the United States, my specimens tracing it from New England along the Atlantic Coast, through North Carolina and Georgia (Mr. Thomas Gentry), and southward to Florida, westward and northward to Minnesota and Wisconsin (Professor Peckham), to Missouri and the American plains.

No. 51. Marxia nobilis (WALCKENAER).   Plate XII, Figs. 4, 5.

1842. *Epeira nobilis*, WALCKENAER .. Ins. Apt., ii., p. 118; ABBOT, G. S., No. 161; Id.,
                            E. cerasiæ, ibid., p. 119; ABBOT, G. S., No. 166;
                            Id., E. iris, ib., p. 120; G. S., No. 336.

Although I have numbered the above as a species, I regard it simply as a variety of M. stellata, from which it differs chiefly in the number of the conical processes on the abdomen. After carefully considering the original descriptions of Baron Walckenaer, it seems to me necessary that "stellata" should be applied to the form having the less number of tubercles (eleven), and "nobilis" to the form having the greater number. In truth, however, neither in the one case nor the other is Walckenaer correct, for he counts but eight tubercles to Stellata and twelve to Nobilis. Occasionally I find specimens with one less or more than the normal number of eleven or thirteen, this asymmetry always occurring in the row across the apical wall, as in Plate XIII., Fig. 9, x. To Epeira cerasiæ and E. iris, which I regard as one with the above, he also assigns twelve tubercles. I have both the male and female of this variety in large numbers from Georgia, collected from substantially the same territory as M. stellata. There is much variety in dorsal patterns among immature specimens, a number having a marked broad band of plumose hairs across the middle, as in Plate XIII., Fig. 9. The adult male corresponds with that of M. stellata, except in the additional two small tubercles in the apical row.

No. 52. Marxia grisea, new species. Plate XIII, Figs. 10, 10a, 10b.

FEMALE: Total length, 8 mm.; cephalothorax, 4 mm. long, 3.5 mm. wide; face, 2 mm. wide; abdomen, 6.5 mm. long, 5 mm. wide. I have seen but one specimen of this interesting species. It closely resembles Marxia stellata, at first glance, in its general appearance, but the clypeus is lower and the arrangement of abdominal tubercles is entirely different. The prominent frontal cone which marks M. stellata is wanting, and the median row of four aligned tubercles along the apex is peculiar to this species. The shoulder cones also are single and not double.

CEPHALOTHORAX: A rounded oval, truncated behind, flattened and sloping from the crest backward; the cephalic suture deep, the corselet grooves distinct, the fosse a wide and rather deep lunette overhung by the abdomen. The head is rather quadrate, square in front, divided into two low ridges by a lateral depression passing just behind the ocular quad, giving the head a lumpy appearance. Two small black circular spots, arranged on either side of the median line, mark both the anterior and posterior part of the caput, the latter being in the suture. The corselet is covered with yellowish hairs and the head more sparsely with the same; color yellowish brown. Sternum shield shaped, with conical elevations in front of coxæ-I, III; elevated in the middle, which is marked by a black pelt shaped figure; the margins are yellowish brown. Labium triangular, wider than long; maxillæ bluntly triangular at the tip, rounded at the sides, and somewhat longer than wide.

EYES: Ocular quad on a high prominence, the front about equal in length to the sides and slightly wider than the rear; MF separated about 1.5 diameter; MR, which are slightly smaller, separated by 1.5 diameter. Side eyes upon decided tubercles; SR as large as SF, from which they are separated by at least a diameter, and are placed behind, well to the side. MF are removed from SF by about 1.5 the area of MF. The front row is slightly procurved, as is also the longer rear row. The clypeus height is about two diameters MF.

LEGS: Order of length, 1, 2, 4, 3; yellow, with brownish bands at the tips of the joints and the middle of the femora. They are heavily clothed with pubescence, with bristles, and numerous white spines with brown bases. The palps are heavily armed with gray bristles, especially upon the last two joints, and colored as the legs. The mandibles are conical, well separated at the tips, where they are dark brown, the bases being yellowish brown, glossy, but provided with yellowish gray bristles.

ABDOMEN: A rounded oval, heavily covered with yellowish gray pubescence and plumose hairs. The dorsal margins are marked by ten tubercles arranged in an arc, containing five on each side. Four others are arranged along the median line of the apical half of the dorsum, passing over the apex one above another, the anterior one being upon the dorsal field. (Fig. 10a.) There are thus fourteen tubercles in all. There is no tubercle upon the middle front of the base as in M. stellata. The color is yellow, with a scalloped or dentate folium lightly edged with brown, and a dark median band passing along the entire dorsal field, which widens at the apex, over which it passes to the spinnerets. The venter is an interrupted yellowish broad band. The epigynum (Fig. 10b) scapus has a wide triangular base, with a long stalk, flat on lower surface, of equal width for most of its length, and terminating in a rounded point.

DISTRIBUTION: Biscayne Bay, Florida. (Marx Collection.)

No. 53. Marxia mœsta (KEYSERLING).   Plate XII, Figs. 11, 11a-c.

1889. *Epeira maesta*, MARX *in litt.* . . . Catalogue, p. 546.
1892. *Epeira môsta*, KEYSERLING . . . . Spinn. Amerk., Epeir., p. 108, Fig. 80.

FEMALE: Total length, 11 mm.; cephalothorax, 5 mm. long, 3.5 mm. wide, 2.5 mm. at the face; abdomen, 9 mm. long, 8 mm. wide across the base, narrowing to 1.5 mm. at the face.

CEPHALOTHORAX: Oval, well rounded at the sides, truncate behind, where it is over-hung by the abdomen; corselet grooves sufficiently distinct; cephalic suture well marked, as is the median fosse. Color dark brown, with lighter band of yellow along the corselet margin; head colored as corselet, and rises from the fosse at a slight inclination; is arched on the posterior and narrower half, and about midway has two knobs, such as appear in other species of the genus; thence the fore part of the caput is depressed to the central eye space, which turns up into a strong prominence, giving from the side view a snouted appearance to the face. In front the face is wide, but the eye space is distinctly separated from the knobby summit of the caput by the depression above alluded to. Sternum shield shape, well rounded at the sides, subtriangular at the apex, glossy, dark brown, or blackish, elevated in the centre and flattened, with sternal cones, sparsely covered with grayish white hairs, which are thickest on the anterior part; skin hard and glossy. Labium widest at the base, where the color is dark brown, the tip obtusely triangular; maxillæ colored like the lip, rounded, at least as wide as long.

EYES: Ocular quad on a high prominence, most elevated behind; the rear slightly narrower than the front, which about equals the sides; MF separated by nearly two diameters, about equal to MR, which are separated by a little greater space. Side eyes on tubercles much less prominent relatively than the median one; separated by a diameter of SR, which is smaller than SF. The space between SF and MF equals about twice the area of MF, or 2.3 times the distance between MF. Height of clypeus equals alignment of MF. Both rows viewed from the front are procurved, the hind row more so, and is much longer on account of SR being placed so far to the side; the margin of the clypeus has yellowish gray bristles.

LEGS: 1, 2, 4, 3; stout; in color dark brown, provided with grayish yellow bristles and spines. The palps are colored and armed as the legs.

ABDOMEN: Subtriangular in form, widest at the base, where it is rounded to the front, and far overhangs the cephalothorax. The dorsum is somewhat flattened, and the abdomen is almost as thick at the apex as in front. Two shoulder tubercles mark the base well back of the anterior middle point, leaving thus the fore part of the abdomen as a wide sub-triangular space sloping toward the front, while the remainder of the dorsum slopes some-what, though but little, toward the rear. The apex is marked by a prominent rounded tubercle, resembling those upon the shoulders, but smaller; on either side of this is a similar smaller tubercle, and beneath it on the apical wall of the abdomen are two others in a row, of similar character, but somewhat flattened. The color is yellow, much broken by irregular and lateral black lines upon the sides. Between the shoulder tubercles and the apical ones extends a folium of dark or darkish brown, mottled with yellow reticula-tions, dentated upon the edges and narrowing toward the apex; an interrupted yellowish band marks the middle of the dorsal field. The basal front has the same black and yellow reticulations that mark the sides. The venter has a broad black band, with lateral mark-ings of yellow at the front; the spinnerets are black, with yellowish spots at the base; the epigynum (Figs. 11b, 11c) has a well arched atriolum, and a scapus whose base is wide and spooned, giving the edges a horseshoe shape; from its central point issues a long lanceo-lated tip.

DISTRIBUTION: Pike's Peak, Colorado; New Mexico. (Marx Collection.) Only the female of this species is known. The localities in which it has been found would indicate a geographical distribution throughout the Southwestern States and Territories, and prob-ably the Pacific Coast.

Genus ORDGARIUS, Keyserling.

The most striking distinctive characteristics of this genus are found in the cephalothorax. The corselet is rounded at the margin, but rises with almost perpendicular walls to the crest. The base is sharply truncated, sloping rapidly upward to the crest. The skin is covered with many warts, particularly numerous and marked upon the caput and face. The summit of the caput is distinguished by two castellated prominences, with strong protuberances on either side. The sternum is scutellate, somewhat longer than wide; the labium wide, rather low; the maxillæ as in Epeira, scarcely as long as wide. The eye rows are both much procurved; divided into two groups, as in Epeira; the central quad upon a low rounded eminence rising at the base of the bossed forehead; the pairs placed close together, so that the side of the quad is much shorter than either the front or rear. The side eyes are placed low down upon the face, on large rounded tubercles; the eyes are all small, not greatly differing in size. The clypeus is very high, as is also the forehead; the middle eyes of the ocular quad are placed a little below the centre of the face. The legs in order of length are 1, 2, 4, 3; stout, provided with bristles and hairs, but without spines. The abdomen is subglobose, as wide as or wider than long, and distinguished by two prominent shoulder humps.

No. 54. Ordgarius cornigerus (Hentz). Plate XII, Figs. 1, 1a–d.

1850. *Epeira cornigera*, Hentz . . . . J. B. S. 20; Id., Sp. U. S., p. 123, xiv., 8.
1879. *Cyrtarachne cornigera*, Keyserling Neue Spin. Amer., i., Verhn. d. z. b. Ges. Wien, p. 300, iv., 4.
- 1879. *Cyrtarachne bicurvata*, Becker . . Ann. Soc. Ent. Belgique, p. 77.
1889. *Cyrtarachne cornigera*, McCook . Amer. Spid. and their Spinningwork, Vol. II., 97.
1889. *Ordgarius cornigerus*, Marx . . . Catalogue, p. 541.
1892. *Ordgarius cornigerus*, Keyserling. Spinn. Amerk., Epeir., p. 40, ii., 34.

Female: Total length, 12 mm.; abdomen, 8 mm. long, 8 mm. wide; cephalothorax, 5 mm. long, 4.5 mm. wide. The general colors of the fore part are red or reddish brown and yellow, with dark brown markings. The abdomen is yellow, with dark or brownish markings upon the front.

Cephalothorax: Rounded at the margin, and rises with almost perpendicular walls to the crest; the base sharply truncated; the skin covered with many warts, particularly numerous and marked on caput and face; summit of the caput marked by two castellated prominences, each with two strong tubercles on either side. The entire front of the creature presents a peculiar appearance on account of the knobby or warted condition. (Fig. 1b.) Sternum shield shape, somewhat longer than wide, has sternal cones, is rounded in the centre, depressed at the base towards the labium, and sparsely covered with hairs; in some species the color is yellow or dark brown; in others dark brown. The labium is triangular, rather low, but wide; the maxillæ scarcely as long as wide; obtusely triangular at the base; color of labium and maxillæ brown, with yellow tips.

Eyes: The ocular quad (Fig. 1b) is upon a greatly elevated prominence; front shorter than rear, and about equal to sides; MF somewhat larger than MR, separated by about 2.5 diameters or more; MR separated by about three diameters, and divided by a longitudinal notch; side eyes separated by about a radius, not greatly differing in size; set on tubercles, with long bases that extend to the clypeus margin; both rows procurved; rear row longer; clypeus high, twice or more the area of MF; on either side above the mandibles is a rounded hump, much warted.

Legs: 1, 2, 4, 3; very stout, the femora and patella covered with yellowish bristles and hairs, but without spines; color yellow, with dark brown apical and median annuli; palps similarly colored and provided, quite wide and stout; mandibles conical, wide at the base, much narrowed at the tips, parallel curved on the anterior surface, receding; dark brown at the base, and yellowish brown beyond.

ABDOMEN: As wide as or wider than long, forming a well rounded oval or irregular hemisphere, which is thickest at the base, where it rises up into marked height. (Fig. 1a, side view.) The crest is surmounted by two large conical tubercles. The front overhangs the cephalothorax, is wide and high; the apical half is arched towards the spinnerets, which are placed somewhat beneath the apex. The dorsal field is without a folium, is bright yellow, with longitudinal lines passing backward along the muscular pits, which are prominent; in some specimens the lines traverse both sides, widening to the venter, giving this part a striped appearance. The front is olive or blackish brown, mottled with yellowish, irregular spots; surface glossy, extremely rugose, marked with numerous black circular pits, arranged laterally in semicircles, curved backward; these pits are found along the sides and apex, though rather small. The venter has a broad yellow patch, with four or five black circular spots arranged longitudinally, and a wide, brownish, median band, which is wanting in some specimens; the epigynum (Fig. 1b) shows a long atriolum, with a tonguelike scapus, short, rounded, and wide at the tip, folded down flat against the genital cleft, almost as though it were attached thereto; the scapus is brown, the front of the atriolum on either side of the portulæ yellow. In one specimen (Fig. 1c) in the Marx collection the abdomen is yellowish brown, smooth, showing only two indentations, and the shoulder humps, instead of being well in the front, as above described, are set off at the sides, a result, probably, of the gravid condition of this female.

DISTRIBUTION: I have received a number of this species, females, from California, where it appears to be common (Mr. S. R. Orcutt, Mrs. Eigenmann, Mrs. Smith, Dr. Davidson, Dr. Blaisdell); and cocoons (Vol. II., page 98) from various localities, ranging from Fort Yukon, Alaska, to San Diego. Becker described it from Louisiana, and Hentz's original description is from Alabama. Dr. Marx records it in the District of Columbia and Virginia. It is, no doubt, distributed throughout the entire Southern States, and along the Pacific Coast in California. Have collected it in Pennsylvania.

## No. 55. Ordgarius bisaccatus (EMERTON).			Plate XII, Figs. 2, 3.

1884. *Cyrtarachne bisaccatus,* EMERTON . N. E. Sp., p. 325, pl. xxxiv., Fig. 11.
1889. *Cyrtarachne bisaccata,* McCOOK . . Am. Spid. and their Spinningwork, Vol. II., p. 95.
1889. *Ordgarius bisaccatus,* MARX . . . Catalogue, p. 541.
1892. *Ordgarius bisaccatus,* KEYSERLING. Spinn. Amerik., Ep., p. 42, ii., 35.

FEMALE: Total length, 10 mm.; cephalothorax, 3 mm. long, 3 mm. wide in the middle, and 2 mm. in front; abdomen, 6 mm. long, 7 mm. wide; in front more than half as broad as in the middle. It may at once be distinguished from O. cornigerus by the absence of shoulder humps.

CEPHALOTHORAX: Rounded at margin; fosse concealed behind the face, which rises from the eye space into an exceedingly high subtriangular vertex, terminated on either side of the summit by two rectangular or castellated prominences, cleft at the top into two obtusely pointed cones, the inner ones the longer. The entire face is covered with numerous warts, which extend to the summit of the vertical prominences; corselet grooves distinct; cephalic suture indistinct; color of corselet behind, yellow; around cephalic suture, brown; forehead and head again yellow. Sternum shield shape, somewhat longer than wide; sternal cones in front of coxæ-I and III; color yellow; slightly pubescent; base with a semicircular depression next the labium, which is subtriangular, half as long as the maxillæ, which are gibbous, apparently a little longer than wide, and colored as the sternum.

EYES: Ocular quad on a rounded eminence; rear slightly wider than front, and the sides shorter than either; MF larger than MR, separated by about 2.5 to 3 diameters, while SR are separated by three or more; rear eyes on tubercles, almost contingent, not greatly differing in size; the space between SF and MF 1.3 area of latter, or about twice or less the intervening space; both rows procurved; the clypeus is high, the margin separated from MF by more than the area of the latter; an oval tubercle marks either side above the articulation of the mandibles, and the whole surface is granulated, the upper part of the eye space containing a few warts.

LEGS: 1, 2, 4, 3; stout, especially at the femora, and narrowing at the tips; color yellow, armed with yellowish short hairs and bristles, but without spines; patella wide, rather flattened; palps colored and armed as legs; mandibles conical, parallel, yellow, covered with short hairs, and abundantly with stiff bristles upon the inside.

ABDOMEN: Ovate, or irregular hemisphere, wider than long, much thickened at the base; dorsum arched, without shoulder humps; no special folium, but flecks of brown around the dark muscular plates; color yellow, reticulated lateral stripes on the sides; ventral pattern, a yellow band extending around the bases of the spinnerets; the epigynum has a wide atriolum, with a very short semicircular scapus extending between the portulæ, like a short flap. . (Fig. 2a.)

MALE: Plate XII., Figs. 3, 3a. 3 mm. long; presents in general the characteristics of the female, in the heart shaped form of the abdomen, wider than long, and in the elongated vertex of the face. The forehead, indeed, is relatively higher in the female, and the crest is divided into two simple cones. The legs are yellowish brown; the abdomen pearly yellow, with a double row of black spots, like muscular pits, arranged longitudinally on either side of the median. The corselet is yellowish brown, and the face and forehead yellow. The palpal digit is globular, brownish yellow. The legs are without spines, but are provided with long bristles, and appear to have no special clasping apparatus.

DISTRIBUTION: New England; District of Columbia. (Marx Collection.)

GENUS VERRUCOSA McCOOK, 1888.

In this genus the cephalic suture is deeply marked; the caput rather shortened and much rounded at the sides, narrowed at the base, elevated above the corselet; the face wide and full; the sternum cordate, somewhat longer than wide; the labium, maxillæ, and eyes as in Epeira. The abdomen is triangular ovate, flattened upon top, and the apical wall marked by rounded protuberances; the skin is hard and glossy; the epigynum, in the typical species, with a long, narrow scapus extending nearly to the spinnerets. The legs in order of length are 1, 2, 4, 3; stout; the spines long and bristlelike, with the exception of a few at the articulation of the joints.

Verrucosa differs from Epeira chiefly in the shape and elevation of the head, and the peculiar character of the abdomen, with its flattened dorsum, tuberculated apical wall, somewhat hardened skin, and the aculeate and lengthened form of the epigynal scapus. The acute spinous armature of the legs also differs from the stout, rather stubby character of Epeïroid spines. The male has the general characteristics of the female as to the form of the caput and abdomen. The tibia is curved, much swollen at the apex, and provided with a long, strong spur, whose point is armed with two spines. The palpal digit is ovate, and the cubital joint is curved. The spines are longer and with stouter bases than those of the female, and the sternum is relatively wider.

I have felt justified in retaining for this genus the name published early in 1888,[1] in my first studies of the manuscript drawings of John Abbot, revising the nomenclature of some of Hentz's species. These results were accepted and embodied in his "Catalogue" by Dr. George Marx. In this Catalogue (page 541) he for the first time made public the generic name "Mahadeva," which he attributes to Keyserling (in litt.), and gives Hentz's Epeira verrucosa as the type. In July, 1889, Cambridge, in his Biologia Centrali-Americana, page 53, adopts this name, credited to Keyserling, but changing the spelling to "Mahadiva." In Keyserling's Spinnen Amerikas, part IV., Epeiridæ, page 67, which was not issued until 1892, the genus is published as new, under the name used by Marx, and for the first time

---

[1] Proceedings Academy Natural Sciences, Philadelphia.

a diagnosis thereof is given. In no one of these works is any reference made to the fact that I had already established for the typical species a new genus.

It may perhaps be a question whether the mere announcement and tabulating of a genus in connection with a fairly typical species by which it can easily be identified, yet with no further diagnosis, is entitled to credit by araneologists. In point of fact, however, such scant description has frequently been recognized, and there is no reason why the rule of courtesy should be broken in the present instance. If, however, the name Mahadeva has right of priority, it must be conceded to Dr. Marx, who first made it known on his own responsibility, although he generously, as in so many other cases, attributes it to Keyserling. Next to him, Cambridge would certainly have priority over Keyserling, whose publication unhappily did not appear until 1892. In view of the above circumstances I have felt bound, under the laws of priority, to give precedence to Verrucosa as the generic name of this interesting group of spiders, of which I still hold Verrucosa arenata (Hentz) to be the type.

No. 56. Verrucosa arenata (Walckenaer).   Plate XII, Figs. 6, 7.

1837. *Epeira arenata*, Walckenaer . . Ins. Apt., ii., p. 133; Abbot, G. S., Nos. 165, 181, 182, 183, 360.
1844. *Epeira verrucosa*, Hentz . . . . B. J. S., 19; Sp. U. S., 121, xiv., 2.
1888. *Verrucosa arenata,* McCook . . . Proceed. Acad. Nat. Sci., Phila., p. 5.
1889. *Epeira verrucosa*, McCook . . . . . Amer. Spiders and their Spinningwork.
1889. *Mahadeva arenata*, Marx . . . . Catalogue, p. 541.
1889. *Mahadiva verrucosa*, Cambridge . Biolog. Cent.-Amer., p. 553.
1892. *Mahadeva verrucosa*, Keyserling . Spinn. Amerik., Epeir., p. 72, pl. 3, Fig. 56.

FEMALE: Total length, 9 mm.; cephalothorax, 3.3 mm. long, 2.75 mm. wide, 1.5 mm. at the face; abdomen, 6.5 mm. long, 6 mm. wide across the base, narrowing to 1 mm. at the apex of the dorsum and 3 mm. across the apical wall. The snare is a round orbweb of the ordinary Epeïroid type.

CEPHALOTHORAX: A somewhat elongated oval, truncate at the base, rounded at the sides, the color dull yellow, flecked with brown, varying from thence to brownish, with the head usually brown or ruddy brown; corselet grooves not distinct; cephalic suture deeply marked; fosse a circular depression from which the corselet slopes sharply to the base; the head rather shortened, much rounded at the sides and narrowed at the base; face wide and full, is glossy brown, with lighter shades around the ocular quad. Sternum shield shape, wide at the base, though somewhat longer than wide, and narrowed at the apex; with sternal cones; glossy blackish brown; somewhat pubescent; labium large, wide at the base, which is brown, subtriangular at the tip, which is yellow; maxillæ as wide as long, rounded at the tips, brown, glossy.

EYES: Fig. 6b. Ocular quad on a squarish eminence, most prominent in front; front wider than rear and about equal in length to the sides; MF somewhat larger than MR, separated by about 1.5 diameter, on separate tubercles, the front of the quad being divided by a notch; MR separated by about one diameter. Side eyes on tubercles; SF larger than SR, divided by about or less than a radius; SF separated from MF by about 1.3 the area of the latter, or more than twice the intervening space thereof; clypeus height about 1.5 MF, the ocular eminence approaching almost to its margin; front row recurved; rear row slightly procurved, resembling thus the eyes of Epeira.

LEGS: 1, 2, 4, 3; stout, yellow, glossy, with bright brown or orange apical and median annuli, abundantly provided with strong yellowish bristles and hair, and with acute yellowish spines; palps colored and armored as legs; mandibles glossy brown, conical, arched, widely divergent at the tips, slightly pubescent.

ABDOMEN: Triangular ovate; almost as wide at the base as long; the dorsum flat, except in gravid specimens, when it becomes slightly arched; is marked along either side by shoulder humps, presents a triangular folium that covers the entire surface, being ordinarily a bright yellow, with beautiful rose-colored or red reticulations. On some of the specimens a brownish line, with curved side branches, passes along the median; the front overhangs the cephalothorax, is rounded, brown, mottled with yellow, showing often rosy hues. The side (Fig. 6a) in some specimens is very beautiful, a reticulated yellow, with stripes of vermilion or lake. The dorsum terminates in a rounded cone, and just underneath this on the apical wall is a second rounded tubercle, and beneath this a third, rather flattened; on either side of this median row are two similar rounded tubercles; the sides in some specimens are roughened or indented; the whole surface of the skin is glossy, and is covered rather scantily with long whitish yellow hairs. The venter is a blackish brown patch, which encompasses the spinnerets, which are black and much overhung by the abdomen; the epigynum (Fig. 6c) has a long needlelike scapus, glossy brown in color, narrow at the base, pointed at the tip, which extends from the narrow atriolum entirely to the spinnerets; the portulæ show on either side as compressed openings into the genital cleft.

MALE: 6 mm. long; cephalothorax yellowish brown, face much projecting beyond the mandibles, which are decidedly weaker than in the female, retreating backward and widely divergent at the tips; legs yellow, with lighter brown annuli; provided with numerous long yellow spines, especially formidable underneath femora-I, II; tibia-II is curved, and provided at the apex with a most remarkable series of clasping spines, one of which is a strong, long spur, thick at the base, which is covered with curved bristles, pointed at the tip into two dark pointed spines, one longer than the other; beyond this spur on either side is a low process which contains two shortened curved brown spines; the apex itself is much thickened, and has one strong curved brown spine; besides these are a number of long acute brownish spines. The abdomen is triangular ovate, and has the characteristic markings and tubercles of the female, but less decidedly. The palps (Fig. 7a) are brownish yellow in color; the digit an elongated oval, the cubital joint much curved, longer than the radial joint, which is quite short.

DISTRIBUTION: I have collected this species in the neighborhood of Philadelphia, where, however, it is not common; in New Jersey and Ohio; and have it from as far west as the American Plains. Abbot took it in Georgia, Hentz in Alabama and North Carolina. Marx reports it from Florida (Cresson City), Utah (Spring Lake), California (Occidental). The species is thus widely distributed over the United States. I have specimens collected by the late Mr. W. H. Gabb from San Domingo varying in but slight particulars from those above described. The species is doubtless well distributed throughout the West Indies; and the States of Central and South America contain species closely resembling it.

No. 57. Verrucosa unistriata, new species. Plate V, Figs. 3, 3a-c.

FEMALE: Body length, 11 mm.; abdomen, 9 mm. long, 7 mm. wide; cephalothorax, 5 mm. long, 5 mm. wide. The general colors are, for the fore part of the body yellow, with orange or rose tints on the legs, and for the abdomen grayish yellow, with a prominent median band of a bright yellow or yellowish white.

CEPHALOTHORAX: A rounded oval, the corselet margin almost circular; color yellowish brown, with a dark or median band of brown, which passes through two large patches of brown, apparently on a depressed spot in the centre of the caput. The head more than half as wide as the corselet, sloping gradually to the face; moderately pubescent, with long, gray, bristlelike hairs scattered over the surface; corselet grooves distinct; cephalic suture well marked. Sternum little longer than wide, attenuated at the apex, raised in the middle, with prominent sternal cones, especially before coxæ-III and in front of the labium; color dark brown, with a yellowish brown median patch. Labium obtusely triangular, thickened at the base, more than half as long as the maxillæ; maxillæ longer than broad, subtriangular at the tip, and tipped with yellowish white.

EYES: Ocular quad elevated, a little wider in front than rear, the length greater than width; MF separated about 1.5 to 2 diameters; MR smaller than MF, and separated one diameter; side eyes on tubercles; tufts of strong gray bristles, like eyebrows, behind them; SF about equal in size to MR; SR decidedly smaller than SF, and not quite contingent; MF separated from SF by about 1.3 to 1.5 their area; MR from SR 2.3 their alignment. The front row scarcely procurved, rear row procurved and longer; long, gray bristles mark the margin of the clypeus, which is about 1.5 diameter of MF in height.

LEGS: 1, 2, 4, 3; the fourth leg scarcely as strong and long relatively as in Epeira; color yellow, with bright rose tints on the femora-I, II, intermingled with orange hues; annuli at joints, and lighter color median annuli underneath tibia and metatarsus; well clothed with long, gray bristles and hairs, and with numerous long, light colored spines, with brownish bases; the hairs on the feet and metatarsus are strong and bristlelike, and are curved outward; palps colored and armored as the legs; mandibles conical, arched at the bases, which are provided with gray bristles; color yellow, diffused with orange brown at the tips.

ABDOMEN: Subtriangular, slightly overhangs the cephalothorax, broad at the base, slightly tapering towards the apex, flat on the dorsum; spinnerets beneath the apical wall; color grayish yellow, marked by lateral stripes on the side, and punctuated by dots of brownish color on the dorsum. The median line is strongly characterized by an interrupted band, consisting of three parts, extending about two-thirds of the length; its upper section is rounded, and has projections at either end, the central part is rectangular, and the third division is a round spot; the median is marked by a few spots and lines. The ventral pattern is black, an irregular quadrilateral, margined by a rather narrow and reticulated yellow band. The epigynum is bright yellow and dark brown, with a remarkably long scapus, broad at the base and tapering to a point, the length extending almost from the insertion to base of the spinnerets. Although the head of this species is more depressed than in the typical Verrucosa and the spinous armature is stronger, in both these particulars resembling Epeira, yet other characters seem to justify placing it in this genus.

DISTRIBUTION: Fort Yuma, Arizona. (Marx Collection.)

GENUS KAIRA CAMBRIDGE, 1889.

In Kaira the cephalothorax is triangular ovate, widest near the base, sharply shelving to the crest, where the corselet is highest, thence sloping forward with considerable inclination toward the face, the caput thus being decidedly depressed. The sternum is longer than wide, and the labium wider than long, as are also the maxillæ. The rear eyes of the ocular quad are little elevated, are less than the front; the relative spaces between the side eyes and the middle group are as in Epeira, as is the curvature of the rows, the front being recurved and the rear procurved. The clypeus is of moderate height, but the high elevation of the quad separates the midfront well from the margin thereof. The legs are especially distinguished by closely set rows of short, stout, spinous bristles, curved towards the apex, and which are especially numerous on the exterior surface of the metatarsus. This joint is particularly stubby, and all the legs are stout. The extremely hirsute character of the legs appears upon the entire body, for stout, spinelike bristles are spread thickly upon the surface of the corselet, and, though somewhat shorter, upon the abdomen. The abdomen is broadly shield shape, high, and thick at the base, slightly narrowing to the spinnerets, which are distal. The base is distinguished by two castellated prominences, which are subdivided into a number of conical protuberances.

No. 58. Kaira alba (HENTZ).　　　　　　　　　Plate XII, Figs. 3, 3a-c.

1850. *Epeira alba*, HENTZ . . . . J. B. S., vi., 20; Id., Sp. U. S., p. 122, xiv., 7.
1884. *Epeira alba*, KEYSERLING . . . . Neu. Spinn. Amer., vi., p. 531, vi., 20.
1890. *Kaira alba*, KEYSERLING . . . . Spinn. Amerik., Epeir., iv., p. 64, iii., 50.

FEMALE: Total length, 7 mm.; cephalothorax, 3.5 mm. long, 3 mm. wide, 1.5 mm. at the face; abdomen, 6 mm. long, 5.5 mm. wide below the shoulder humps; across the shoulder humps, 6.3 mm.

CEPHALOTHORAX: Triangular ovate, widest near the base, where it is truncated, and sharply shelving to the crest, where the corselet is highest; thence it slopes forward again with considerable inclination toward the face; margins rounded; cephalic suture well marked; corselet grooves not distinct; foss a semicircular depression, overcovered by the cephalothorax; color yellowish brown, glossy, rather sparsely covered with stout, gray, bristlelike hairs, whose bases are distinctly marked with slight rugosities; caput much depressed, sloping to the face, which would appear contracted except for the strong tubercles on which the side eyes are set. The sternum is shield shaped, and longer than wide, on account of the projecting apex; with sternal cones; the sides deeply indented; the centre covered with bristles; color yellow, with rather lighter shade in the middle. Labium sub-triangular, wider than long, and about half the height of maxillæ, which are rounded at the sides and tip, and but little wider than long. Both labium and maxillæ yellowish brown; slight elevations mark the sternum opposite the coxæ.

EYES: Ocular quad on a high prominence, which is elevated above the side tubercles, and is higher behind than in front. The quad front is slightly wider than rear, and about equals the sides; MF separated by about or more than 1.5 diameter, and a little larger than MR, which are separated by about the same space. Side eyes on prominent tubercles; barely contingent; the front somewhat larger than the rear; SR well behind SF. MF are separated from SF by about 1.3 their area, and from the margin of the clypeus about 2 to 2.5 diameter of MF. The front row is recurved, the rear row slightly procurved, and the longer. (Fig. 3b.)

LEGS: 1, 2, 4, 3; stout, heavily covered with gray pubescence and bristles, with numerous white spines with dark bases. The front and inner sides of the metatarsi, particularly of the first, second, and third pairs, are marked by thickly set rows of yellowish spines, with brown bases, curved toward the front and inner sides, which extend also to the apical half of the tibia. This hairy and spinous armature gives the legs a particularly hirsute appearance. The color is a warm yellow, marked with narrow, dark brown annuli at the tips of and between the joints; palps colored and armed as the legs; mandibles cylindrical, not projecting beyond the clypeus.

ABDOMEN: Subglobose, arched in front and along the dorsum; the base with two castellated tubercles, whose summits are marked by numerous conical spurs or warts, a few of which appear upon the front, just below the base of the castle. The color is cretaceous, mottled with blackish spots; and a folium of indistinct outlines, in the specimen in hand, marks the dorsal field, with wavy lines of black or blackish brown. The surface is covered with numerous short, gray and black bristles, which extend also to the sides of the tubercles and the warts thereon. The venter is a broad, blackish band, marked with median lines of yellow; spinnerets distal, dark brown in color; epigynum (Fig. 3c) is without a prolonged scapus, having a short triangular flap, somewhat pointed, which does not extend beyond the posterior margin of the genital cleft. The atriolum is widened and arched, hairy and yellow.

DISTRIBUTION: Kentucky, North Carolina, and Florida. Hentz found specimens in the mud daub nests of Sphex cyanea, and supposed the species to be rare; but it is probably well distributed throughout the Southern States. (Marx Collection.)

GENUS WAGNERIA, NEW.

I propose this genus, in honor of Professor Waldemar Wagner, of Moscow, to receive Cambridge's species, Epeira tauricornis. It is distinguished by a corselet high at the summit, descending abruptly to the truncated base; cephalic suture so plainly marked that the caput is sharply differenced from the corselet, is squarely truncate at the base, and rises thence arched to the vertex, whence it slopes to the face, which is broad. The labium is compressed at the base, triangular at the tip, wider than long. The maxillæ are gibbous, rounded upon the outer margin, inclined toward each other, rather longer than broad. The eyes, as in Epeira, have a wide space between the central and sidefront groups; but the intervening spaces are relatively greater, the sidefront being separated from the midfront by about four times their intervening space, the siderear from the midrear eyes by even a greater distance. Both rows are procurved, the rear row the longer. The legs, in order of length, are 1, 2, 4, 3, the difference between 1, 2, and 4 being slight; the tibiæ are curved, as in Gasteracantha; the patella is wide, somewhat flattened upon the exterior surface; the ordinary spinous armature is wanting in the female, but the legs are clothed with rows of stout spinelike bristles. The abdomen is cylindrical-ovate or rectangular-ovate in form, about one-third longer than wide. The dorsal margins of the abdomen are characterized by a number of conical tubercles, of which the two forward ones in the typical species are thornlike, sharpened, and curved forward, resembling miniature cow's horns. The dorsal apex is also tuberculated and overhangs the spinnerets, the ventral part of the body having a somewhat conical form. The male resembles the female in general form and pattern; the palps are short, the palpal bulb large, and the copulatory organs prominent, large, and complex. The tibiæ-I and II are armed with a few short spines. Leg-IV has a series of strong, long spines, arranged the entire length underneath; while the femora have a row of short, black, denticulate spines. The cephalothorax is not as highly peaked as in the female, nor the head as high.

No. 59. Wagneria tauricornis (CAMBRIDGE).   Plate XIII, Figs. 1, 2.

1889. *Epeira tauricornis*, MARX *in litt.* . Catalogue, p. 548.
1892. *Epeira tauricornis*, CAMBRIDGE . . Biol. Centr. Amer. Aran., p. 44, vi., 2, 3; viii., 1, 2.
1892. *Epeira tauricornis*, KEYSERLING . Spin. Amer., Epeir., p. 90, iv., 68.

FEMALE: Total length, 7.8 mm.; cephalothorax, 2.7 mm. long, 1.9 mm. wide; facial width, 1.3 mm.; abdomen, 4.5 mm. long, 3 mm. wide.

CEPHALOTHORAX: Corselet rounded at the edges, high and peaked at the summit, descending abruptly to the truncated base; corselet grooves distinct; cephalic suture marked so that the head is distinctly differenced from the corselet. Color brown to brownish black, caput dark brown, covered with hair; fossa a deep indentation; caput squarely truncate at the base, rising thence and arched to the vertex, whence it slopes to the face, which is broad and rather quadrate, well rounded on the forehead. Sternum cordate, squarely truncate at the base, where it is tufted with yellowish bristlelike hairs that are sparsely distributed along the margins; sternal cones distinct, flattened in the middle; color black to blackish brown. The labium is compressed at the base, yellow; maxillæ scarcely as wide as long, gibbous, yellow. (Fig. 1d.)

EYES: Ocular quad on a rounded and projecting eminence, more prominent before than behind; front somewhat wider than rear, the sides a little longer; MF but slightly, if any, larger than MR, and separated by 1.5 to 2 diameters; MR by about 1.3. Side eyes are on low tubercles; somewhat smaller than those of the middle group. SF separated from MF by 1.5 the area of the latter, or about three times their intervening space; SR from MR by even a greater distance. The clypeus height about 1.5 diameter MF; both rows are procurved, the rear the longer.

LEGS: Order, 1, 2, 4, 3; color yellow to yellowish brown, with dark apical and median annuli, with golden yellow bristles and hairs, with rows of stout spinous bristles on the inter surface, curved toward the apex, and almost comblike, on the metatarsi especially; the ordinary spinous armature is otherwise wanting; patellæ wide, somewhat flattened on the exterior surface; tibiæ short and curved, as in Gasteracantha (Fig. 1e); palps colored and armed as the legs; mandibles rather short, strong, conical, wide at the base, covered rather evenly with short, yellow, bristlelike hairs.

ABDOMEN: About one third longer than wide, rectangular-ovate, the form difficult to describe. The dorsum appears to be flattened or concave, and provided on the margins with a series of conical protuberances, which terminate in points more or less blunted. The shoulder humps are bifid, one of the projections terminating in a long, curved, corneous point, something like the horns of a cow. The apex is about as thick as the base, the spinnerets distal, but overhung by the projecting and tuberculated apex. The dorsal surface is thickly covered with short golden yellow hairs; about the middle on each side rises one pair of horns, beneath which, upon the sides and a little posterior, is another process projecting laterally. The apex of the dorsum is divided into at least four mammal-like projections, and in the middle of the apical wall is a fifth. The sides are covered with golden yellow pubescence. At the base of the cones are flecks of brownish hue. The venter is golden yellow, covered with pubescence; the spinnerets blackish brown. The epigynum (Figs. 1a, 1b) has a brief columnar atriolum of dark brown, from which projects a scapus very wide at the base, from the middle of which issues a short rounded point.

MALE: Total length, 4.9 mm.; cephalothorax, 2.5 mm. long, 2 mm. thick; facial width, 1.1 mm.; abdomen, 2.4 mm. long, 1.8 mm. wide. The male resembles the female in general characters and colors, but is smaller. The face lacks the dark transverse band observed upon the female. The abdomen differs from that of the female in having the shoulder tubercles single instead of bifid. The legs are all provided with short strong spines; order of length, 1, 2=4, 3, as follows: 7.9, 6.2, 6.2, 4.6. Tibia-II is thicker than tibia-I, and armed with more and stronger spines. Coxa-I has an obtuse spur at its under side, and the fourth coxa a smaller and sharply pointed one.

DISTRIBUTION: Florida, Louisiana, Alabama, and probably throughout the entire Southern States. According to Cambridge, this interesting and curious spider has been collected in Guatemala and Panama, and is probably distributed throughout Central America. (Marx Collection.)

GENUS WIXIA, CAMBRIDGE, 1882.

The genus Wixia is especially differenced from Epeira: (1) By the shape of the caput, which is not narrowed in front, but is wide, or even a little wider than the base; (2) by the strong eye eminence upon which the ocular quad is placed, the lower part of which towers above and overhangs the face; (3) by the excessive size of the rear eyes of the middle group; (4) by the fact that both eye rows are procurved, and that the separating distance between the side group and the middle group is greater than the proportionate height of the clypeus, which is even greater than that of the length of the ocular quad, which (5) also differs from that of Epeira in having the rear width greater even than the length; (6) in the fact that the abdomen, which is cylindrical-ovate, is carried by the spider in a plane almost at right angles to that of the cephalothorax.

The male of the genus is also distinguished by the above characteristics, though in a modified form. The legs are especially marked by blackish denticulate spines placed underneath femora-II and IV, and similar spines placed upon both their trochanter and coxæ, also upon coxæ-II. Corneous spurs mark the articulation of coxæ-I.

No. 60. Wixia ectypa (WALCKENAER).   Plate XIII, Figs. 4, 5.

1842. *Epeira ectypa*, WALCKENAER . . . Ins. Apt., ii., p. 129.
1850. *Epeira infumata*, HENTZ . . . . . J. B. S., p. 19; Sp. U. S., p. 122, xiv., 4.
1863. *Epeira ectypa*, KEYSERLING . . . Beschr. n. Orbit., Isis, p. 135, vi., Figs., 13-16.
1884. *Epeira infumata*, EMERTON . . . N. E. Ep., 319, xxxvii., 11, 12, 13.
1890. *Epeira infumata*, MARX . . . . . Catalogue, p. 544.

FEMALE: Total length, 7 mm.; cephalothorax, 3.5 mm. long, 2.5 mm. wide, 2 mm. at the face; abdomen, 6.5 mm. long, 5 mm. wide, narrowing to 2.5 mm. at the base and 1.5 mm. at the apex.

CEPHALOTHORAX: Corselet a long oval; high, truncated, and indented at the base; fosse deep, circular; corselet grooves distinct; cephalic suture decidedly marked; head quadrate in front, flat on top, with a slight bulge on either side at the middle part, which lifts it a little above the corselet level; color yellow, lighter on the sides, and darker, tending to brown, upon the caput, which deepens into blackish at the face; covered closely with gray and yellow pubescence, and two blackish spots upon the knob of the caput. The sternum (Fig. 4c) is shield shaped, with pointed apex and sternal cones, somewhat longer than wide; flattened in the middle; yellow, covered with pubescence, which is quite thick and long at the margins. Labium triangular, but narrowed into a squarish stock at the base; more than half the length of the maxillæ, which are gibbous, inclined toward each other, obtusely triangular at the tips, and colored as the sternum.

EYES: Ocular quad (Fig. 4b) on a high prominence, which projects much at the rear, where the eyes are placed on two elevated, rounded, distinct tubercles, and are located well to the sides of the same, thus giving the summit a notched appearance; the quad much narrower in front than rear, which is wider even than the sides; MF are black, separated about 1.5 their diameter, much smaller than MR, which are light colored, and separated by at least two diameters or more; but they are so much diverted to the side of their tubercles that it is difficult to estimate the space. Side eyes on low tubercles; SF somewhat larger than SR, both much smaller than MF, and greatly smaller than MR; they are separated by about the radius of SF; MF are removed from SF by about 1.5 their area, and as SR are placed much to the side of SF, the space between them and MR is much greater. Both eye rows are procurved, the front but little, the rear row, which is longer, much more so; clypeus height about the area of MF. The face is yellowish brown, with a blackish area in front of the ocular quad. The forehead, by reason of the strongly projecting rear eyes, is hidden when the face is viewed from the front.

LEGS: Stout, rather short for the length of the species, yellow, with dark bands on tips of femora, and slight annuli at other joints; well clothed with hairs, and sparingly with short yellowish spines; palps thick, colored as the legs, hairy; mandibles conical, much separated at the tip, covered with hair at the arching base, yellow, with brownish at the fang.

ABDOMEN: Ovate (Fig. 4a), the base carried at a wide angle with the cephalothorax; contracted in front into a compressed prominence, which is bifid at the top; color yellow, or grayish yellow, covered thickly with pubescence; an indistinct folium upon the dorsal field, which in some specimens forms an elongated W about the middle thereof. On the sides beneath are bands of reticulated white and yellow, marked out by lines of brown or brownish black. The surface is marked by numerous circular rows of impressed dots, brown, smooth, which extend along the sides to the basal front, and to the summit of the frontal tubercle. The spinnerets are prominent, distal. The venter has a dark, brownish yellow band. The epigynum (4d) has a dark brown, glossy, corneous, arched atriolum, which terminates in a short, wide, blackish flap, with two notches on either side; viewed well from beneath it is hollow, and presents a shelllike appearance.

MALE: The mature specimen in hand (Figs. 5, 5a) is 6.5 mm. long, being equal in size to, or even greater than, the female. The color is darker, the cephalothorax being brownish yellow, the caput lacking the blackish coloring of the female; the legs less strongly annulated, and darker than the female; abdomen about the same color. The notched compression or tubercle on the middle base of the abdomen is even more distinct than in the female, and the irregular folium upon the dorsal field somewhat better marked; the skin heavily pubescent, and muscular pits, particularly the median ones, deep. The femora-I, II are wide, the latter particularly so; patella-I is long, and much thinner than the femur; tibia about the thickness of patella, and armed on the inside with rows of long, thick, yellowish brown spines, with a pair on either side of the joint. Patella-II is wider and shorter than I, the tibia thicker, and armed on the inside with a double row of strong, thick, clasping spines, stouter but shorter than those on the tibia-I. Coxæ-I are marked by strong spurs at the inside of articulation; coxæ-II, IV have short blackish brown spines, stronger on coxæ-IV. The fourth trochanter is also provided with a short toothlike spine. The palp is highly developed in the digital bulb. The eyes resemble substantially those of the female.

DISTRIBUTION: I have an immature male and female, collected from the neighborhood of Philadelphia. The female above described is from the District of Columbia, the male from Selma, Ala. (Marx Collection.) The spider as originally described was collected and figured by Abbot in Georgia, and it has further been observed in North Carolina, Florida, Virginia, and as far north as Connecticut. It thus appears to inhabit the entire Atlantic Coast from New England to Florida, showing a large elasticity of organism. It probably has a wide distribution throughout the United States.

Genus CAREPALXIS L. Koch, 1871.

This genus is distinguished by a high and rounded corselet, steeply arched at the sides. The caput is a lofty conical elevation, rounded at the top, which towers above the crest of the corselet, the summit occupying about the middle point of the cephalothorax, whence it rolls downward to the fosse by a sharp declivity, and with a somewhat arched incline forward to the eye space. The face is distinguished by a forehead which projects far above the eye space, and is itself considerably higher than in the typical species, twice as high as the lower part of the face. Both eye rows are procurved, and the side eyes placed low down upon the outer margins of the clypeus. The midfront eyes are separated from the sidefront by a space much greater than that which divides themselves, in this respect resembling Epeira. The legs are in order of length 1, 2, 4, 3, the first and second pairs almost equally long, and are without spines. The length of the maxillæ equals the width, thus substantially resembling Epeira. The abdomen is cylindrical; about one-third longer than broad; clothed with soft skin; carried in a position almost perpendicular; the base has tubercles, whose summits in the typical species are divided into several peaks. The spinnerets are distal.

No. 61. Carepalxis tuberculifera CAMBRIDGE. Plate XIII, Fig. 6.

1889. *Carepalxis tuberculifera*, CAMBRIDGE . . Biolog. Cent. Amer., Aran., p. 48, iv., 9.
1890. *Carepalxis tuberculifera*, MARX in litt. . . Catalogue, p. 542. (Keyserling in litt.)
1892. *Carepalxis tuberculifera*, KEYSERLING . . Spinn. Amerik., Epeir., p. 50, ii., 40.

FEMALE: Total length, 3.5 mm., but from the peculiar manner in which the animal carries its abdomen, in a plane almost perpendicular to the cephalothorax, this length is deceptive; cephalothorax, 2 mm. long, 1.5 mm. wide; abdomen, 3.3 mm. long, 2 mm. wide.

CEPHALOTHORAX: The corselet is hidden by the overhanging abdomen, somewhat rounded on the sides, truncated behind, smooth and sharply sloping to the base; the caput is extremely high, the forehead towering far above the eyes, having a height nearly twice as great as that of the face from the midrear eyes to the margin of the clypeus (Fig. 6b, 6c), giving the front thus a triangular appearance; the color is dull brown, covered with golden yellow hairs; the forehead has two knobs about midway between MR and the crest. The sternum is broadly shield shaped, rounded at the sides, bluntly rounded at the apex, covered with grayish yellow pubescence, dark brown, with lighter shade in the centre; the labium is bluntly triangular; the maxillæ gibbous, curved toward each other, somewhat longer than wide.

EYES: Ocular quad on a high squarish prominence; the rear width greater than both the front and sides; eyes about equal in size; MF separated by 1.5 diameter; MR by at least two diameters. The side eyes are on slight tubercles placed far down on the side margin of the clypeus; they are equal in size, somewhat less than the middle eyes, scarcely contingent; SR placed well below and to the side of SF; the clypeus is high, more than the area of MF. Both rows are greatly procurved, the rear row the longer.

LEGS: Order, 1, 2, 4, 3; short, stout, yellow to orange yellow in color, covered with grayish yellow bristles abundantly, but apparently without spines; the palps are colored and armed as the legs; the mandibles conical, well compressed at the tips, yellowish brown, covered with hairs at the base and inside.

ABDOMEN: Carried in a plane almost at right angles to that of the corselet, is cylindrical-ovate, broadened at the base, where there are two thick cones, whose summits are bifid; at the anterior bases of these cones and from the middle point of the abdominal front projects forward a smaller similar cone; the whole base of the abdomen is thus much roughened; the color is dull yellow, with a blackish band between the shoulder humps and an irregular semicircular patch of yellow upon the dorsal field; the whole surface is thickly covered with gray hairs; the spinnerets are distal; the venter a brownish band; the epigynum (Fig. 6a) has a short bowllike scapus extending from the atriolum between the portulæ.

DISTRIBUTION: Florida. (Marx Collection.) Cambridge describes the species from Panama, and it will probably be found along the entire Gulf Coast of the United States.

GENUS GEA, C. KOCH, 1843.

In Gea the base of the head is flattened and marked by slight tubercles on either side of the median line. The caput is depressed at the base, elevated at the face. The sternum is cordate, about as wide at the base as long, with sternal cones. The labium is subtriangular; the maxillæ are gibbous, somewhat longer than wide. The space between the midfront and sidefront eyes is but little, if any, greater than that between the midfront; the space between the midrear and the siderear eyes, on the contrary, is much greater than between the midrear eyes. The front row is slightly procurved, almost aligned; the rear row, which is considerably longer, is very much procurved. The clypeus is moderately high. The legs are, in order of length, 1, 2, 4, 3; stout, and well provided with bristles and spines; the mandibles are rather long, somewhat conical, slightly retreating backwards. The abdomen is a long oval, truncate at the base, but little arched upon the dorsum, which is marked by prominent shoulder humps and smaller tubercles. The spinnerets are distal; the skin soft and with metallic white or silvery lustre.

No. 62. Gea heptagon (HENTZ).   Plate XII, Figs. 8, 8a–d.

1850. *Epeira heptagon,* HENTZ . . . . . J. B. S., vi., p. 20; Sp. U. S., p. 122, xiv., 4, 5, 6; xviii., 52, 72.
1890. *Gea heptagon,* MARX . . . . . . Catalogue, p. 541.
1892. *Gea heptagon,* KEYSERLING . . . . Spinn. Amerik., Epeir., p. 76, iii., 58.

FEMALE: Total length, 5 mm.; cephalothorax, 2 mm. long, 1.75 mm. wide; abdomen, 3.5 mm. long, 3 mm. wide. General colors for the fore part of the body are dark brown, with yellowish and blackish, the abdomen being dark, with silvery marks. I have no mature male, but an immature one from Georgia, nearly full grown, closely resembles the female in size and markings.

CEPHALOTHORAX: Oval, well rounded at the margin; fossa a longitudinal depression; corselet grooves and cephalic suture distinct; color brown, with a light margin of yellow, and yellow flecks along the corselet grooves; base of the head flattened and marked by very slight tubercles on either side of the median line, the color yellow, with brown patches on the sides and towards the vertex; the caput is depressed at the base, and elevated at the face. Sternum (Fig. 8c) cordate, nearly as wide at the base as long, with sternal cones; color brown on the edges, with a broad median band of yellow, with lateral branches to the sternal cones; slightly pubescent; lip subtriangular, yellowish; maxillæ gibbous, somewhat longer than wide, obtusely triangular at the tips, which are yellow; the bases a yellowish brown.

EYES: Fig. 8b. Ocular quad narrower in front than behind, sides decidedly longer than rear; MF on tubercles, which project over the face; separated by about 1.5 diameter; less in size than MR, which are separated by at least 1.5 diameter. Side eyes on tubercles; SF, which are smaller, are at the base thereof, and well in front. The space between MF and SF is but little, if any, greater than that between MF; the space between MR and SR, on the contrary, is much greater than between MR; the front row is slightly procurved, almost aligned; the hind row, which is considerably longer, is much procurved; clypeus height at least 1.5 diameter MF. The eyes are on black bases; the color of the face and maxillæ is yellow.

LEGS: 1, 2, 4, 3; stout for such a small species, well provided with bristles and spines; color yellow, with dark annuli at tips and middle of joints; palps lighter yellow, with less decided annuli; mandibles' yellow, mottled with brown, rather long, somewhat conical, slightly retreating backward.

ABDOMEN: A long oval, truncate at the base, but little arched upon the dorsum, except with gravid females; dorsum marked by prominent shoulder humps, and two smaller tubercles on each side. The dorsal field is brown, relieved with metallic white, and has a black shield shaped folium, scalloped upon the edges, and tipped with white marks in the middle of the apical half; this part of the abdomen is arched to the spinnerets, which are distal and blackish brown; the venter is blackish, mottled with silvery white or yellow; the epigynum has a wide, shortened scapus (Fig. 8d), hollowed beneath like the half of a bowl.

DISTRIBUTION: The Southern United States; collected in the District of Columbia, Virginia, Georgia, Alabama, southward to Florida, and so along the Gulf Coast through Georgia, Alabama, and New Orleans.

GENUS GASTERACANTHA, LATREILLE, 1831.

This strongly marked genus is at once distinguished by the peculiar character of the abdomen, which is broader than long, with a hard, glossy skin, punctuated with many dimplelike depressions. The venter is subconical, and the spinnerets placed at the apex thereof. In the female many of the species have a dorsal field prolonged into various corneous, more or less pointed, tubercles. The head is wide, almost as wide as the cephalothorax, steeply elevated above the flattened corselet, and with hard, glossy skin. The eyes are divided into three groups, as in Epeira, but the intervening spaces are relatively much greater. The fourth leg is relatively longer than the others; the tibiæ slightly curved; ordinary spines are lacking, but the legs are armored with long, aculeate, spinelike bristles. The males are very small, but roughly resemble the female, the abdomen commonly lacking the pointed dorsal cones.

No. 63. Gasteracantha pallida C. KOCH. Plate XIV, Fig. 8.

1850. *Gasteracantha pallida,* KOCH C. . Die Arachn., xi., p. 60, pl. 374, Fig. 881.
1889. *Gasteracantha pallida,* MARX . . Catalogue, p. 539.

FEMALE: Total length, 6 mm.; cephalothorax, 2.75 mm. long, 2.5 wide across the face; abdomen, 5.5 mm. long, 8.5 mm. wide across the middle, 4 mm. across the front, 3.5 mm. across the dorsal apex.

CEPHALOTHORAX: Corselet quite concealed by the overhanging base of the abdomen, is rounded at the sides, flat and smooth on the dorsum, which slopes downward from the caput to the base. The head rises so prominently as to hide the corselet when viewed in front; the head is almost as wide in front as the cephalothorax; the forehead is high, the distance from the sides to the vertex being considerably greater than from the margins of the clypeus to MR; the forehead is divided at the median into two rounded humps; color glossy brown, with short grayish yellow pubescence. Sternum an elongated shield shape, decidedly longer than wide; surface somewhat rugose, blackish to blackish brown, with lighter touches of yellow in the middle; slight pubescence, glossy. Labium subtriangular, about half the height of maxillæ; maxillæ as wide as or wider than long, obtusely triangular at the tips, which are inclined toward one another.

EYES: Ocular quad on a rounded eminence; blackish brown in color, much wider behind than in front, the length not greater than the rear width; MF slightly larger than MR, separated by about one diameter, MR by about 1.5. Side eyes on blackish tubercles; about equal in size, scarcely contingent; distance between SF and MF very great, about three times the area of the latter, or four to five times the intervening space; clypeus low, the ocular quad touching its margin; MF removed by scarcely a diameter therefrom; front row slightly recurved, rear row is almost aligned, scarcely procurved.

LEGS: 1, 4, 2, 3; short, stout, bright glossy brown in color, with dark apical annuli which almost cover some of the joints; patellæ wide; tibiæ short; pubescence yellowish and blackish bristles and hair, but no spines; palps armored as legs, almost black; mandibles conical, wide at the base, where they are much arched, relatively much attenuated at the apex; color glossy bright brown.

ABDOMEN: An elongated oval, wider than long, the dorsum and dorsal field uniform yellow, the four ordinary muscular pits forming a prominent quadrilateral in the centre; around the margin are rows of dimples resembling the central ones, being nine in number, on either side of a strong spine which extends laterally and upward from the middle of the margin; this spine is covered with short blackish hairs, whose strong bases give it a rugose appearance; it is tipped with bright yellowish brown; sides and apex brown, strongly contrasting with the pale yellow of the dorsal field, mottled here and there with round yellow spots and with dimples scattered over the rugose surface. From the apex and below the dorsum project two spinous cones like those upon the sides, and rather flatter; the spinnerets are near the middle of the venter, are black, as is the surrounding space, but the base encompassed with circular yellow spots. In front of the epigynum and midway between it and the spinnerets is a strong black cone, projecting downward. The epigynum (Fig. 8a) shows a subtriangular atriolum, with a short flaplike projection serving as a scapus, which droops down towards the genital cleft.

DISTRIBUTION: The habitat of Koch's type was unknown. Dr. Marx locates the species in California. (Marx Collection.)

No. 64. Gasteracantha maura, new species.   Plate XIII, Fig. 12.

FEMALE: Total length, 8 mm.; cephalothorax, 4.5 mm. long, 3.5 mm. wide, 3 mm. wide at the face; abdomen, 7.5 mm. long, 11.5 mm. wide.

CEPHALOTHORAX: Rounded at the sides; corselet wider than long, truncate and indented at the base, concave on the summit; fosse a semicircular pit, with concavity directed forward; the surface roughened, scantily pubescent on the margin; color dark glossy brown; cephalic suture marked, the head rising in a steep slope from the corselet, smooth, glossy black on the posterior, slightly notched in the middle, where it is highest; the forehead as wide as the eye space, leaving a triangular depression in the middle; color glossy blackish to brown, almost black. Sternum shield shape, compressed and pointed at the apex, longer than wide; sternal cones distinct; indented at the margin and flattened in the middle; glossy black at the margins, with a median band of brownish yellow covered with bristlelike hairs with strong bases; the labium long and obtusely triangular, well rounded at the sides; the maxillæ as wide as long, rounded at the tips and sides, color, as the labium, glossy blackish brown.

EYES: Ocular quad on a prominence, rounded and more projecting in front, where it is narrower than behind; sides somewhat longer than rear. The eyes are small, of about equal size; MF separated by 1.5 diameters, MR by at least two. Side eyes on tubercles strongly projecting sidewise, of about equal size, separated by about the diameter of SR; they are distant from MF by three times or more the area of the latter; of clypeus about two diameters MF high, the margin touching the quad eminence. The front row is very slightly recurved, almost aligned; the rear row is slightly procurved, almost aligned.

Legs: 1, 4, 3, 2; in all specimens a uniform blackish brown; freely provided with strong spinelike bristles, but with no ordinary spines; palps colored and armed as legs; mandibles arched at the base, conical, glossy, blackish brown.

Abdomen: Considerably wider than long; semicircular in front and behind; the sides somewhat square and truncate, caused by the presence on each of two blunted cones, of which the posterior is somewhat larger; the dorsal field is yellow, but deeply marked with black patches, which environ the dimples, and in front are merged into two bands. There are twelve dimples in the front row and nine in the posterior. From the apex of the dorsum project two short, black, conical spines; the sides and apical wall are mottled yellow and black, covered profusely with short hairs set in a circular cavity; a row of dimples encompasses the spinnerets, which are jet black and set at the apex of the conical venter. The epigynum (Fig. 12a) is an obtusely triangular or semicircular flap, slightly prolonged at the apex into a short rounded scapus; the atriolum is wide, and, looked at from beneath, shows deep hollows on either side of the base of the scapus; color black, glossy, except at the extreme tip, which is yellowish brown.

Distribution: Numerous specimens, young and old, of this species have been obtained from California, particularly the southern part (Dr. Blaisdell, Mr. Orcutt), and from islands off the coast (Mr. C. H. Townsend).

No. 65. Gasteracantha preciosa, new species. Plate XIV, Fig. 7.

1889. *Gasteracantha preciosa*, MARX *in litt.* . . Catalogue, p. 539.

Female: Total length, 6 mm.; cephalothorax, 2.5 mm. long, width at the face 2.3 mm.; the abdomen 6 mm. long, 8 mm. across the middle. The description of Gasteracantha pallida will apply to this spider in almost every detail except the general color. The cephalothorax, mouth parts, and legs are extremely glossy, and blackish to blackish brown, except upon the terminal joints of the legs. The dorsal field of the abdomen, instead of being a pale yellow, as with G. pallida, is jet glossy black, mottled with large irregular patches of bright yellow placed in front and along the margins, and near the middle part of the margins present a striking appearance. The number of circular pits or dimples around the margins is also greater than in G. pallida, numbering eleven on the base and ten on the apical half of the dorsal circle. The spines projecting from the middle part are yellow at the base, succeeded by a bright rosy red, the point being jet black, all covered with circular rows of spines with roughened bases. There is a decided conical process on coxa-II which is wanting in the example of G. pallida before me. The epigynum shows simply a low atriolum, with a hard, glossy black, semicircular edge without a scapus.

Distribution: Mohave Desert, California.

No. 66. Gasteracantha cancriformis (LINNÆUS). Plate XIV, Fig. 9.

1767. *Aranea cancriformis*, LINNÆUS . . Syst. Nat. Ed., xi., p. 1037.
1837. *Plectana cancriformis*, WALCK. . . Ins. Apt., ii., p. 151.
1837. *Plectana elipsoides*, WALCKENAER . Ins. Apt., ii., p. 155; ABBOT, Ga. Spid.
1850. *Epeira cancer*, HENTZ . . . . . . J. B. S., vi., p. 23. Sp. U. S., p. 126, xiv., 13.
1889. *Gasteracantha cancriformis*, McCOOK. Amer. Spid. and their Spinningwork, Vols. I, II.
1889. *Gasteracantha cancriformis*, MARX. Catalogue, p. 539.

Female: Total length, 7 mm.; width at the face, 3 mm.; abdomen, 5 mm. long, 8 mm. wide.

The cephalothorax and caput are of the typical form, the color glossy black, with a tint of brown. The sternum and mouth parts are typical, as are also the eyes. The legs are glossy brownish black, the pubescence gray bristles, without spines. The abdomen differs from that of G. maura in that the six thornlike spines are much more pointed, and are a bright rosy, reddish brown color, though the tips are blackish, and entirely covered with bristles, rising out of strong black bases. The anterior thorns are directed somewhat forward; the posterior are larger, and decidedly directed backward; two similar cones project from the apical part. The abdomen is yellow, reticulated, and covered sparsely with soft, short, grayish bristles, the dimpled pits surrounded by black patches, in some cases blended together. There are ten of these dimples upon the basal row and nine upon the posterior row. The latter are arranged in three groups, the two larger at the base of the posterior side thorn, the posterior one of these two forming, with the posterior two of the central quadrilateral, a row whose curvature is backward. The marginal row has five smaller dimples between the bases of the posterior thorns, and slightly curving backward. In the basal row the two dimples near the bases of the anterior spines are larger than the others. The sides and venter are black, mottled with yellow spots, covered with soft yellowish gray bristles. The spinnerets are black, and the epigynum is a strong projecting cone, which is tipped with reddish brown. The epigynum (Fig. 9a) has a long subtriangular atriolum, whose edge is deeply indented on either side of the base of a short scapus, which is little more than a rounded and roughened hump.

DISTRIBUTION: I have collected this species in Texas, have specimens from North Carolina, and a number from the islands of the Caribbean Sea, collected by Mr. C. H. Townsend. Dr. Marx locates the species in Florida, Alabama, Texas, Arizona, New Mexico, and California. It is therefore distributed along the entire Gulf Coast of the United States, and thence across the continent to the Pacific Coast, where it abounds, at least in the southern portion thereof.

GENUS ACROSOMA, PERTY, 1834.

In Acrosoma the cephalothorax is a rather lengthened oval, truncated behind, but low in front. The fosse is not deep, the cephalic suture distinct, the head arched and somewhat erect, quadrate and wide in front; the skin hard, glossy, and comparatively smooth. The sternum is longer than wide, marked by decided sternal cones. The labium and maxillæ are as in Epeira. The fourth pair of legs the longest, the order being 4, 1, 2, 3; well clothed with pubescence, bristles, and short, stubby, bristlelike spines, but lacking the ordinary spinous armature. The eyes are substantially as in Epeira. The abdomen is an elongated rectangular ovate, upon a dorsum the skin of which is hard, glossy, covered with dimples, and marked along the margin by conical tubercles, more or less pointed. The venter is conical, the spinnerets placed at the apex, nearly underneath the middle of the abdomen. The characteristics of this genus are so distinct that it is difficult to mistake the identity, at least of the known American species.

No. 67. Acrosoma gracile (WALCKENAER).   Plate XXI, Figs. 1, 4.

1837. *Plectana gracilis*, WALCKENAER . . Ins. Apt., ii., p. 193; ABBOT, G. S., No. —.
1845. *Acrosoma matronale*, KOCH, C. . . Die Arach., xi., p. 68, Fig. 887.
1850. *Epeira rugosa*, HENTZ . . . . . . J. B., vi., p. 21; Sp. U. S., p. 124, xiv., 10.
1884. *Acrosoma rugosa*, EMERTON . . . . N. E. Ep., p. 326, xxxviii., 10.
1888. *Acrosoma gracilis*, McCOOK . . . Proceed. Acad. Nat. Sci., Phila., p. 5.
1889. *Acrosoma gracile*, MARX . . . . . Catalogue, p, 530.

FEMALE: Total length, 10 mm.; abdomen, 7 mm. long (not including the posterior spines), 4.5 mm. wide; cephalothorax, 3.5 mm. long, 2.5 mm. wide. The colors vary in different species, but the fore part of the body is generally a dark orange or reddish brown, while the abdomen is yellow, with brown or red thornlike spines.

CEPHALOTHORAX: Corselet irregular oval, truncated behind; rather flat on the edges, elevated in the centre; fosse a low circular depression; skin glossy, hard, slightly covered with short pubescence; face wide, head raised above the corselet, from which it is separated by a deep suture, which strongly delineates caput from corselet. A bright yellow ribbon girdles the edge of the corselet, which for the rest is a reddish brown, more or less decided in various specimens. The cordiform sternum is longer than wide, has decided sternal cones; is rather heavily pubescent; color dark brown, with median band of yellow, which is broadened into a rhomboid figure at the middle, and in front is yellow; labium large, subtriangular, and has a yellow tip like the maxillæ, which are rounded, and as wide as long, or even wider.

EYES: Fig. 1a. The quad on an elevated prominence; narrower in front than behind, longest at the sides; MF separated by about 1.5 diameter; MR longer, and separated by about one diameter; side eyes on decided tubercles, almost contingent, almost equal in size; MF separated from SF by more than twice their alignment, and about once their alignment from margin of clypeus; front row shorter than rear and recurved, rear row procurved.

LEGS: 1, 4, 3 2; color from dark brown to yellow, the final joints being brighter. The femur has a wide median annulus, and the joints are tipped with darker color; well armed with short hairs and stout bristles, arranged in longitudinal rows. Underneath, on the femora particularly, the skin is extremely rugose, an effect produced by the elevated bases of the stumpy spinelike hairs. Palps colored as the legs; mandibles strong and well rounded at the base, but not projecting beyond the face.

ABDOMEN: Rectangular ovate, flattened or even hollowed on the dorsum, except in gravid females; a series of thornlike conical spines set around the edges, of which the basal are shorter than the middle, and those on the apical corners are double; immediately beneath each of the latter is a shorter spine; color usually bright yellow, with a row of four circular black pits on the base, two large ones at the middle, and others, again, on the posterior; the bases of spines dark brown or red, deepening into black at the sharp points; skin glossy, hard, sparsely pubescent. The sides are rugose, streaked with about four longitudinal rows of yellow, alternating with brown bands, and thickly covered with short hairs, with brown circular pits. The spinnerets are so far underneath the abdomen as to give it a conical or pyramidal shape; are dark brown or blackish; venter pattern yellow, mottled with brown; the epigynum (Fig. 2a) is dark brown in color, tipped with yellow, with a prominent scapus, wide at the base and narrowing to a rounded spooned tip. The specimens of this spider vary much in color, apparently according to age. In some the head and dorsum of the corselet are orange yellow, with a median band of brown, and a broad circular belt of brown next below and adjoining the marginal band of yellow; the legs, in such specimens, become yellow, with brown annuli, not only at the ends of the joints but in the middle thereof. The dorsal spines also differ in length, and the abdomen varies slightly in color. Where specimens have been long in alcohol the circular pits or spots above alluded to appear in marked prominence, and are arranged laterally across the dorsum and between the apical spines, and again in rows' passing from the dorsum to the spinnerets, underneath the venter.

DISTRIBUTION: I have collected along the entire Atlantic Coast of the United States, and in the Middle States. It inhabits the Gulf States, and probably the entire region east of the Rocky Mountains.

No. 68. Acrosoma reduvianum (WALCKENAER). Plate XXIII, Figs. 6, 7.

1837. *Plectana reduviana*, WALCKENAER . Ins. Apt., ii., 201; ABBOT, G. S., No.
1850. *Epeira mitrata*, HENTZ . . . . . J. B. S., vi., 22; Id., Sp. U. S., p. 125, xiv., 11.
1884. *Acrosoma mitrata*, EMERTON . . . N. E. Ep., p. 37, xxxviii., 9.
1888. *Acrosoma reduviana*, McCOOK . . Proceed. Acad. Nat. Sci., Phila., p. 5.
1889. *Acrosoma reduvianum*, MARX . . Catalogue, p. 540.

FEMALE: Body length, 5 mm.; abdomen, 4 mm. long, 3 mm. wide; cephalothorax, 1.5 mm. to 2 mm. long, and 1 to 1.3 mm. wide. Specimens differ in size. The above measurements are taken from one of the largest females.

CEPHALOTHORAX: Corselet irregular oval, truncated behind; the median fosse deep, cephalic suture well marked; corselet grooves distinct; color yellowish brown, with a yellow marginal band; in some specimens lighter streaks of yellow mark corselet grooves and caput; skin glossy and slightly pubescent; head on a level with or slightly above the corselet; color yellowish brown, uniform with corselet and mandibles. The sternum is shield shaped, about as wide as long; blackish or dark brown, slightly pubescent; skin hard and glossy; labium subtriangular, colored like the sternum; the maxillæ wide as long, and rounded at the tip.

EYES: Ocular quad on a black prominence; narrower in front than behind, sides slightly longer than rear; the eyes are set on black rings; MR slightly larger than MF; MF separated by about 1.5, MR by one diameter; side eyes on tubercles, not very pronounced; propinquate; SF somewhat larger than SR. The clypeus height about two diameters MF; front eye row well recurved, the longer rear row slightly procurved or aligned.

LEGS: 4, 1, 2, 3; yellow, with slight brown apical annuli; armed with bristles and short bristlelike spines, and are rugose, particularly upon the femora; palps colored and armed as the legs.

ABDOMEN: Somewhat an irregular oval, wider behind than in front, widest in the middle; the posterior is armed at the dorsal edges with strong but not long spinal processes, and immediately beneath these are similar but smaller processes. The color is yellow, with a folium of blackish color, interrupted in the middle. Several curved rows of circular brown spots cross the dorsum, modifying the color, some bowed to the front and some to the rear; similar spots are on the sides and underneath. The dorsum is reticulated, and is glossy and hard; the spinnerets are placed well underneath, the venter is yellow, the spinnerets brown, as well as the space surrounding the gills and · epigynum. The epigynum (Fig. 6 front view, 6b side view) shows anteriorly a brown chitinous atriolum and a scapus bent and blunt at the tip. Underneath the scapus and dividing the postulæ is what may be called a secondary scapus, wide at the base and somewhat rounded at the tip.

MALE: Figs. 7, 7a, 7b. The male does not differ widely from the female in coloring and markings. The apex of the abdomen is rounded, and is destitute of spinal processes. Its size is a little more than half that of the female, being 3 mm. long.

DISTRIBUTION: The geographical distribution appears to correspond closely with that of its congeners, A. gracile and A. sagittatum. It has been collected throughout the New England States, in New York, Pennsylvania, Maryland, District of Columbia, Virginia, North Carolina, Georgia. I have specimens from Ohio, and as far to the north and west as Wisconsin. Its distribution west of the Mississippi has not been determined, and it is probably limited by the Rocky Mountains.

No. 69.　Acrosoma sagittatum (WALCKENAER).　　　Plate XXIII, Figs. 8, 9.

1837. *Plectana sagittata*, WALCKENAER . Ins. Apt., ii., p. 174; ABBOT, G. S., No.
1850. *Epeira spinea*, HENTZ . . . . . . J. B. S., vi., p. 21; Sp. U. S., p. 123, xiv., 9.
1884. *Acrosoma spinea*, EMERTON . . . N. E. Ep., p. 326, xxxviii., 5–8.
1888. *Acrosoma sagittata*, McCOOK . . . Proceed. Acad. Nat. Sci., Phila., p. 5.
1889. *Acrosoma sagittatum*, MARX . . . Catalogue, p. 540.

FEMALE: Total length, 9 mm.; abdomen, 4.5 mm. long from base to apex at the median line, 7 mm. long from base to tips of posterior spines; width, 4 mm.; cephalothorax, 3 mm. long and 2 mm. wide.

CEPHALOTHORAX: A long oval, rounded, slightly indented at the base, the posterior half raised instead of depressed; the fosse distinct, as are also the corselet grooves; the margin somewhat rimmed or turned up; caput elevated; cephalic sutures distinct; the head rather stumpy and wide at the face. Color yellow or orange brown, with darker streakings on the sides; a marginal rim of yellow surrounds the corselet, with a wider dark brown band above it; the head light yellow or bright brown; skin smooth and hard. Sternum is a long shield, engrailed at the edges, the apex pointed and somewhat lengthened; the anterior or chief extending at both dexter and sinister points into rectangular arms; the centre is raised; slight sternal cones opposite the coxæ and on the apex; skin is hard and smooth like the corselet, with which it corresponds in color, though of a rather lighter shade; labium large, subtriangular, colored like the sternum with a yellowish tip, as also are the maxillæ, which are ovate, somewhat longer than wide, the tips subtriangular.

EYES: Ocular quad on a slightly raised prominence, the rear considerably wider than the front, and equal to sides or even a little longer; MF somewhat smaller than MR, and separated by about one diameter; MR by about 1.5 diameter; eyes on a black ring. Side eyes on decided tubercles, and each upon a slight secondary tubercle; SF larger than SR, from which they are separated by about or more than the diameter of SR; MF are removed from SF by about twice their alignment, and from clypeus margin by about 1.5 diameter; front row recurved, the longer rear row slightly procurved.

LEGS: 4, 1, 2, 3; the fourth leg being 9 mm. long, the first leg 8 mm.; clothed with hairs and bristles, with decided spines, and the skin roughened partly by the lumpy bases of bristles; color uniform yellow, or yellowish brown, or orange brown, varying with individuals. They are tolerably stout, the two first pair more so than the others; the third leg, instead of extending backward like.the fourth leg, extends forward like the two first pair; palps colored like legs; mandibles yellowish brown, rather short, rounded at the base, where they project a little beyond the face, tapering at the tip.

ABDOMEN: An irregular oval, truncated at the base, widened at the apex, where on either side it is elongated into two long thornlike spines; the dorsal front has two decided spines, but shorter than the posterior, and midway between these a much abbreviated pair; the skin is hard and smooth, slightly pubescent, and covered with rows of brown circular pits, which encompass the abdomen. The color varies somewhat, in some specimens being more brilliant than in others, but typical specimens have the dorsum bright yellow, the bases of the spines crimson or orange, toward the point deepening into brown or blackish brown or black. The sides are ridged with alternate rows of yellow or brown. The spinnerets are on a conical projection well underneath the abdomen; venter colored like the sides, the posterior parts around the gills and epigynum being orange brown. The epigynum (Figs. 8b, 8c) has a well elevated and rounded atriolum of the same chitinous texture as the abdomen, slightly pubescent, and at the middle elongated into a scapus which is wide and triangular at the base, with an attenuated shank, enlarged at the tip into a rounded spoon; the terminal parts are light yellow, the base brown. The portulæ are prominent, and on either side and in front the wall is elevated into a subtriangular partition, lying just beneath the scapus.

MALE: Figs. 9, 9a, 9b. In form and general characteristics the male differs widely from the female. Its length is about 4.5 mm.; the color is black or deep blackish brown; the cephalothorax dark brown, or sometimes dark yellow. The fore legs are black or blackish brown, except the two terminal joints; the third and fourth legs are usually lighter in color. The cephalothorax is a long oval. The eyes are arranged substantially as in the female; the palp distinguished as at 9a, 9b. The abdomen widens toward the apex, where it is nearly twice as wide as at the base, and is rounded at the corners. It is destitute of the spinous processes which characterize the female, but in some species there are slightly developed humps on either side, though the posterior spines are wanting; the spinnerets are placed underneath the abdomen as in the female; the skin is tough, smooth, and shining.

DISTRIBUTION: This species reminds one of tropical fauna in its general appearance, but is widely distributed throughout the Northern and Middle as well as the Southern United States. It has been traced from New England through New York, Pennsylvania, New Jersey, along the Atlantic Coast, to Florida, along the Gulf Coast; it has been found in Alabama, Mississippi, and Louisiana. I have received no specimens from the Pacific Slope; the distribution is probably limited by the Rocky Mountains.

Genus CERCIDIA, Thorell, 1869.

In Cercidia the cephalothorax is oval, the head elevated and somewhat quadrate. The sternum is somewhat longer than wide, and marked with distinct sternal cones. The labium and maxillæ are as in Epeira. The eyes are substantially as in Epeira, except that the siderear eye is larger. The fourth leg is longest of all. The abdomen is a regular oval, but is distinguished from Epeira, Zilla, and Singa by the glossy hard skin. The individuals of Cercidian species are small. The typical American species is C. funebris. The cocoon is a small egg shaped sac.[1]

No. 70. Cercidia funebris Keyserling.   Plate XIX, Fig. 9, 9a–c.

1889. *Cercidia funebris*, Marx *in litt.* . . Catalogue, p. 540 (Keyserling *in litt.*).
1892. *Cercidia funebris*, Keyserling . . Spinnen Amerikas, Epeir., p. 37, ii., 32, 32a.

Female: Total length, 4 mm.; abdomen, 3 mm. long by 2.5 mm. wide across the shoulders; cephalothorax, 1.75 mm. long by 1.3 mm. wide. The general colors in front, dark brown to blackish; the abdomen glossy black, with longitudinal stripes of cretaceous or yellow.

Cephalothorax: A rounded oval; head prominent; cephalic suture strongly marked; fosse entirely overhung by the abdomen; caput high; color blackish brown. Sternum shield shape, somewhat longer than wide, but rather rounded at the apex; sternal cones distinct; flat in the middle; slightly pubescent; color jet black. Labium triangular, half the height of the maxillæ, which are as broad as long; cut square at the tips, and, like the labium, glossy black, in contrast with the yellow coxæ.

Eyes: Eye space black or blackish brown; ocular quad on a rounded eminence; manifestly wider in front than behind, and the sides longest; MF decidedly smaller than MR, separated by about one and a half to two diameters, while MR are separated by about or a little more than one; side eyes on tubercles, subequal, contingent, removed from MF by a space equal to the area of the latter, and at least half greater than the intervening space of MF; front eye row recurved, the longer rear row procurved; clypeus two and a half to three diameters MF in height.

Legs: 4, 1, 2, 3, as follows: 4.4, 4.2, 3.9, 2.6 mm.; color dark brown, lighter at the bases of the femora, the coxæ underneath yellow; stout rows of strong, short, yellow bristles both above and beneath along the joints, and apparently without spines; palps colored and armed as legs; mandibles strong, conical, parallel.

Abdomen: An oval longer than broad; narrower at the apex than at the base; the dorsum arched; the skin hard and bright, glossy; the folium a black to blackish brown figure, undulating at the margin, and a broad interrupted ribbon of cretaceous or yellow along the median band, which is divided in the middle by an irregular thread of blackish brown color, except upon the base. A broad waving band of yellow passes longitudinally along each side beyond the folium; and below that again on the sides are broad bands of glossy brown, which terminate in a narrow longitudinal strip of cretaceous that borders the blackish quadrilateral venter. This black ventral band quite encompasses the spinnerets, which are of a light color, and also the epigynum. The latter organ (Fig. 9b) appears to be a crescent shaped cup, with the opening toward the front, and without a distinct scapus. In the figure, however, upon the plate the artist has represented a short scapus, widening into a spoon shaped atriolum.

Distribution: Crescent City, Florida. (Marx Collection.)

Genus ARGIOPE, Savigny and Audouin, 1825.

Argiope is distinguished by a cephalothorax which is flattened upon the summit of the corselet, giving this organ a comparatively low and weak appearance. The caput also is flattened along the entire length, and placed upon, or very nearly upon, a level with the

---

[1] Thorell, "On European Spiders," page 59.

corselet. The corselet is oval, somewhat longer than wide. It is ordinarily closely furred with metallic white hairs, that give it a silvery appearance. The sternum is somewhat longer than wide, and is often marked with prominent sternal cones. The labium and maxillæ are substantially as in Epeira, though the latter are rather longer than wide. The eyes are divided into three groups, usually placed upon decided eminences. . The eye rows are both procurved, the anterior row slightly, the posterior row decidedly. The legs are in order of length 1, 2, 4, 3; sufficiently stout, armored with hair, bristles, and spines, the two terminating joints being usually attenuated in size, and the front metatarsal joint of greater relative length. The abdomen is commonly longer than wide, cylindrical or subcylindrical, or a lengthened oval; in some species the dorsal base is marked by shoulder humps, in others the posterior margin and sides by tubercles. The colors are brilliant, sometimes metallic. The skin is soft and thickly covered with hairs. Some of the largest species of both American and exotic spider fauna are found within this genus. The males usually differ much from the females, and are relatively small. The orbicular snare is frequently decorated with ribbons of white silk, symmetrically arranged.

No. 71. Argiope cophinaria (WALCKENAER). Plate XV, Figs. 1-6; Pl. XVI, Figs. 5, 6; Pl. XVI, Figs. 1, 2.

1837. *Epeira cophinaria*, WALCKENAER . Ins. Apt., ii., 109; ABBOT's Ga. S., No. 151.
1837. *Epeira ambitoria*, WALCKENAER . Ins. Apt., ii., p. 112.
1839. *Nephila vestita*, KOCH . . . . . . Die Arachniden, v., p. 35, pl. 153, Fig. 358.
1847. *Epeira riparia*, HENTZ . . . . . J. B. S., v., 468; Sp. U. S., p. 106, xii., 5.
1882. *Argiope riparia*, McCOOK . . . . Proceed. Acad. Nat. Sci., Phila., p. 256.
1884. *Argiope riparia*, EMERTON . . . . N. E. Ep., p. 329.
1888. *Argiope cophinaria*, McCOOK . . . Proceed. Acad. Nat. Sci., Phila., p. 1; and Amer. Spiders and their Spinningwork throughout.
1889. *Argiope cophinaria*, MARX . . . . Catalogue Described Araneæ, p. 541.

FEMALE: Total length, 20 to 28 mm.; cephalothorax, 11 mm. long, 8 mm. wide; abdomen, 18 mm. long, 11 mm. wide. The size of the adult female varies greatly, but the above measurement gives one of the largest, although at times a gravid female is found with even a much larger abdomen.

CEPHALOTHORAX: A long oval, thin, flat on top; fosse a deep rounded pit in the centre thereof; corselet covered uniformly with grayish or silvery white hairs. A lighter band of yellowish hue girdles the margin; the caput is low at the base, flat on top, the skin beneath, like that of the corselet, yellowish brown, mottled with yellow, but so thickly covered with silvery white hair as to disguise the color thereof. These hairs extend to the eyes, and entirely encompass the forehead surrounding the upper part of the eyes. Sternum shield shape, much wider at the base, and pointed at the apex; strong sternal cones, a particularly large one at the apex; flattened in the middle; color dark brown to blackish, with a broad bright yellow median band the entire length; covered, particularly at the margins, with white pubescence, intermingled with long, dark, spinous bristles. Labium subtriangular, less than half the height of the maxillæ, which are gibbous, somewhat longer than wide, brown, with yellowish tips, and provided with strong, black, curved bristles.

EYES: Ocular quad on a squarish elevation, which much projects in front, the rear eyes at the base thereof; length much greater than width, the front narrower than rear; eyes not greatly different in size, but MF are somewhat larger, situated on triangular projections from the corner of the quad; separated by about 2.5 diameter; MR separated by 1.5 to 2. Side eyes on strong black tubercles, separated by about one diameter of SR, which is slightly larger than SF; the space between SF and MF is about 1.3 the area of the latter, or not more than 1.5 the separating space; the distance between SR and MR is 1.3 to 1.5 times greater than that which separates SF and MF; the clypeus is rather high, the margin separated from MF by 2.5 to 3 times the diameter of the latter, which, however, are much elevated above it; the eye space is yellowish brown, the tubercles and ocular quad glossy, black, corneous, with a few curved bristles placed along the margin of the clypeus, strongest at the base of the quad; front row slightly procurved, rear row greatly procurved.

LEGS: 1, 2, 4, 3, as follows: 52, 49, 42, 31.5 mm.; stout at the thighs, but gradually tapering to the tarsus, which is thin, as is also the metatarsus; in many adult females the last two joints, and frequently the last three joints, also including the patella, are glossy black, the thighs alone showing at the bases a bright yellow hue; in the younger specimens the leg is striped with brown and yellow (see Fig. 4); the whole surface is covered with short hairs, yellow on the femora and dark on the other joints; femora–I are usually darker, and specimens will be found in which they are almost black, a peculiarity which extends to the third leg; surface thickly covered with bristles, and less abundantly with short dark spines; the palps are colored and armed like the legs, but in some specimens are entirely yellow; mandibles conical, dark brown, and glossy with yellowish tips.

ABDOMEN: An elongated oval, narrowed or broadened according to the maturity or age of the individual; its width is usually about two-fifths its length, broadest about the middle, slightly tapering towards the apex, which overhangs the spinnerets; it also narrows somewhat towards the base, which is marked by two well developed conical processes, which project rather forwards than sideways, as in the case of Epeira. They show decidedly in half grown specimens, and more or less in adults not gravid, when the abdomen is distended with eggs; the dorsal field is marked by a broad irregular median band of blackish brown, flanked on either side by an irregular broken band of yellow spots; the median blackish brown band is marked by four to six, usually the latter, patches or spots of yellow, arranged symmetrically along the median line; four longitudinal brownish stripes, which are often indistinct, pass from the middle of the dorsal field along the folium to the apex; the surface is somewhat freely covered with short, silvery white hairs, and the skin is reticulated; the venter has a broad brown band, compressed towards the spinnerets, with eight circular yellow spots arranged symmetrically on the median line, and a broad ribbon of bright yellow along either margin, which passes to the base of the brown spinnerets, encircles the same in a broken band. The epigynum (Plate XVI., Figs. 5, 5b, 5c) has a low, wide, yellow atriolum, subtriangular in form, but little separated from the venter; the scapus is brown, glossy, corneous, widest at the base, and curved along the lower side, short, terminating in a wide, notched tip, which is hollowed on either side; when removed from the venter, and the under part is viewed, this organ is seen to be quite hollow; it is covered with short, curved, black bristles.

MALE: Plate X, Figs. 5 and 6. The male of this species is about one-fourth the size of the female, being 6 to 8 mm. in body length. The cephalothorax is yellowish brown, covered, but less densely than the female, with white and yellowish white hairs, which also sparsely clothe the caput, and are found within the eye space. The eyes are arranged nearly as in the female. The MF are about 1 to 1.5 diameter removed from the margin of the clypeus, and are separated from SF by a space a little greater than that which divides themselves. The rear eyes are on tubercles, propinquate, SF slightly the smaller. The front row is arranged in a nearly straight line, the rear row much procurved. The legs are brownish yellow, with faintly marked annuli, and well provided with spines, bristles, and hairs. The abdomen is a long oval, widest at the top, where at each corner the base presents a slight conical process, and is marked by a cluster of long bristles. Along the middle of the dorsum extends a wide dentated band of brown color, interspersed with yellow, on either margin of which along the sides is a zigzag band of white, sprinkled with silver hairs, corresponding with the interrupted zigzag bands on the sides of the female. The mandibles do not extend beyond the border of the eye space, and the head on either side projects beyond them. The humeral, cubital, and radial joints of the palp are yellowish white, the two last named being short and irregularly rounded, the last one particularly having a number of radiating bristles. The digital joint is bright brown, intermingled with yellow color, and may most readily be distinguished by three projections extending from the outer side, of which two are relatively long, and the third, or shorter one, is palm shaped, and strongly toothed upon the edge. (See Plate II., Fig. 6.) The males are found hanging upon the outer borders of the female's snare during August and September, sometimes several individuals being in attendance upon one female.

DISTRIBUTION: This fine spider, whose habits are so fully given in Volumes I. and II. of this work, is well known to all frequenters of our fields, and is familiar to even the most careless observer. It inhabits grasses, bushes, and generally low positions, on which it spins a strong web, the central portion of which is covered thickly with a white shield-like patch, from either extremity of which proceeds a broad zigzag ribbon. It is able to capture the strongest insects in its mature stages; and feeds largely upon grasshoppers and locusts. Its geographical distribution is almost coterminous with the United States. I have collected specimens in New England, as far north as Massachusetts, Connecticut, and Northern Illinois (Chicago), and southward along the Atlantic Coast as far as Florida. To the southwest I have taken specimens in Texas, and have them from the Pacific Coast as far south as San Diego, California. I also have specimens from the Rocky Mountains, and from various points along the great plains and prairies, both east and west of the Mississippi River. Northward specimens have been obtained from Minnesota and Wisconsin. It may therefore be considered as distributed throughout the entire United States. In all locations it appears to preserve the same habits, spins the same sort of web and cocoon, and, whether in Southern California or Northern New England, remains in shape, color, and size substantially the same.

No. 72. Argiope argyraspis (WALCKENAER). Plate XV, Fig. 8; Pl. XVI, Figs. 3, 4.

1837. *Epeira argyraspides,*WALCKENAER.[1] Ins. Apt., ii., p. 110; ABBOT, G. S., No.
1847. *Epeira fasciata,* HENTZ . . . . . J. B. S., v., p. 468; Sp. U. S., p. 107, xii., 8.
1873. *Argiope fasciata,* L. KOCH . . . . Arachniden Australiens, p. 133; pl. 10, Fig. 5.
1882. *Argiope fasciata,* McCOOK . . . . Proceed. Acad. Nat. Sci., Phila., p. 256.
1884. *Argiope transvera,* EMERTON . . . N. E. Ep., p. 330, pl. 24, Fig. 20.
1888. *Argiope argyraspides,* McCOOK . . Proceed. Acad. Nat. Sci., Phila., p. 1.
1889. *Argiope argyraspis,*[1] McCOOK . . Amer. Spiders and their Spinningwork, Vol. I.
1890. *Argiope argyraspides,* MARX . . . Catalogue, p. 541.

FEMALE: Total length, 19 mm.; abdomen, 12 mm. long, 9 mm. wide; cephalothorax, 6.5 mm. long, 6 mm. wide, 3 mm. at the face. This spider is distinguished at once from its congener, A. cophinaria, by the color and shape of its abdomen, which is truncated at the base and pointed at the apex, in the younger specimens quite decidedly so, though the young of cophinaria have the same form. The spinnerets are located rather further underneath, the abdomen being placed at least one-third of the distance from apex to base. (Plate XV., Fig. 8.) Adult specimens differ much in size, from 16 mm. or less in length to 25 or 30 mm., as with some specimens from Southern California.

CEPHALOTHORAX: A long oval; the corselet rounded at the margin, broadly truncated, and indented at the base, low and flat; the fosse a conical pit; skin yellow, with wide brownish patches on either side of the broad yellow median line; the margin yellow, with strong grayish yellow bristles; the whole surface so covered with silvery white hairs that it has a silvery gray lustre; caput narrow at the base, compressed at the middle, colored and armored as the corselet, somewhat contracted towards the face. Sternum shield shape, wide at base, rather rounded at margins, obtusely triangular at the apex, the centre a broad yellow band; strong yellow sternal cones; color of margins brown; the surface covered thickly with strong hairs and stiff brownish bristles. Labium subtriangular, yellow, with but a fleck of brown at the base, as are also the maxillæ, which are somewhat longer than wide, and sparsely covered with brown, curved, strong, spinous bristles.

---

[1] The proper term is undoubtedly "argyraspis," and I use it in accordance with the author's liberty to correct obvious errors in spelling.

EYES: Ocular quad on a rounded eminence, projecting in front; longer than wide, front narrower than rear; eyes on black bases; MF somewhat larger than MR, separated by about 1.3 diameter; MR separated about 1.5 diameter. Side eyes on high tubercles; SR larger than SF, propinquate, but not contingent; the space between SF and MF is equal to about 1.3 the area of the latter; space between SR and MR is at least one-third greater than that between SF and MF; clypeus height about three diameters MF; front row slightly procurved, rear row much procurved.

LEGS: 1, 2, 4, 3; not so stout at the thighs as A. cophinaria, and not tapering quite so sharply to the feet; color yellow, strongly annulated at the tips and between the joints with brown; covered with yellowish white hair and bristles, and with rather short brown spines; palps yellow, with little brown at the tips, strongly armed with brown spines and spinous bristles; mandibles yellow, with flecks of brown; conical, but not greatly tapering.

ABDOMEN: A long oval, rather truncated at the base (in young specimens subtriangular) and narrow at the apex, which much overhangs the brown spinnerets; color silvery white, broken by lateral bands of black and yellow, alternating from base to apex; dorsum not much arched, except in the female when gravid. A dark branching median line, which divides into four longitudinal lines, passes along the centre of the dorsal field; the muscular pits are strongly marked, and the surface is punctuated irregularly with dots; the sides are marked by longitudinal stripes of brown and silver, the colors produced chiefly by the hues of the pubescence; the brown stripes contain many dark brown curved bristles; the ventral pattern is a broad brown patch, with eight circular silvery spots symmetrically arranged on either side of the median; on the margins a ribbon of yellow or whitish yellow, which encompasses the base of the brown spinnerets in an interrupted band. The epigynum has a bowl shaped atriolum, brown, corneous, which is spanned in the middle by the scapus, that arches over the bowl from side to side, like the clasp of a padlock; the apex is not free, but attached to the posterior margin of the atriolum. (Plate XVI., Figs. 3a, 3b.) It is hollow, and has an opening at the point where the curve touches the venter.

MALE: Plate I., 10, 11. About the same size as that of A. cophinaria, which it resembles in many respects. The cephalothorax and legs are uniform yellow or yellowish brown, the legs well provided with numerous spines, bristles, and hairs. The front eye of the lateral pair, as in the female, is decidedly smaller than the rear eye. Tufts of silvery white hairs are found along the margins of the cephalothorax, extending to the eye space. The abdomen has a brownish, scalloped band along the median line, with longitudinal rows of silver white hairs on either side. It is a long oval in shape, and, unlike the male of Cophinaria, is not bifid at the base. The venter is silvery white, with a long rectangle of blackish color drawn between the spinnerets and the epigynum. Instead of the three flattened processes projecting from the external margins of the palps, as in the case of the male of A. cophinaria, there is a curled or crescent shaped process, strongly marked with black chitine along the edge. (Plate II., 4.) The habits of the male, as far as known, are precisely those of the male of A. cophinaria.

DISTRIBUTION: The distribution of this species is coterminous with that of A. cophinaria, at least I have rarely failed to get the two in the same general bound. I have taken it all along the Atlantic Coast, in the Middle States, and have specimens from the Pacific Coast. L. Koch reports it in New Grenada, Madeira, and Australia.

No. 73. Argiope argentata (FABRICIUS). Plate XVI, Figs. 1, 2.

1775. *Aranea argentata*, FABRICIUS . . Entom. System, ii., p. 414.
1837. *Epeira argentata*, WALCKENAER . Ins. Apt., ii., p. 110; ABBOT, G. S., 551; 117, vol. 15.
1839. *Argiopes argentatus*, KOCH, C. . . Die Arachn., v., p. 38, Fig. 360.
1839. *Argiope fenestrinus*, KOCH, C. . . Ibid., p. 155, Fig. 361.
1889. *Argiope argenteola*, McCOOK . . . Amer. Spiders and their Spinningwork.
1889. *Argiope argentata*, McCOOK . . . Ibid., Vol. I., p. 108.
1889. *Argiope argentata*, MARX . . . . Catalogue, p. 541.

FEMALE: Total length, 16 mm.; cephalothorax, 6 mm. long, 4.5 mm. wide; abdomen, 11 mm. long, 8 mm. wide at the middle, 4 mm. at the base, narrowing to 2 mm. and less at the apex. The spider is easily distinguished from its two largest indigenous congeners by the peculiar serrated form of the abdomen, and by the brilliant silvery and beautiful yellow markings on the basal half of the dorsum. The metallic lustre is caused by masses of silvery hairs which closely thatch the cephalothorax and abdomen. This sheen does not show in alcohol, and the specimen requires to be dried in order to bring it out in any degree.

CEPHALOTHORAX: A long oval; corselet rounded at margins, indented at base, flat, with a deep circular fosse; corselet grooves distinct; cephalic suture well marked; ground color blackish brown, with a yellow band along the margin of the corselet and a yellowish median strip both on corselet and caput; the whole is covered thickly with silvery hairs, which give a shining metallic appearance; the caput is quite pointed at the base, flattened in front, and much widened at the face, covered also with silvery hairs; sternum shield shaped, wide across the front, obtusely triangular at the point; the field dark brown, glossy, but covered thickly along the margins with yellowish or silvery white hair; on the anterior portion strong black bristles; raised in the middle and in front; with decided sternal cones. Labium high, subtriangular; maxillæ longer than wide, gibbous.

EYES: Ocular quad on an elevation distinctly depressed in the middle, thus separating the front and rear parts into two eminences on which the eyes are seated; the quad much longer than wide, decidedly wider behind than in front; eyes not greatly different in size, but MF slightly larger, separated by about 1.75 to 2 diameters; MR by 2.5; side eyes on high tubercles; MR much larger than MF, the two barely contingent; MF separated from SF by a little more than their area, or 1.5 the interval of MF; MR from SR by a space much greater than divides MF and SF. Clypeus height 3.5 to 4 diameters MF; front row slightly, rear row much procurved.

LEGS: 1, 2, 4, 3; yellow, with broad apical and median dark brown annuli; legs-I, III in many specimens darker than II, giving indeed the impression of dark brown legs with yellow annuli; freely provided with bristles and long, thin, yellowish spines, and thickly covered with silvery white pubescence; palps yellow, with a touch of brown on the terminal joints, which are tolerably well provided with dark spines and gray bristles; mandibles conical, dark brown, flecked with yellow, glossy and slightly pubescent.

ABDOMEN: Rectangular ovate widest about the middle, narrowing towards the base, which is cut squarely across, and narrowing also towards the apex, which terminates in a pointed tubercle like a caudal part that extends quite beyond the spinnerets. The dorsum rather flat, but the surface much broken; divided into two well marked parts about the middle by a thick armor of silvery hairs, with which the base is clothed; at the narrowed anterior of the base are two short, distinct, rounded tubercles; the line of separation above noted is marked on either side by a similar but smaller tubercle, and there is another in the middle of the dividing line; the apical part is divided along its edge on each side into two rounded projections or lobes, leaving the apex extended posteriorly from the middle as a fifth lobe. The silver armor extends downward along the median line by a narrowing band somewhat rounded before it terminates; on either side of this the space is a yellow and blackish brown folium, through which branching lines extend towards the apex. The venter is beautifully marked by a truncated pyramid of brownish yellow, thickly studded with short blackish bristles, and bordered on both sides and behind by a silvery band, the posterior part being the widest; between this and the spinnerets is a patch of brown flanked on either side by a blackish brown belt, which entirely encompasses the spinnerets, at whose base there is a somewhat broken band of yellow; the sides along either flank of the ventral folium are black, flecked with brown, in the midst of which are many bristles, whose bases are glossy, chitinous, well raised above the surface, and when the shafts are broken off add to the glistening appearance of the surface. The epigynum (Figs. 1c–d) has a marked resemblance to that of A. argyraspis; the atriolum is rather narrow, the scapus rising from its base in a wide yellowish belt, which is indented at the base, and projects at the middle into a scapus, which is depressed in the middle but highly arched, and widened towards the end, which bends under, leaving the base within the atriolum and the portulæ exposed on either side; it thus presents the appearance of the clasp of a padlock when shut down; the color of this scapus is amber.

MALE: I have but one specimen which I regard with little doubt as the male of this species, and which is perfect with the exception of the two hind legs. (Fig. 2b.) The specimen is relatively even smaller than the males of Argiope cophinaria and argyraspis, being 3.7 mm. in body length. The cephalothorax and legs are a uniform yellowish brown; the abdomen has the same silvery white appearance as that of the female, but to a less degree. Longitudinal stripes of yellow pass on either side of the median line or the dorsum. The basal processes are distinctly marked, and along the sides of the abdomen appear what I take to be indications of the lateral projections or lobes. In a specimen in the Marx Collection the lobes show plainly. (Fig. 2.) The face, mandibles, and eyes, in general construction and arrangement, resemble males of A. argyraspis. The eyes have the front row slightly procurved, the rear row decidedly so; SF decidedly smaller than SR and contingent. (Fig. 2c, Plate II., 2b.) The palp has a single strong spinal bristle projecting from the tip of the cubital joint, and the radial joint is well provided with long bristles. The boatlike cymbium is of yellowish color, and strongly covered with bristles. The inlying convolutions (Figs. 2d, 2c) present characteristics something like a combination of the palpal digits of both A. cophinaria and A. argyraspis. A strong black spiral process, bluntly terminated, is a marked characteristic by which it may be easily determined.

DISTRIBUTION: This beautiful species inhabits Southern California, from which I have received numerous specimens (Mr. Orcutt, Mrs. Eigemann); from Magdalena Bay, off the coast of Lower California (Mr. C. H. Townsend). I have also specimens from Florida, and Dr. Marx reports it in his Catalogue from Texas and Arizona. Abbot collected it in Georgia and the Southern States, and has two fine figures in his manuscript drawings. The species is well distributed throughout South America, and inhabits also the West Indies (Costa Rica, Santo Domingo), whence I have several specimens (Mr. William Gabb). It may be regarded as tropical and subtropical in its habitat.

No. 74. Argiope avara THORELL.   Plate XIV, Figs. 1, 1a.

1868. *Argiope avara*, THORELL . . . . . Fregatten Eugenies Resa., Araneæ, Vetensk. Akad. Handling, p. 27.

FEMALE: Total length, 18 mm.; abdomen, 14 mm. long, 13 mm. wide; cephalothorax, 6.5 mm. long, 5.5 mm. wide; face, 2.5 mm. wide.

CEPHALOTHORAX: Corselet rounded at the margin, very flat and thin, though somewhat elevated in the centre; fosse deep, head depressed, and flattish on top; color brown, with yellowish white on the margin, covered with silvery white hairs; sternum dark brown, with a broad median band of bright yellow, narrowing towards the apex, and with radiate points at the corners, presenting the appearance of a pelt of a vertebrate animal; somewhat longer than wide, pointed at the apex, with well developed sternal cones; clothed with strong bristles and hairs. Labium is subtriangular; maxillæ as wide as long, and both brown, with yellowish margins.

EYES: Ocular quad elevated; widest behind, somewhat longer than wide; MF separated by about 1.5 diameter; MR about equal MF in size, and separated at least or more than two diameters; side eyes on strong tubercles; SR much larger than SF, separated by about or less than radius SF; MF are removed from SF by about their alignment; height of clypeus about two diameters MF; front row somewhat the longer, rear row decidedly procurved. The side tubercles have a band of yellow at the base, which passes across the middle of the ocular quad.

LEGS: 1, 2, 4, 3, as follows: 35, 33, 28, 16 mm. Although probably darkened much by the alcohol, appear to be a rich dark brown, with dark annuli at the tips of the joints, which on the metatarsi and tarsi are black; they are not stout, rather thin, indeed, for such a large species; are well clothed with spines, bristles, and hairs, the spines, particularly, grouped underneath the femora and tibia in longitudinal rows, and are long, strong, brownish yellow. Mandibles brown; slightly tapering at the fangs, projecting at the base; palps light yellow, the digital joint well armed with spines and bristles.

ABDOMEN: Almost as wide as long in the specimen in hand, which appears to be a gravid female. In shape a rounded oval, narrowing but little at the base, where it overhangs the cephalothorax, and somewhat more at the apex, which well overhangs the spinnerets. The dorsum beautifully reticulated, the base and front a bright yellow, which extends along the sides, deepening into yellowish brown towards the apex; dorsal field silvery white, with flecks of yellow. The apical half is traversed by lateral bands of dark or blackish brown, extending in semicircles around the sides to the venter; also by longitudinal lines of the same color, one median, with side branches, more or less curved, which cross the lateral bands above described, passing to the apex. The venter has a blackish brown, wide band, girdled on either side by an interrupted ribbon of yellowish white, which encompasses the brown spinnerets at the base, at which point it is broken into spots. Within this ribbon, and symmetrically arranged upon the central band, are four yellowish spots, the anterior ones closer together than the posterior. In the epigynum (Plate XIV., Fig. 1a) the scapus is without a free end, which, like the curved clasp in a padlock, bends over from the posterior to the anterior side of the atriolum, thus dividing the portulæ or seminal chambers. In this respect it resembles the epigynum of A. argyraspis and A. argentata.

DISTRIBUTION: In the United States, Fort Yuma, Arizona. (Marx Collection.) Originally described by Dr. Thorell from a specimen collected on the voyage of the frigate "Eugenies."

No. 75. Argiope Marxii, new species. Plate I, Fig. 5.

1890. *Epeira americana* MARX, *in litt.*

FEMALE: Total length, 17 mm.; cephalothorax, 6 mm. long by 4.5 mm. wide; abdomen, 12 mm. long by 7 mm. wide. This fine large spider has a strong resemblance to A. argyraspis, from which, however, it may be at once distinguished by the oval contour of the abdomen and the absence of the strong annular markings on the legs.

CEPHALOTHORAX: A rounded oval, widest towards the base, which is truncated and shelving, thin, flat upon the summit, the fosse a deep rounded pit; cephalic suture deep; corselet grooves sufficiently distinct; color brownish yellow, with darker bands along the grooves and on the median of the caput, covered scantily with silvery gray hairs. The sternum is shield shaped, somewhat longer than wide, indented at the edges; with sternal cones; rounded in the centre, the color blackish brown and yellow. Labium wide at the base, subtriangular at the tip; maxillæ rounded, nearly as wide as long, and, like the labium, brown, with yellowish tips, and covered, like the sternum, with numerous black bristles.

EYES: Ocular quad on a prominence, most decided in front; rear slightly wider than front, sides decidedly longer; MF separated by about one diameter, MR by 1.5 to 2 diameters, smaller than MF; side eyes on tubercles, slightly separated, about equal in size; MF removed from SF by about 1.3 their area, and from the margin of the clypeus by about 1.5 to 2 diameters; front row recurved, rear row slightly procurved.

LEGS: Long, stout, freely armored with hair and black bristles, the latter arranged in thick rows along the upper surface; spines numerous, long, blackish, arranged in rows underneath the joints; beneath the femora especially are three rows of long spines, with elevated bases, of which the largest is directly in the middle of the joint. The other femora are similarly, but less abundantly, provided, and there are spines also beneath the tibia. The color is yellow (in the alcoholic specimen), with no indication of annuli, but darker bands upon the femora beneath; the palps are of like color and armature; the mandibles are cylindrical, arched at the base, glossy, and slightly pubescent; yellow, with darker shades at the base.

ABDOMEN: Well arched, rounded to the spinnerets, which are situated well under the apex. The dorsum is marked by a broad black median band, varying but little in width from base to apex, but with contractions at two points near the middle; whitish yellow marginal bands mark the outlines, and double rows of whitish spots symmetrically arranged are grouped along the median band on each side the median line. The sides are marked by whitish yellow irregular patches, running longitudinally along the base, but on the sides and towards the apex sloping to the spinnerets. Beneath the abdomen is dark brown, the venter marked by two whitish yellow broken lines, well separated, running from the epigynum to the spinnerets; within these, symmetrically arranged on either side of the median line, are buttonlike rows of whitish yellow spots, unequal in size, and somewhat narrowing toward the spinnerets. The epigynum is without a prolonged scapus, and shows simply a cuplike atriolum, separated into two chambers, within which the opening of the oviducts appear. (Plate I., Fig. 5.)

This beautiful specimen, by the flat form of its cephalothorax and the location of the spinnerets beneath the abdomen, approaches Argiope more closely than the typical Epeira. The eyes, however, more closely resemble those of Epeira, the front row being recurved, as in Epeira, instead of procurved, as usual in Argiope. The rear row also is but slightly procurved, as in Epeira, instead of greatly procurved, as in the typical Argiope. For this reason I place it here rather than create for it a new genus, which might perhaps properly be done, as a connecting link between Argiope and Epeira.

DISTRIBUTION: Fort Yuma, Arizona. (Marx Collection.) Only a single specimen collected.

GENUS CYCLOSA, MENGE, 1876.

In Cyclosa the cephalothorax is cordate or oval, the head arched and. elevated above the corselet. The sternum is wide at the base, somewhat longer than broad. The maxillæ are as in Epeira. The labium is relatively smaller. The eyes are divided into three groups; the front eyes of the central group upon a projecting eminence; the quad longer than wide; the clypeus is moderately high; the sidefront eyes are divided from the midfront by a space greater than that which separates themselves; the front row is decidedly recurved, the rear row aligned, or nearly so. The legs are, in order of length, 1, 2, 4, 3; stout, rather scantily pubescent, the spines being long, thin, and aculeate. The abdomen is an irregular oval, thickest at the apex, marked in some species by shoulder humps, and in others by a conical projection from the dorsal median that has the appearance of a tail. The apical wall is high, the venter conical, and the spinnerets situated well beneath the middle part thereof.

No. 76. Cyclosa turbinata (WALCKENAER). Plate XVII, Figs. 5, 6.

1842. *Epeira turbinata*, WALCKENAER . Ins. Apt., ii., p. 140.
1850. *Epeira caudata*, HENTZ . . . . . J. B. S., vi., p. 23; Sp. U. S., p. 126.
1888. *Cyclosa turbinata*, McCOOK . . . Revised Nomenclature, Proceed. Acad. N. S., Phila.
1889. *Cyclosa caudata*, McCOOK . . . . Amer. Spiders and their Spinningwork, I., II.
1890. *Cyclosa turbinata*, MARX . . . . Catalogue, p. 549.

FEMALE: Total length, 5 mm.; cephalothorax, 1.7 mm. long, middle width 1.2 mm., front width 0.7 mm.; abdomen, 3.4 mm. long, 2 mm. wide.

CEPHALOTHORAX: Corselet a rounded oval; caput and the somewhat higher anterior part of the corselet are roundly arched; cephalic suture deep; the fosse small; sternum flat, longer than broad; maxillæ and labium broader than long; color, as also sternum and mouth parts, dark brown or black.

EYES: Ocular quad somewhat longer than broad, wider in front than behind; MF separated by about 1.3 diameter; MR smaller than MF, and separated by less than their diameter; the distance between MF and SF is a little greater than the area of MF; lateral eyes contingent, on a small tubercle, and are about as large as MR. Front eye row decidedly recurved, rear row slightly recurved or aligned, and longer; the clypeus projects over the mandibles, and is scarcely the width of the diameter of MF.

LEGS: 1, 2=4, 3, as follows: 4.6, 4.1, 4.1, 2.8 mm.; color light yellow, with narrow, dark, apical annuli, somewhat wider on femora; palps colored and marked as legs, and mandibles retreating; somewhat shorter than the patella, and not quite as thick as femur-I.

ABDOMEN: A long ovate, narrowed at the apex into a taillike termination, directed more or less upward, and extending beyond the spinnerets; it is longer than broad, highly arched on the dorsum, which has two shoulder humps. (Fig. 5d.) The field is brown or yellow; the sides are marked with a few blackish stripes; the dorsal pattern somewhat resembles that of C. conica, being an irregular blackish arrow shaped patch, connected by a narrow band to an irregular rectangular apical patch of like color; the caudal part is yellowish, with blackish borders. The venter has a broad black band which surrounds the spinnerets, bearing white spots at the base; the epigynum (Fig. 5e) has a well marked scapus of nearly equal width throughout, but obtusely triangular at the tip; the portulæ are open and rounded, the atriolum arched, and has several small white spots of color on each side.

MALE: Fig. 6, 6a, 6b. Total length, 3 mm.; cephalothorax 1.5 mm. long, middle width 1.1 mm., front width 0.5 mm.; abdomen, 1.5 mm. long, 1.2 mm. broad; cephalothorax a little longer than femur-I, not quite one-third longer than broad, in front not quite half as broad as in the middle; dorsally highly arched. The face at the eye space projects considerably over the clypeus, which is rather high; eyes grouped as in the female; mouth parts and sternum dark brown, the latter sometimes with traces of yellow spots. The legs have distinct spines on all joints; tibia-II has no special thickening or clasping spines; the coxæ are without tubercles; the distal joints are of darker color; the order of length is 1, 2, 4, 3, as follows: 3.9, 3.3, 3.2, 2.1 mm. The abdomen is but little longer than broad, is dorsally highly arched, rounded at both ends, and apparently destitute of the caudal projection which marks the female; spinnerets distal; color brownish yellow or gray, with a broad scalloped longitudinal band, and two white spots on the basal half; the venter black, with two round white spots on each side close to the spinnerets. The palpal digit is marked as at Fig. 6b.

DISTRIBUTION: I have collected this interesting spider from Canada and the Northern New England States, southward along the Atlantic Coast as far as Florida, as well as in Ohio and some of the Western States. Professor Peckham sends it from Wisconsin. Dr. Marx reports it in Louisiana and Alabama; Dr. Blaisdell sends it from California. It is probably distributed throughout the greater portion of the United States. Its habits are fully given in Vols. I., II.

No. 77. Cyclosa conica (PALLAS). Plate XVII, Figs. 3, 4.

1772. Aranea conica, PALLAS . . . . . Spicil. Zool., 1, 9, p. 48, pl. i., 16.
1776. Aranea triquetra, SULZER . . . . Abgeck. Gesch., etc., p. 254, xxx., 3.
1778. Aranea conica, DE GEER . . . . Mem., vii., p. 231, xiii., 16–20.
1805. Epeira conica, WALCKENAER . . . Tab. d. Aran., p. 64.
1837. Epeira conica, WALCKENAER . . . Ins. Apt., ii., p. 138.
1837. Singa conica, KOCH, C. . . . . . Ubers. der Arachn. Syst., i., p. 6.
1842. Singa conica, KOCH, C. . . . . . Archn., xi., p. 145, pl. 392, Figs. 943–945.
1864. Epeira conica, BLACKWALL . . . Spid. G. B. & I., ii., p. 363, xxvii., 261.
1866. Cyclosa conica, MENGE . . . . . Preuss. Spinn., i., p. 74, xii., tab. 18.
1869. Cyrtophora conica, THORELL . . . European Spiders, p. 57.
1870. Cyrtophora conica, THORELL . . . Synonyms, p. 18.
1874. Cyclosa conica, SIMON . . . . . . Arachn. d. France, i., p. 38.
1884. Cyclosa conica, EMERTON[1] . . . . N. E. Ep., p. 321, xxxiv., 3.

[1] Emerton probably refers in his description to this species, at least his drawing shows no shoulder humps; but he erroneously confounds it with C. turbinata.

FEMALE: Total length, 7 mm.; cephalothorax, 2.2 mm. long; width in middle, 1.6 mm.; width in front, 0.9 mm.; abdomen, 5.1 mm. long, 3.6 mm. wide. The species differs from C. turbinata in its genital organ, but may be at once easily distinguished by the absence of the two shoulder humps.

CEPHALOTHORAX: Corselet, mouth parts, and sternum uniformly brown, or blackish brown; caput sometimes a little lighter color; the latter is covered freely, the former sparsely, with gray hairs; dorsally flat; caput not higher than corselet; median fosse small; corselet grooves not deep; cephalic suture distinct. Labium obtusely pointed, and maxillæ broader than long.

EYES: Ocular quad elevated; wider in front than behind, and decidedly longer than wide; MF separated by about 1.5 diameter, larger than MR, which are separated by about or less than one diameter. SF removed from MF by about the area of MF; side eyes not contingent; SF larger than SR, and not much less than MF. Clypeus margin separated from MF by 1.5 to 2 diameters; front row recurved, rear row longer, nearly aligned, or even a little recurved.

LEGS: 1, 2, 4, 3, as follows: 7.6, 6.7, 6.3, 4.5 mm.; color yellow, with decided brown apical and median annuli; armed with hairs, bristles, and long, thin, acute spines. The mandibles retreat backward?

ABDOMEN: A rounded oval; at the apex a more or less conical protuberance, which is directed upwards; color yellowish white, or whitish, with brownish side, and a dark longitudinal folium, which is sometimes indistinct, compressed about the centre, and irregularly indented on the edges. The venter black, with a large angulated long white spot on each side; four white small spots near the base of the spinnerets.

MALE: Total length, 3.6 mm.; cephalothorax, 0.2 mm. long, 1.6 wide in the middle, 0.7 mm. in front; abdomen, 1.8 mm. long, 1.4 mm. wide. The cephalothorax is quite pointed in front, overhanging the clypeus, dorsally flat, without grooves. Mandibles much shorter and thinner than patella, one directed backward. The legs have long spines; tibia-II not thickened, and has no special clasping armature. Coxa-I has a small toothlike tubercle; coxa-IV has two vertical, sharp, and strong spines. On the outside of the maxillæ is a pointed, spinelike tubercle.

DISTRIBUTION: Numbers of this species have been collected in California, where they appear to be abundant, especially in the southern parts. It is found in Utah (Professor Orson Howard), and I have taken it in Florida. Emerton appears to have found it in New England. The facts of distribution, as thus far reported, would seem to indicate the introduction of the species by immigration from Europe.

No. 78. Cyclosa Walckenaerii (KEYSERLING).　　　　Plate XVII, Figs. 1, 1a–d.

1892. *Epeira Walckenaerii*, KEYSERLING . Spinnen Amerikas, Epeiradæ, p. 98, v., 85.

FEMALE: Total length, 6.5 mm.; cephalothorax, 2.5 mm. long, 1.7 mm. wide; abdomen, 4 mm. long, 2.5 mm. wide.

CEPHALOTHORAX: Oval, squarely truncate at the base, high in the middle, the sides rounding to the crest, in which is located the fosse, a deep circular pit; color glossy brown, with grayish pubescence; corselet grooves indistinct; the cephalic suture, on the contrary, is so strongly marked that it presents almost the appearance of a neck, the suture passing entirely around the corselet. (Fig. 1b.) The head is thus quite divided from the central portion of the corselet, the anterior being flattened into the necklike connection above referred to. The head is rounded, somewhat quadrate or bluntly triangular; the vertex is nearly as high as the crest of the corselet, slightly depressed at the face, which is broad; the color is brown, interspersed with gray hairs. The sternum is an elongated shield, longer than wide, the apex prolonged and rounded, the sternal cones strongly marked, rising into high rounded knobs before coxæ–I, with a smaller cone between these two; these are all colored yellow, as is the apex, and a connected median band in the centre of the shield, giving thus a curious lumpy appearance to the organ. (Fig. 1a.) Outside of these cones the color is brown, and there is a slight gray pubescence. The labium is wide, triangular at the tip, which is yellow, the base being somewhat compressed, and brown; maxillæ as wide as or wider than long, the tips obtusely triangular, directed toward one another; colored as the labium.

EYES: The ocular quad has the front eyes upon a semicircular prominence, is higher in front, where it projects over the clypeus; the quad is decidedly longer than wide, wider in front than behind; MF larger than MR, separated by about 1.3 diameter, while MR are almost contingent. The side eyes are on elevated bases rather than tubercles; scarcely contingent; SF larger than SR; MF are separated from SF by about 1.3 their area, or at least twice the intervening space of MF; clypeus height about one diameter MF; front eye row much recurved; the longer rear row slightly procurved or aligned.

LEGS: 1, 4=2, 3; stout, yellow, with blackish brown apical annuli, and lighter median ones; provided freely with bristles and hairs, and rather sparsely with stout black spines; the palps have lighter color than the legs; the mandibles are long, parallel, as wide at the tips as at the base, where they are well arched; somewhat receding; color glossy brown, flecked with yellow.

ABDOMEN: Oval, widest at the base, where it is marked by two conical shoulder humps; the front is rounded, somewhat narrowed, and overhangs the cephalothorax; the apex is marked by a wide conical hump, which is cleft at the top, and has a similar smaller cone on either side; the dorsum is yellow, interspersed with brown, covered with silvery white hairs; an indistinct folium, with a yellowish border inclosing a blackish rectangular pattern, lies between the shoulder humps and the apical cones; the venter is conical, the spinnerets being placed at the apex, underneath the middle part of the abdomen; the color (Fig. 1a) is brown, with a circular ring of yellow spots around the base of the spinnerets; the epigynum (Fig. 1b) presents a wide, semicircular, arched atriolum, yellow, with grayish hairs, from the centre of which issues the scapus, which is wide at the base, narrowed and rounded toward the tips, which slightly overhang the genital cleft. It apparently lacks one moult of maturity.

MALE: Plate XVII., Fig. 2. Smaller than the female. The cephalothorax has the same general characteristics, but the interspace between the head and the corselet is not so strongly marked. The head is well elevated. The abdomen is a rectangular ovate; the dorsal cones smaller than in the female; the color is black, with cretaceous spots at the base and near the middle. The legs are yellow, annulated with a light brown; provided with long, aculeate spines; without any special clasping apparatus upon tibia-II. I cannot positively identify the individual here figured as the male of Cyclosa cervicula, but consider the identity as highly probable.

DISTRIBUTION: Two specimens, female, from California, one of them from Magdalena Bay (Mr. C. H. Townsend); a male, San Diego, California, in the Marx Collection. Keyserling described the species from Guatemala specimens.

No. 79. Cyclosa bifurca McCook.   Plate XVII, Figs. 9, 10.

1887. *Cyrtophora bifurca,* McCook . . . Proceed. Acad. Nat. Sci., Phila., p. 342.
1889. *Cyclosa bifurca,* McCook . . . . Amer. Spiders and their Spinningwork.
1889. *Cyclosa bifurca,* MARX . . . . . Catalogue.

FEMALE: Total length, 7.5 mm.; cephalothorax, 3 mm. long, 2.3 mm. broad; abdomen, 6 mm. long, 3 mm. broad.

CEPHALOTHORAX: The corselet high and well rounded; the fosse deep, the grooves well defined; cephalic suture deep and decided, sharply defining the head from the corselet; the head is on the level of the corselet, and covered with grayish white hairs. The color is green, fading out in alcohol to a light yellow. The sternum is shield shaped, darkish brown along the margin, with irregular greenish median and radiating bands; flattened; covered sparsely with gray hairs, pointed at the apex; almost as wide as long. The labium large, triangular; the maxillæ rounded, about as wide as long; color of both the last named organs green.

EYES: The ocular quad (Fig. 9c) is slightly wider in front, and is longer than wide; MF are separated by 1.7 diameter, and are larger than MR, which are separated by less than a diameter; the space between MF and SF about equals the space between MF; SF are slightly smaller than SR, and the two eyes are scarcely contingent; SR placed well to the side of SF, making with the front row of eyes a continuous curve. The space between MR and SR is at least three times greater than between MR; front eye row recurved; rear row slightly procurved; MF are placed close to the margin of the clypeus, and project over it.

LEGS: 1, 4, 3, 2; the color in some specimens is dark green, with strong blackish brown markings at tips of joints, which embrace half or more of the length of the femora; in others these annular bands are less decided, leaving the legs, like the cephalothorax and abdomen, prominently green; they are rather sparingly armed with long brown spines and with stout short bristles on the two terminal joints. The mandibles are conical, on the same plane with the face, by which they are overhung.

ABDOMEN: A long oval, widest in front; color green, with a lighter band passing from the base to the apex, in some individuals marked on the edges of the dorsum by scalloped bands of light brownish hue. A median line of brown passes along the entire dorsum, with curved radiating lines on either side. The dorsum is rounded at the base, curved downward in the middle, and terminates in two decided points, separated by a deep triangular bifurcation or notch, giving the apex the form of a fish's tail. Four conical tubercles are placed on the base of the dorsum, two at the shoulders and two farther back more widely separated than the former. The venter is a brownish patch, with median spots of green, and a green marginal band between spinnerets and epigynum. The spinnerets are placed nearly midway of the abdomen, caused by the greatly projecting caudal part; they are surrounded by a broad white band, which extends along the venter as far as the epigynum, which is a subtriangular flap, without prolonged scapus. (Fig. 9b.)

MALE: The only male which I have is immature, apparently lacking one moult of maturity. It is about 4.5 mm. in length, and resembles the female. (Fig. 10.)

DISTRIBUTION: I found this species on Merritt's Island, Indian River, Florida, nested in great numbers on the porch of a rustic hotel. Its general habits, snare, and method of preserving its cocoons resemble that of Cyclosa conica and C. caudata.

No. 80. Cyclosa Thorelli, new species.   Plate XIX, Fig. 11.

FEMALE: Not quite mature. Total length, 11 mm.; abdomen, 8 mm. long, 2 wide; cephalothorax, 3.5 mm. long, 1.7 mm. wide.

CEPHALOTHORAX: A long oval, truncated and scalloped at the base, high and peaked; the fossa a deep slit along the summit of the corselet, which decidedly slopes both backward and forward; the head much depressed; almost as wide at the face as at the cephalic suture, which is indistinct. Color dark yellowish brown, with yellowish irregular streakings upon summit of corselet and base of caput. Both corselet and caput sparsely furnished with short, white, bristlelike hairs, which continue around the face. The sternum (Fig. 11b) shield shaped, with scalloped edges pointed at the apex; decidedly longer than wide; the sternal cones well marked; rounded in the middle; color yellowish brown. The labium is triangular; not more than one-third the height of the maxillæ, which are rounded, of a distinct Epeiroid type, but slightly longer than wide.

LEGS: 1, 4, 2, 3; stout; not heavily armored, although in the specimen under description the spines and bristles have been lost in the alcohol. Color yellowish, without annuli; but with traces of longitudinal stripes on some of the joints. The mandibles are conical; quite divergent at the tips, and sloping backward a little toward the sternum.

EYES: Ocular quad (Fig. 10c) longer than wide, and the front slightly longer than the rear; MF are black, and separated by at least 2.5 times their diameter; and are smaller than MR, which are amber color, and separated by about 1.5 their diameter. The quad is placed upon a smooth rounded prominence. The side eyes are distinctly separated, but approximated; SF larger than SR; the latter placed well to the side; the two upon a slight tubercle. SR are widely separated from MR by at least twice the area of the latter; MF are separated from SF by a little more than their area. The front row is slightly recurved; the rear row strongly procurved. The clypeus margin is distant from MF by two or more diameters.

ABDOMEN: Elongated oval, much longer than wide; thickest at the base, where it overhangs the cephalothorax, narrowing at the apex to a caudal part. The spinnerets are placed a little short of the middle point of the venter (Fig. 11a), which is drawn downward somewhat into a subtriangular shape. The color and pattern of the folium are indistinct in the specimen, but the latter appears to be a median ribbon of brown color extending quite from the base to the apex, with a narrow border of yellow on either side. Beyond this again upon the dorsum is a belt of brown, succeeded by a mottled band of yellow, and still lower upon the belly a broader band of blackish brown. The ventral pattern is a long brownish ribbon extending between the spinnerets and the epigynum, bordered by the yellow band already referred to, along the lower part of the sides. The caudal part beneath is mottled with yellow and brown, principally the latter. The specimen appears to be scarcely mature; the epigynum (Fig. 10b) is without scapus, and indicates a subtriangular hood, widest in front, where it is apparently a little notched, with lateral slits on either side opening underneath the hooded atriolum.

DISTRIBUTION: Key West, Florida. Specimen received from Dr. Marx.

GENUS SINGA, C. KOCH, 1837.

This genus is composed almost exclusively of small species, which in general appearance suggest certain species of the Retitelariæ, with which the female especially may easily be confounded. It approximates Epeira in the form of the cephalothorax and mouth organs, and in the general stoutness and brevity of the legs. The cephalothorax is constructed as in Epeira, but the caput is more evidently arched, and not so much depressed toward the face; the surface is less pubescent, and is glossy. The siderear eyes are removed from the midrear usually by a space only a little greater than that which separates between the midrear and the midfront. The midfront eyes are separated from the sidefront by a space which is usually not much greater than, or nearly equal to, the space which separates the midfront eyes. Westring in his diagnosis states that in one species only (S. melanocephala) do the midrear eyes approximate one another more closely than the midfront eyes. But in American species considerable difference exists among the various species in the relation, size, and position of the midfront and midrear eyes. Indeed, the genus may easily be divided into several groups, according as the midrear are larger than or smaller than the midfront eyes, or are arranged in a line shorter or longer. · The clypeus in some species, especially in the living examples of the male, is not much narrower than the space between the midfront and midrear eyes. The mandibles and maxillæ are shaped as in Epeira. The legs are stout, and rather sparingly armed with stubby spines. The palps of the male are short; the digital bulb almost wider than the femur of the front legs; and the cubital is usually clothed with the normal two exceedingly long robust bristles. Coxa-I in the male is usually without any apical denticulation, but there are some exceptions to this. The abdomen is obtusely oval or shortly eliptical; but little pubescent; the skin glossy. Its markings frequently consist of longitudinal fillets of cretaceous or yellowish black color. The snare is like that of Epeira in its general characteristics.

No. 81. Singa Mollybyrnæ, new species.   Plate XIX, Fig. 1.

1880. *Singa Mollybyrnæ*, MARX *in litt.* . Catalogue, p. 549 (KEYSERLING *in litt.*).

FEMALE: Total length, 5 mm.; abdomen, 3.5 mm. long, 3 mm. wide; cephalothorax, 1.8 mm. long, 1.6 mm. wide. This species resembles certain forms of S. variabilis, but is at once distinguished by the ocular quad being wider in front, instead of the reverse, as in S. variabilis.

CEPHALOTHORAX: A rounded oval, almost as broad as long; corselet highest in the centre, sloping sharply behind from the caput to the face; corselet grooves not distinct; cephalic suture marked, thus sharply delineating the caput; fosse, a longitudinal pit; color orange yellow, darker on the centre of the caput to the face, which is blackish. Sternum shield shaped, manifestly longer than wide; sternal cones well marked, slightly pubescent; color orange yellow to yellow, like the surrounding coxæ; labium subtriangular, widest at the base; maxillæ as wide as long; rather obtusely triangular at the tip, and, like the labium, yellow in color, with darker shades at the base.

EYES: Ocular quad on a blackish eminence, wider in front than behind, the sides somewhat longer than front; MF somewhat larger than MR, and separated by a little more than one diameter; MR by not more than a radius; SR on slight tubercles; SF longer than SR, and separated by about a diameter of the latter, the dividing space being greater than that between SR. SF removed from MF by a space but little greater than that which separates the latter; SR removed from MR by a space manifestly greater than that which separates SF and MF. Front row recurved, rear row slightly procurved, or aligned; clypeus height about one diameter MF.

LEGS: 1, 2, 4, 3; stout; yellow, with slight tinge of orange; the tips of the joints annulated with brown; the armature is more decided than with many species of Singa, the bristles passing into light, long, yellowish spines. The palps are colored as the legs; the maxillæ conical, strong, brownish yellow color, as the sternum; curved at the base, but not extending beyond the eyes, which, however, project slightly over the face.

ABDOMEN: A short oval, thickest at the base, and diminishing somewhat toward the apex; the entire dorsal field is a reticulated cretaceous color, with an indistinct marginal band, which is strongest on the front, and which continues along the side. A black irregular line extends along the median, with branches to the dark impressed dots, which are increased to four at and along the apex. The sides are yellowish white, merging into brown, the whole beautifully reticulated. The ventral pattern a quadrilateral patch of yellowish or testaceous color, extending to the spinnerets, which are light brown. The epigynum (Fig. 1c) has a short spooned scapus, very wide at the tip, extending to the margin of the atriolum, leaving the portulæ well displayed on either side.

DISTRIBUTION: Biscayne Bay, Florida; District of Columbia, where it was taken in June by Dr. Marx, to whom I am indebted for the single specimen here described. I have adopted the name suggested by him, which was given in honor of one of his lady friends.

No. 82. Singa Keyserlingi, new species.   Plate XIX, Figs. 2, 2a-c.

1889. *Singa rubella*, MARX . . . . Catalogue, p. 549 (Epeira rubella, HENTZ).
1893. *Singa rubella*, KEYSERLING . . . . Spinn. Amerik. Epeir., p. 234, xiv., 209.

FEMALE: Total length, 5.6 mm.; cephalothorax, 2+ mm. long, 1.5 mm. median width; at the face, nearly 1 mm.; abdomen, 4 mm. long, 2 mm. broad. Dr. Marx in his Catalogue considers this species identical with Hentz's Epeira rubella, Sp. U. S., page 120, pl. 31, Fig. 22; but I find it impossible to agree with this opinion. The eyes are not unlike, and the broad, median, longitudinal band which traverses the abdomen at least suggests the folial stripe of this species. But apart from difference in general form, the characters of the legs, and especially of the maxillæ, in Hentz's drawings are so wholly unlike those of the specimens in hand that they cannot be attributed to the same species. I have, therefore, felt it necessary to propose a new name.

CEPHALOTHORAX: Oval, corselet rather high, sloping behind; base of caput raised above level of corselet, but sloping to the face; fosse a lateral pit; corselet grooves sufficiently distinct; cephalic suture decided, giving the head a marked delineation; color reddish brown; skin glossy, with slight pubescence; the caput a blackish brown to blackish, smooth, glossy, a little pubescence. Sternum shield shaped; somewhat longer than broad, with sternal cones before; cut square in front; a broad, yellow, median band, with a margin of blackish brown; slightly pubescent. Labium triangular, color dark brown, less than half the height of maxillæ; maxillæ as wide as, or wider than, long; squarish at the tip.

EYES: Ocular quad on a rounded prominence, which projects in front, leaving MF on separate tubercles; decidedly wider in front than behind, and length greater than front width. MR separated by about 1.5 diameters, MF by scarcely more than a radius; space between MF and SF but little greater than that between MF; side eyes on tubercles, separated by about a radius; not greatly differing in size, but SF somewhat larger. Front row recurved, rear row, which is decidedly longer, slightly procurved, almost aligned. The space between SR and MR is about twice the diameter of the latter; the clypeus about the height of one diameter MF.

LEGS: 1, 2, 4, 3; stout, short, yellow, with dark shades upon the femora; well provided with yellowish bristles and gray pubescence; sufficiently provided with stout, brownish spines; palps colored and armed as the legs; mandibles blackish brown; conical; parallel; slightly inclined backward.

ABDOMEN: A long oval, not greatly differing in width throughout, but widest at the middle, and in specimens in hand rather narrowed at the base, which much overhangs the cephalothorax. The skin is glossy, covered with a yellowish white pubescence, which, being removed, leaves numerous minute pits. The folium is outlined on either side by a broad brown band, somewhat undulating at the margin; while a reticulated yellow ribbon occupies the median, narrowing both at the front and the apex, on either side of which are symmetrically placed five impressed dots, of which the basal pair are the largest. Both on the base and apex the brown marginal bands deepen into black. On the sides are reticulated ribbons of yellow, like that upon the dorsal median, and below this again mottled bands of brown, which quite encompass the spinnerets and the apex of the abdomen, which projects beyond the spinnerets about quarter the length of the venter. The latter is a broad band of brownish color, with interrupted and reticulated yellowish marginal ribbons, which pass as far as the spinnerets. The epigynum (Fig. 2b) has a short, subrectangular, spooned scapus, which does not extend beyond the apical margin of the atriolum, which presents towards the venter a deeply indented parm.

DISTRIBUTION: St. Louis, Missouri; District of Columbia. (Marx Collection.) I am compelled here to differ with the opinion of Dr. Marx (expressed both in his Catalogue and in the Synonyma of Keyserling's Epeiridæ) that this species is Hentz's Epeira rubella. The thin legs and characters of the cephalothorax and abdomen, as shown in Hentz's drawings, appear to me to be too unlike the spider here described to justify the supposed identity.

No. 83. Singa Listerii, new species.   Plate XIX, Figs. 3, 4.

FEMALE: Total length, 5 mm.; abdomen, 3.5 mm. long, 2 mm. wide. This species closely resembles S. Keyserlingi; but in a large number of specimens from Georgia and North Carolina the differences in the shape and markings of the abdomen and color of the head are persistent. The midfront eyes are smaller than, or at least equal to, the midrear, while in S. Keyserlingi the midfront eyes are the larger. The ocular quad in S. Keyserlingi is relatively wider in front than in S. Listerii, wherein the quad is slightly narrower than, or at least equal to, the midrear area. Moreover, in S. Keyserlingi the side eyes are well separated, while in S. Listerii they are nearly contingent.

CEPHALOTHORAX: Oval; the corselet high, well rounded, cordiform; the fosse a lateral pit; sloping to the base, which is notched; base of caput not lower than summit of corselet; the head slightly elevated in the middle, but depressed at the face; cephalic suture well marked; corselet grooves distinct; color glossy; slightly pubescent; yellow, with dark streaks along the corselet grooves and caput; sternum shield shaped, slightly longer than broad; sternal cones distinct before all the coxæ; dark brown, with a lighter yellowish band in the centre, where it is elevated; labium subtriangular, color of sternum, not half the height of the maxillæ, which are colored as the sternum, as wide as long.

EYES: Fig. 3b. Ocular quad on a rounded eminence, the centre of which is blackish; the width in front slightly narrower than behind, or at least not wider; the sides of greatest length; MF somewhat smaller than MR; MF separated by about or a little more than a diameter; MR separated a little less than a diameter; side eyes on tubercles, about equal in size; almost contingent; SF removed from MF by about the area of the latter, the intervening space being about 1.5 MF; front row recurved, longer rear row slightly procurved; clypeus height about two diameters MF; the front eyes project somewhat beyond the bases of the mandibles.

Legs: 1, 2, 4, 3; stout, of uniform yellow color, well provided with yellowish hairs, and a few short, feeble, dark spines, with a few stout bristles; the palps are colored as the legs; the mandibles conical, yellow, thick at the bases, and slightly retreating towards the sternum.

Abdomen: Oval, considerably longer than broad; slightly thicker across the shoulders, and diminishing but little towards the apex, which is high above the spinnerets; the latter placed immediately beneath the apical wall; the base extends well over the corselet. The dorsum has a broad, median, yellow ribbon, in the centre of which runs longitudinally, the entire length, an irregular thread of brownish hue; on either side are longitudinal bands of brown extending to the base in front, and quite to the spinnerets posteriorly. These are followed on each side by a band of yellow similar to those on the dorsum, and this again lower down upon the sides by a broader longitudinal band of brown, which extends to the venter. These markings give to the dorsum the appearance of being longitudinally striped. The skin is glossy, pubescent, and reticulated over the entire surface. The original color of the specimens is lost in the alcohol, but appears to be brown and yellow. The venter has an oblong band of yellowish brown, with a bright marginal ribbon of yellow on either side, the latter surrounding the dark spinnerets in interrupted lines. The epigynum (Fig. 3a) has a rather wide atriolum, with a short, rounded, subtriangular scapus extending a little beyond the margins of the portulæ, which are slits under the atriolum.

The male (Fig. 4) is somewhat smaller than the female, being 4 mm. in length. In marking and color it quite closely resembles the female, but the hues appear to be somewhat lighter. The palpal digits are distinguished as in Fig. 4a.

Distribution: Georgia and North Carolina. Specimens presented by Mr. Thomas Gentry.

No. 84. Singa nigripes Keyserling.   Plate XIX, Figs. 5, 6.

1863. *Singa nigripes*, Keyserling . . . Neue Spinn. Amer., v., p. 655, pl. 21., Fig. 7.
1889. *Singa nigripes*, Marx . . . . . Catalogue, p. 549.
1893. *Singa nigripes*, Keyserling . . . Spinnen. Amerik. Epeir., p. 290, xv., 214.

Female: Total length, 3.5 mm.; abdomen, 2+ mm. long, 1.75 mm. wide. The general colors are orange and yellow, and on the abdomen a light reddish yellow or ash color, except on the legs, where the last three members in most species are black or blackish. This peculiarity of the feet casually marks the species, in connection with the black spots on the abdominal dorsum.

Cephalothorax: A well rounded oval, elevated in the centre, sharply sloping behind the caput; prominent, squarish, slightly pubescent; corselet and head a uniform orange, with a slightly darker shade on the marked cephalic suture; the sternum heart shaped, almost triangular, wide in front, much rounded at the apex; sternal cones distinct, flattened in the middle, very slightly pubescent; color orange yellow, uniform with the surrounding coxæ and the mouth parts; the width in front almost equal to the length. Labium triangular; maxillæ as broad as long, cut square at the tips.

Eyes: Ocular quad on a rounded eminence, whose interior is blackish, thus sharply contrasted with the orange color of the face; front somewhat narrower than rear, the sides somewhat longest; MF slightly smaller than MR, separated by about one diameter, MR by a less space. MF from SF by about their area, the intervening space about that between MF and MR. Side eyes on a slight tubercle; contiguous; SR separated from MR by a space about equal to the area of the latter; clypeus high, the margin separated from MF by at least three or more diameters. Front eye row recurved, rear row slightly procurved, nearly aligned. The caput somewhat contracted at the face, which projects over the mandibles.

Legs: 1, 2, 4, 3; sufficiently pubescent upon the apical joints, which are blackish, or in some specimens a darkish yellow; without spines, but with long yellowish bristles; palps colored and armored as legs; mandibles uniform in color with the head, conical, strong, slightly retreating, and overhung by the frontal eye space.

Abdomen: An even oval; dorsum arched; color ashen gray; folium a light square of reticulated yellow or cretaceous, occupying the middle of the dorsum, marked at the corners by four black spots, of which the two basal are more widely separated. In some specimens an indistinct, interrupted double line of yellow passes along the middle of this folium, and shapes itself upon the front in an indistinct lance headed figure. The impressed spots are prominent upon the dorsum; the skin is covered with soft, short pubescence. The venter, like the sides, is a light ashen color, the spinnerets, which are just beneath the apical wall, being somewhat lighter in color. The epigynum (Fig. 5b) has a very slight, short scapus, of uniform length.

Male: Somewhat smaller than the female, 2.5 mm. long, and in general color and markings resembles it. Cephalothorax bright orange; the head somewhat more pointed at the face than the female; the legs without any special clasping apparatus, but with bristle-like spines, more numerous and longer; the palps (Fig. 6a) rather stout, joints orange color, except the digital, which is greatly distended, and colored as the feet.

Distribution: Specimens have been taken as far southeast as Florida, southwest in Texas, and as far to the northwest as Nebraska. (Marx Collection.) This would indicate a wide distribution over the United States.

No. 85. Singa variabilis Emerton. Plate XX, Figs. 11, 12, 13; Pl. XIX, Fig. 7.

1884. *Singa variabilis*, Emerton . . . . N. E. Ep., p. 332, xxxiv., 16.
1889. *Singa variabilis*, Marx . . . . . . Catalogue, p. 550.

Female: Total length, 4 mm.; abdomen, 2.75 mm. long by 1.75 mm. wide; cephalothorax, 1.5 mm. long by 1+ mm. wide. The individuals of this species, as may be observed from the drawings on the plate, differ so greatly in color that it is hard to determine the typical colors. The specimen of which description is made (Fig. 12) is, in front, yellowish brown; and for the abdomen a drab or cretaceous hue for the field, with longitudinal markings of silvery white. Some of the specimens (Fig. 11) have the abdomen a uniform glossy black. These changes of color do not depend upon moulting stages, but appear in adult individuals of both sexes.

Cephalothorax: A rounded oval, high in the centre, from which the corselet slopes sharply backward; the fosse overhung by the abdomen; the head rounded and erect above the corselet; cephalic suture strongly marked; corselet grooves indistinct; color yellow, mottled with splotches of brown; sternum shield shape, glossy black; sternal cones distinct; flat in the centre, which is raised; very little pubescence. Labium triangular, wide at the base, more than half as high as the maxillæ, which are rather rounded at the tip, as broad as long; color, blackish brown at the bases, and lighter brown at the tips.

Eyes: The entire eye space glossy black; the ocular quad on a high eminence, narrower in front than behind; longest at the sides; MF somewhat smaller than MR, and separated by about 1.5 diameters, MR by about one diameter. Side eyes on tubercles; subequal, but MF apparently somewhat smaller than MR, divided by about or less than a radius; separated from MF by a little more than the space which divides the latter; the distance between SR and MR is greater than that between SF and MF; front row recurved, rear row procurved; clypeus rather high, measured from MF to the margin, but the black central eminence extends very close to the margin.

Legs: 1, 2, 4, 3; short, stout, yellowish color, in harmony with the corselet; abundant yellowish hairs, which are blackish upon the feet, and a few bristlelike spines; the palps colored and armed as the legs; the mandibles stout, parallel, and not retreating backward.

Abdomen: An oval, longer than broad; the dorsum arched; the base overhanging the cephalothorax; the spinnerets placed beneath the apical wall; color greatly varying, from silvery white, drab, and yellow to black, with brown longitudinal markings, or blackish brown. The ventral pattern is a broad blackish stripe between the dark spinnerets and the epigynum, which is yellow. This organ (Fig. 11a) has a short scapus of nearly equal length throughout, but somewhat widened at the tip, which is directed downward, in some specimens so decidedly as to touch and seem to unite with the margin of the portulæ.

MALE: Fig. 13. Smaller than the female, being 2.5 mm. long by 1+ mm. wide; the cephalothorax is somewhat wider; the head more pointed and less prominent. The legs are longer, relatively, and thinner, without any special clasping apparatus or thickening of tibia-II, but with some longer spinous bristles. The sternum in the specimen under description is orange yellow, corresponding with the coloring of the surrounding coxæ, instead of jet black, as in the female. The palps are stout, and long for so small a species.

DISTRIBUTION: Throughout New England, in Massachusetts, Maine, Connecticut; in the District of Columbia, and as far to the northwest as Hill City, Southern Dakota. (Marx.) These widely separated points of collection indicate a wide distribution.

No. 86. Singa maura (HENTZ).   Plate XIX, Figs. 8, 8a-d.

1847. *Epeira maura*, HENTZ . . . . . . J. S. B., v., p. 474; Sp. U. S., p. 114, xiii., 8.
1889. *Singa maura*, MARX . . . . Catalogue, p. 549.
1893. *Singa maura*, KEYSERLING . . . . Spinn. Amerikas, Epeir., p. 283, xiv., 208.

FEMALE: Total length, 5.5 mm.; abdomen, 4 mm. long, 3 mm. wide; cephalothorax, 2.5 mm. long, 2 mm. wide. General colors of the fore part of the body are a rich brown to yellow, and of the abdomen black, with folium marks of yellow.

CEPHALOTHORAX: A well rounded oval, high; the fosse overhung by the abdomen; the base sharply sloping; corselet grooves deeply marked; cephalic suture distinct; head prominent; the base of the caput as high as the corselet; color a uniform glossy orange brown; slightly pubescent. Sternum shield shaped, slightly longer than wide; sternal cones distinct, but not prominent; slightly raised, but flat in the centre; color a uniform orange brown, in harmony with the legs and mouth organs. Labium subtriangular, not half as high as the maxillæ, which are as broad as long, cut squarely at the tips.

EYES: Ocular quad on a rounded, blackish eminence, especially prominent in front, which makes MF project beyond the mandibles; the front decidedly wider than rear, the side longest; MF manifestly larger than MR, and separated by 1.5 to 2 diameters, while MR are separated by about a radius. Side eyes on separate tubercles, SF most pronounced, and with dark brown base; separated from each other by at least, or more than, a diameter; about equal in size. SF removed from MF by a space not greatly different from that between MF. Front eye row decidedly recurved, the longer rear row aligned, but appears slightly procurved when viewed from in front and beneath on the same line with the front row; the clypeus height is about 1.5 diameter MF.

LEGS: 1, 2, 4, 3; short, stout, uniform orange brown, except tips of feet, which are black; well covered with yellowish hairs, and a few slight yellowish spines or spinous bristles. The palps are armed and colored as the legs, with black tips heavily pubescent; the mandibles conical, strong, colored as the corselet, slightly retreating towards the sternum.

ABDOMEN: A well rounded oval, whose width differs little throughout; the skin is glossy, covered sparingly with soft yellowish hairs; the dorsal folium marked in some species by a whitish or yellow spot upon the base, followed toward the apex by two similar and smaller patches, and still further by a transverse patch, or lunette. In other specimens this is yellow, reticulated, and the terminal patches united by an irregular median band. On either side are interrupted patches of yellow or cretaceous color. The field of the dorsum is glossy, blackish brown, lighter in some specimens. The rounded apex somewhat overhangs the spinnerets. The venter has a glossy, brownish rectangle, outlined on either side by patches of yellow. The epigynum (Fig. 8b) is without an elongated scapus, the median projection of the atriolum being simply marked by a brown chitinous ridge, semicircular in shape, with the concavity directed forward.

DISTRIBUTION: Hentz found the species in Alabama, Marx in the District of Columbia. (Marx Collection.)

No. 87. Singa maculata (EMERTON). Plate XIX, Figs. 9, 9a, 9b.

1884. *Singa maculata*, EMERTON . . . . N. E. Ep., p. 323, xxxvii., 18.
1889. *Singa maculata*, MARX . . . . . Catalogue Described Araneæ, p. 549.
1893. *Singa maculata*, KEYSERLING . . Spinn. Amerik., Epeir., p. 285, xiv., 210.

FEMALE: Total length, 3+ mm.; abdomen, 2 mm. long, 1.5 mm. wide; cephalothorax, 1.4 mm. long, 0.6 mm. wide. This is a small species, and although Emerton's description, as usual, is extremely insufficient, and he gives only a figure of the epigynum, the drawings as above and description as below probably point to the species above named by him.

CEPHALOTHORAX: Oval, with corselet margins well rounded; about one-fifth longer than broad; in front more than half as wide as in the middle; corselet grooves tolerably distinct; median fossa rather flat; cephalic suture well marked; the head prominent, somewhat depressed at the face, where it projects beyond the mandibles. Color of corselet orange brown, with rather lighter shade upon base of caput, which deepens into black or blackish around the ocular area. Sternum shield shaped, about or almost as broad as long; orange brown color; the labium and palps about as in S. Keyserlingi, brownish yellow in color.

EYES: Ocular quad on a rounded eminence; narrower in front than behind; length about equal to, scarcely greater than, rear width; MR decidedly larger than MF; MF separated by about two diameters, and by the same or a little greater distance from MR, which are separated by about 1.5 of their diameter; side eyes amber color, scarcely contingent, about equal in size, upon a slight tubercle; SR apparently larger than SF; MF separated from SF by about their area, the intervening space being somewhat greater than between SF and MF; clypeus high, equal to about two or more diameters MF. The forehead, viewed from the front, is high, and semicircular in outline. The front eye row is slightly recurved, the rear row procurved.

LEGS: Short, stout, brownish or orange yellow; sparsely pubescent; without annuli; with a few stout bristles, or bristlelike spines. Palps as the legs; mandibles slightly inclined backward; orange brown.

ABDOMEN: An oval, longer than broad; dorsum arched, and apex as thick as the base; the spinnerets distal, or placed immediately beneath the rounded apical wall; skin glossy, and rather inclined to silvery in alcoholic specimens; covered with grayish pubescence. The dorsum is marked by a broad folium, with blackish outlines, limited by a dark transverse band at the rounded part of the apex; four prominent impressed spots mark the central portion of the dorsum. The sides, along the base, are marked by a little longitudinal patch of yellow or cretaceous. At each shoulder of the dorsum is a circular spot of black, almost resembling an eye; the folium is cut off squarely at the apical part of the dorsum, leaving a blackish, semicircular, lateral band, beyond which is a reticulated narrow band of yellow. The ventral pattern is a black quadrilateral, bordered on all sides by a band of yellow, including the spinnerets, which are of like color. The epigynum (Fig 9b) in the specimen in my possession is much damaged, and I give the organ as figured by Emerton.

DISTRIBUTION: New England (Emerton); District of Columbia. (Marx Collection.)

GENUS CYRTOPHORA, SIMON, 1864.

The chief distinguishing characteristics of Cyrtophora are the eyes and abdomen. The eight eyes are arranged in three distinct groups, as in Epeira, but the side eyes, instead of being contingent or approximated, are, relatively, widely separated, thus resembling some of the species of Tetragnatha. The space between the sidefront and midfront eyes is but little, if any, greater than that between the two midfront eyes; the eye rows are both slightly recurved. The cephalothorax is short and rather feeble, the head, relatively, long and powerful. The sternum is triangular, the labium low, the maxillæ wide at the tips and compressed at the shank, and somewhat longer than wide. The legs are in order of length 1, 2, 4, 3, and, for so small a species, quite stout, but rather scantily armored. The abdomen is conical, tuberculated, and carried in a position nearly perpendicular to the plane of the cephalothorax.

No. 88. Cyrtophora tuberculata KEYSERLING.   Plate XVII, Fig. 11.

1889. *Cyrtophora tuberculata*, MARX *in litt.* . Catalogue, p. 549. (KEYSERLING *in litt.*)
1893. *Cyrtophora tuberculata*, KEYSERLING . Spinn. Amerikas, Epeir., p. 265, xiv., 197.

FEMALE: Total length, 3 mm.; cephalothorax, 1.4 mm. long, middle width 1 mm., front width 0.7 mm.; abdomen, 1.9 mm. long, 1.4 mm. broad.

CEPHALOTHORAX: One-third longer than broad, in front two-thirds as broad as in the middle; corselet grooves deep; median fosse feeble or wanting; sternum triangular, a little longer than broad. The labium rounded at the tip, low, being twice as broad as long, and not quite half as long as the maxillæ, which are as long as, or slightly longer, than broad, in front a little broader and truncate. Color of corselet yellow; sternum and mouth parts spotted brown.

EYES: Ocular quad longer than broad, and narrow behind (Fig. 11c); MF separated by about one diameter, and a little further removed from MR, which are smaller and contingent; MF are distant from SF by little more than the space between themselves. The side eyes are as widely separated as MF, more widely separated than MR, and about as far apart as SF and MF; they are on separate elevations. The clypeus is low, about as high as half the diameter of MF. Both eye rows are about equally and moderately recurved, and differ little in length, the rear row being slightly longer.

LEGS: 1, 2, 4, 3, as follows: 5.2, 4.4, 3.3, and 2.4 mm. They are sparsely pubescent, with a few isolated spines on the upper sides of the joints; color yellow, with brown annuli. The palps are similarly marked. The mandibles are longer than the patella, and as thick as femur–I.

ABDOMEN: One-fourth longer than broad, rounded at the base, attenuated towards the spinnerets, giving the dorsum a shieldlike shape. Two roundish tubercles are situated upon the shoulders, two smaller ones, more closely approximated, near the middle region on either side of the median band. (Fig. 11b.) The dorsum is arched; reticulated; the color white, but marked at the base, apex, and sides with blackish spots. The dorsal pattern is a rather broad, irregular, scalloped, whitish band passing down the median line, widened into a narrow shape on the apical part, beyond the posterior tubercles. On either margin of this median band are broad irregular margins of black or blackish. The venter is dark, with a white spot in the middle; the epigynum (Fig. 11a) a flaplike atriolum, without a scapus.

DISTRIBUTION: Florida. The description has been made from two female specimens in the collection of Dr. George Marx.

GENUS ZILLA, C. KOCH, 1837.

In Zilla the cephalothorax is constructed substantially as in Epeira, rather small and weak, but is little pubescent; it is moderately convex, the cephalic part, in females at least, sufficiently large. The siderear eyes are separated from the midrear eyes by a space not greater, or, in the male, but little greater, than that between the midfront and midrear eyes. The midfront eyes are removed from the sidefront by a space not greater than that which divides themselves. The midfront eyes are a little larger than the midrear as a rule. The rear row of eyes (seen from above) is nearly aligned or slightly recurved. The face does not project beyond the margin of the mandibles. The sternum, mandibles, and maxillæ are formed as in Epeira. The palps in the male are tolerably long, the humeral joint being either equal to or but little longer than the patella of the first pair of legs. The legs, unlike Epeira, are provided with few spines, but are sufficiently armored with hairs and bristles. The femora about the middle have only one to two spines, and, as in many species of Epeira, about the apex itself are two to three spines. Abdomen is a short oval, subelliptical, depressed, sufficiently pubescent, and with a slight silvery lustre. The dorsal area is marked, as in many Epeira, with a folium included within a sinuate margin. The snare is a vertical orb, with an open sector above the median, by which the viscid spirals are looped backward and forward across the radii. (See Vol. I., page 140.) Through the open sector a trapline connects the centre of the web with a bell shaped nest, within which the spider dwells. This trapline and nest are placed above the horizontal centre of the orb, but indifferently to the right or left, or directly above the same.

No. 89. Zilla x-notata (CLERCK).　　　　　　　Plate XVIII, Figs. 1, 2.

1757. *Araneus x-notatus*, CLERCK . . . . Svenska Spindl., p. 46, pl. 2, tab. 5.
1757. *Aranea x-notata*, CLERCK . . . . Ibid., p. 154.
1789. *Aranea literata*, OLLIVIER . . . . Encyclopédie Methodique, iv., p. 206.
1802. *Aranea calophylla*, WALCKENAER . Faune Par., ii., p. 200. (In part.)
1805. *Epeira calophylla*, WALCKENAER . Tabl. d. Aran., p. 62. (In part.)
1832. *Epeira calophylla*, SUNDEVALL . . Svenska Spindlarness.
1834. *Zygia calophylla*, KOCH, C. L. . . Herr.-Schaeff., Deutschl. Ins., 123.
1839. *Zilla calophylla*, KOCH, C. L. . . Die Arachn., vi., p. 148, Taf. ccxvi., 538, 539.
1844. *Epeira similis*, BLACKWALL . . . Ann. & Mag. Nat. Hist., xiii., 186.
1858. *Zilla x-notata*, THORELL . . . . . Om Clercks Original Spindel-samling, p. 146.
1861. *Zilla x-notata*, WESTRING . . . . Araneæ Sveciæ, p. 71.
1864. *Epeira similis*, BLACKWALL. . . . Sp. G. B. & I., ii., 337, xxv., 244.
1884. *Zilla x-notata*, EMERTON . . . . . N. E. Ep., p. 324, pls. 34, 37, 40.
1889. *Zilla x-notata*, McCOOK . . . . . Amer. Spiders and their Spinningwork.
1889. *Zilla x-notata*, MARX . . . . . . Catalogue Described Araneæ.

FEMALE: Total length, 8 mm.; abdomen, 6 mm. long by 5 mm. wide; cephalothorax, 4 mm. long by 2.5 mm. wide.

CEPHALOTHORAX: A shortened oval; the fosse a longitudinal depression; corselet grooves sufficiently distinct; cephalic suture distinct; caput slightly depressed; color pale yellow, with a brownish patch on the summit; glossy, slightly pubescent. The sternum longer than broad, rounded at the apex; shield shaped; a wide, rounded, yellow band in the middle; sternal cones distinct. Labium subtriangular; yellow, as are also the maxillæ, which are subtriangular at the tip.

EYES: Ocular quad slightly wider behind; a little longer at the sides; MF separated by a little more than one diameter; MR, which are smaller, by about 1.5 diameter. Side eyes barely contingent; SR somewhat the larger, but not greatly differing in size; removed from MF by a space about equal to that between MF. Front row recurved, rear row longer and procurved; clypeus height about one diameter MF.

LEGS: 1, 2, 4, 3; rather stout; sparsely armored with spines; color pale yellow, with slight annuli at the joints, and median indistinct annuli on the femora beneath.

ABDOMEN: Oval; of nearly equal width throughout; the dorsum scarcely, or but little, arched; color yellow; the folium outlined with black, the edges being scalloped; skin with a subdued sheen. The epigynum is without manifest scapus, and consists of a hood shaped atriolum, penetrated by portulæ on either side; color dark brown.

MALE: Fig. 2a. Length, 6.5 mm. Resembles the female in markings. The legs are not so stout, relatively; tibia-II without any special clasping apparatus; not thickened at the apex. Color yellowish, with brown annuli at the tips of the joints, and median annuli on the femora underneath. The palps (Fig. 2a) are yellow, without annuli; the palpal digit rather pointed, and the joints comparatively short. The male Z. x-notata is easily distinguished from the male Z. atrica by the brevity of the palps, and by the absence of the row of ten long vertical hairs on the external side of the front metatarsi. Neither the palps nor the feet are so distinctly tinged with rufous-brown color; the hairs on radial joints of palps are less thick, and are not of equal length; some are bent downward. In Z. atrica all the hairs are vertical and equally long, and the joint itself is very long (8a), four times longer than wide.

DISTRIBUTION: New England, New York, California. The species has not been reported in the interior States, and may have been introduced to the Atlantic and Pacific Coasts by commercial intercommunication or by emigration. It is widely distributed throughout the continent of Europe, where I have collected it as far north as the Highlands of Scotland. It is one of the oldest and best known of the Orbweaving species.

No. 90. Zilla atrica C. Koch.   Plate XVIII, Figs. 7, 8.

1802. *Aranea calophylla*, Walckenaer . Faune Par., ii., p. 200. (In part.)
1805. *Epeira calophylla*, Walckenaer . Tabl. d. Aran., p. 62. (In part.)
1834. *Zygia calophylla*, Koch, C. . . . Herr.-Schaeff, Deutschl. Ins., 123, 17. (In part.)
1844. *Eucharia atrica*, Koch, C. . . . . Die Arachn., xii., p. 103, pl. 419, Figs. 1030, 1031.
1851. *Epeira atrica*, Westring . . . . Forteckning, etc., p. 35
1856. *Zilla atrica*, Thorell . . . . . . Recensio Critica, p. 107.
1861. *Zilla atrica*, Westring . . . . . Araneæ Svecicæ, p. 69.
1864. *Epeira calophylla*, Blackwall . . Sp. G. B. & I., ii., p. 338, xxv., 245.
1866. *Zygia atrica*, Menge . . . . . . Preuss. Spinn., i., p. 78, pl. 12, tab. 20.
1867. *Zygia calophylla*, Ohlert . . . . Aran. d. Prov. Preuss., p. 30.
1889. *Zilla atrica*, McCook . . . . . . Amer. Spiders and their Spinningwork.
1889. *Zilla atrica*, Marx . . . . Catalogue, p. 550.

FEMALE: Total length, 8 mm.; abdomen, 5.5 mm. long, 3.5 mm. wide across the base; cephalothorax, 4 mm. long, 2.5 mm. wide; the face, 1.5 mm. The general colors of the fore part are yellowish brown, and of the abdomen yellow, white, and blackish.

CEPHALOTHORAX: A long oval; black on the summit; the sides high; fosse a longitudinal slit; corselet grooves not distinct; cephalic suture distinct; color brownish yellow, lighter on top, with a patch blackish at the caput base, and a dark band passing thence to the eye space; slightly pubescent, especially around the caput base; skin glossy along lower margin of corselet; a row of bristles curved along the front. Sternum longer than wide, shield shaped; glossy at the margin, with a broad yellow median band; sternal cones distinct, indented on the edges, raised in the centre. Labium triangular; maxillæ as broad as long; subtriangular at tip; light brown, shading to yellow at tip.

EYES: Ocular quad slightly wider behind than in front, and the side somewhat longer. Eyes about equal in size. MF separated by about one diameter; MR by about 1.5 diameter. Side eyes on tubercles; contingent; about equal in size, and not greatly different in size from MR. SR removed from MR by a space but little greater than the distance between MR. The space between SR and MR at least twice the distance between the latter. The front eye row recurved; the rear row longer, and procurved. The clypeus is about the height of one diameter MF.

LEGS: 1, 2, 4, 3; stout, particularly the femora; yellow, with dark brown annuli at tips of joints, and median annuli on femora and tibia, especially underneath; spines short, black, few in number; the joints not heavily but sufficiently clothed with yellow hair and bristles. Mandibles uniform brown, slightly pubescent, extending almost directly from the face, and sharply bent downward; conical; quite thick at the apex; parallel.

ABDOMEN: Oval, but slightly diminishing toward the apex, where it is nearly as thick as at the base. The folium is a broad, oval figure, with undulating margins of black, mottled with yellow, there being about five prominent scallops thereto; it narrows to the apex; the central part is yellow, with branching longitudinal lines from the median point backward. The dorsum is reticulated, the sides well covered with hairs, the whole having a slightly glistening appearance in life. The venter has an irregular rectangular patch of brownish yellow, well covered with yellowish hairs, and a margin reticulated on either side. The epigynum (Fig. 7a) is dark brown, glossy in color, lacking the prominent scapus characteristic of most species of Epeira. The figure on the plate is not a very good representation thereof.

THE MALE: Figs. 8, 8a. Is smaller than the female, being 6 mm. long, and resembles it in general color and markings, although the legs are less decidedly annulated. These members are long, provided with comparatively few spines, with no special clasping spines upon tibia-II; color uniform yellowish brown. The palps (Fig. 8a) are long and strong, and thus are an admirable substitute for the tibial clasping spurs.

Zilla atrica is distinguished from Z. x-notata, which it strongly resembles, (1) by the reddish brown color of the mandibles; (2) it does not have, in the living species, the shining black, and, after death, the blackish, color of the latter; (3) the black bases of the spines and hairs on the feet are less strongly expressed, and (4) the fuscous annuli on the femora and the apices of joints are less deeply colored, so that the feet are rather a pale testaceous; the male is easily distinguished by the remarkable length of the palps. From Z. montana, Z. atrica differs by its greater size; the palps of the two males are differently constructed, and there is also a marked difference in the sexual organs of the females.

DISTRIBUTION: The only specimens of this species in my collection are from Wood's Holl and Annisquam, Massachusetts. Dr. Marx reports it as received from California. It is a well distributed European species, and has probably been imported into the United States.

No. 91. Zilla montana C. KOCH.   Plate XVIII, Figs. 3, 4.

1832. *Epeira calophylla*, SUNDEVALL . . Svenska Spindlarnes Beskrifning, p. 253. (Variety b.)
1834. *Zilla montana*, KOCH, C. . . . . Herr-Schaeff. Deutschl. Ins., 125, 19.
1839. *Zilla montana*, KOCH, C. . . . . Die Arach., vi., p. 146, Figs. 536, 537.
1851. *Epeira montana*, WESTRING . . . Forteckn., etc., p. 35.
1856. *Zilla x-notata*, THORELL . . . . . Recensio Critica, p. 26.
1858. *Zilla montana*, THORELL . . . . . Om Clercks Origin. Spindelsn., p. 148.
1861. *Zilla montana*, WESTRING . . . . Araneæ Svecicæ, p. 75.
1867. *Zilla calophylla*, MENGE . . . . . Preuss. Spinn., i., p. 76, pl. 12, tab. 19.
1870. *Zilla Stroemii*, THORELL . . . . . On European Spiders, p. 235.
1884. *Zilla montana*, EMERTON . . . . . N. E. Ep., p. 323, pls. 34, 37.
1889. *Zilla montana*, McCOOK . . . . . Amer. Spiders and their Spinningwork.
1889. *Zilla montana*, MARX . . . . . . Catalogue Described Araneæ.

FEMALE: Total length, 6.5 mm.; abdomen, 4 mm. long, 3 mm. wide. It resembles Z. x-notata and Z. atrica, but the legs are more strongly annulated, the median annuli being distinctly dark brown upon yellow.

CEPHALOTHORAX: Strongly marked with brown; corselet grooves and cephalic suture distinct; the head not depressed; the eyes as in Z. x-notata; MF distinctly larger than MR. The clypeus has a height of at least one diameter MF.

LEGS: Abundantly armed with long, yellowish bristles, which stand in rows on the femora, both above and beneath; the spines are not numerous. The venter has an elongated blackish band, flanked on either side by a triangular belt of yellow, which extends to the spinnerets. The epigynum is without a scapus, dark brown or blackish, corneous, and is simply an opening underneath the atriolum.

ABDOMEN: A short oval; color shining cretaceous or yellowish, with a folium whose margins are marked black within a wide median figure of shining cretaceous. The folium is broad at the base and narrowest towards the apex and the front.

MALE: The male does not differ greatly from the female in size; is lighter in color; the legs not so distinctly annulated. The palp is much shorter than that of Z. x-notata. The digital bulb (4a) is prominent, terminated by a strong hook, with a long black fang; the cubital joint short, subtriangular; the humeral joint but little longer than the cubital; the radial joint the longest of all. The abdomen is marked as in the female, the color shining cretaceous, and the folium outlined in black; the sides reticulated, cretaceous next the folium, and beyond that a belt of black. The legs are well provided with long yellowish spines, with dark bases, which, however, are almost wanting on the femora. Tibia–II has no special clasping organs.

DISTRIBUTION: This is also a European species, and is widely distributed from Sweden southward. In this country it has been taken in New York and New England.

GENUS ABBOTIA, new.

With some hesitation I have separated Epeira gibberosa of Hentz and E. maculata of Keyserling from Epeira, and established for them a new genus, in recognition of Mr. John Abbot, the earliest worker in the field of American Araneology. The whole general appearance of these spiders is different from that of the typical Epeira. In their more dainty and delicate formation and colors, the shape of the cephalothorax and abdomen, and, in part, the structure and armature of the legs, they at once strike the eye as quite different. A special difference is noted in the structure of the **cephalothorax**, which in Epeira is rather low, of almost equal thickness throughout, rising gradually from the margin of the corselet to the fossa. In Abbotia, on the contrary, the corselet, as viewed from the side, is a truncated cone in shape (Plate XX., Fig. 17), rising rather abruptly from the margin to the crest, which is comparatively high, and cleft into two parts by the longitudinal fosse. The **head**, again, is more depressed than in the typical Epeira, sloping forward from the fosse, with an incline somewhat less than that of the base, which slopes downward rather abruptly. The legs are relatively not so short and stout as in Epeira (except at the femora, which are strong), are less heavily clothed with pubescence; the spines comparatively few, and longer and thinner than in typical Epeira. The face is somewhat narrower than that of the typical Epeira, but not widely different. The **eyes** resemble those of Epeira, but the intervening space between the sidefront and the midfront eyes is relatively not quite so great, in the typical Epeira being from 2 to 2.5 times that of the intervening space of the midfront eyes, while in Abbotia the space is not greater than 1.5 that interval. The **abdomen** (Figs. 7, 7a), instead of having the subtriangular, subglobose, or oval form of Epeira, is a decided ovate, contracted at the base and widening toward the apex, where it is again somewhat narrowed. The known species make horizontal, orbicular webs, but occupy an adjacent shelter of retitelarian lines, underneath which they hang, back downward, clasping a trapline that unites the shelter to the orb.

No. 92. Abbotia gibberosa (Hentz). Plate XX, Figs. 7, 8; Pl. XXIV, Figs. 4, 4a, 4b.

1847. *Epeira gibberosa*, Hentz   . . . . J. B. S., v., p. 457; Sp. U. S., 119, xiii., 20.
1884. *Epeira gibberosa*, Emerton   . . . N. E. Ep., p. 317, xxxiv., 17.
1889. *Epeira gibberosa*, McCook . . . . Amer. Spiders and their Spinningwork, Vol. I.
1889. *Epeira gibberosa*, Marx . . . . . Catalogue, p. 545.

Female: Total length, 4 mm.; cephalothorax, 1.75 mm. long, 1 mm. wide; abdomen, 3.3 mm. long, 2 mm. wide.

Cephalothorax: High, peaked; the centre cleft with a longitudinal fosse; sloping sharply downward to the roundly truncated base, which is slightly indented, sloping also forward at a less angle to the face; the cephalothorax, from the side, presents the view of a truncated cone; corselet grooves distinct; cephalic suture well marked; color green or greenish yellow (which in alcohol becomes yellow), with a brownish crest; caput depressed below the summit of corselet, and slopes well to the eye space; color rather lighter than the corselet, covered with yellowish pubescence, thicker at the face, where the head is narrowed. The sternum is broadly shield shaped, rather squarely truncate at the base, not sharply pointed at the apex, almost as wide as long; with sternal cones, especially in front of coxa-I; rather elevated and rounded at the centre; sparsely covered with brownish bristles; color in alcohol yellow; the labium is rather small, subtriangular, scarcely half as long as the maxillæ, which are obtusely triangular at the tips, somewhat longer than wide, colored as the sternum and labium.

Eyes: Ocular quad on a low eminence, projecting in front, thus giving the face a somewhat compressed appearance; the quad is longer than wide, the front scarcely, or but little, wider than the rear; the eyes on black bases, and not greatly differing in size; MF and MR both separated by about one diameter; side eyes without decided tubercles, but on black bases; barely contingent; about equal in size, smaller than the middle eyes; the space between SF and MF about one and a half to two times the intervening space of MF; the space between SR and MR is somewhat greater than the above; clypeus height about one diameter MF; front eye row decidedly recurved, rear row procurved.

Legs: 1, 2, 4, 3; stout at the femora, and narrowing toward the feet; provided with a few short and a number of long, acute, brownish spines, set in double rows upon femora and tibia, and with short pubescence; a dark longitudinal band of color extends beneath femora–I, II the entire length, along which are placed a number of long, aculeate, epine-like bristles; the palps are colored and armed as the legs, but a lighter hue, and the spines curved; the mandibles are conical, colored as the face, receding at the tips.

ABDOMEN: Ovate, rounded at the base, which overhangs the corselet, somewhat narrowing at the spinnerets, which are distal; the dorsum highly arched; color green, without folium, but the apical half of dorsum and sides marked strongly by blackish, longitudinal stripes, passing downward to the spinnerets and along the sides of the venter; color green, the surface beautifully reticulated, and covered with short, yellowish pubes-cence; the venter yellowish green, surrounded by a broken band of yellowish green or white, which encompasses the spinnerets, with dark anterior corners; the spinnerets brown; the epigynum has a wide subtriangular atriolum, with a short scapus (but rather longer than shown in Fig. 7b), rounded at the tip.

MALE: Fig. 8. Somewhat smaller than the female, which it resembles closely in general form and color. The legs are without any special clasping armature, but the few spines are long and strong. Femora–I, II have the same black longitudinal stripes as the female on the under side; a few aculeate bristles are placed underneath femora–IV, and at least one underneath femora–III; coxæ–I have spurs on the articulation with the trochanter. The sternum is decidedly cordate, and the labium even feebler, relatively, than in the female. The digital joint of the palp is large, subglobose; the embolus and associated parts dark brown, glossy, corneous. (Fig. 10a.)

DISTRIBUTION: This beautiful spider is common in the woods surrounding Philadelphia, and is widely distributed throughout the United States, my collections ranging from New England southward to the Carolinas (Gentry) and to Florida, and as far to the northwest as Wisconsin (Professor Peckham).

No. 93. Abbotia maculata (KEYSERLING). Plate XX, Figs. 9, 10.

1865. *Epeira maculata*, KEYSERLING . . Verh. Zool. Bot. Gesellsch. Wien., p. 827.
1884. *Epeira gibberosa*, EMERTON . . . N. E. Ep., p. 317.
1889. *Epeira maculata*, MARX . . . . . Catalogue, p. 546.

A. maculata in its general characteristics so closely resembles A. gibberosa that detailed description would be mere repetition. The species might, indeed, with much propriety be classified as a variety of the latter. The differences, however, are striking, although chiefly in color markings. The abdomen of A. maculata is, relatively, rather longer and a more even oval than A. gibberosa, which is somewhat wider and thicker at the apex; the latter is marked by several distinct, blackish stripes, drawn from the dorsum along the sides downward and backward. (Fig. 7a.) In A. maculata these stripes are lacking, and instead thereof, on the median apical part of the dorsum, are six black circular spots, arranged symmetrically three on each side. Again, A. maculata is distinguished by lacking the distinct blackish longitudinal stripe beneath the femora of legs–I, II of A. gibberosa.

DISTRIBUTION: The species has substantially the same geographical distribution as A. gibberosa along the entire Atlantic Coast, and inward perhaps to the Mississippi River, at least. The habits of the two congeners are also the same.

GENUS ARGYROEPEIRA, EMERTON, 1884.

The genus Argyroepeira includes a number of spiders intermediate between the typical Meta and Tetragnatha. The legs are long and slender; their order of length 1, 2, 4, 3; the first two much the longest, and not greatly different in length. The maxillæ are longer than those of Meta, but less in length than those of Tetragnatha; broad at the extremities, and divergent. The sternum is subtriangular. The falces are powerful, but not developed to the remarkable extent usual in Tetragnatha. The abdomen is sub-cylindrical, stouter, and shorter than in Tetragnatha; it is often rather humped before, and its hinder extremity is sometimes prolonged in a quasicaudal.form. All the known species are more or less ornamented with patches, streaks, stripes, lines, or sprinklings of brilliant silver, sometimes varied with black and reddish on a whitish or yellowish, or often, in life, a greenish colored ground; occasionally the abdomen is short, stout, or subglobular. The true Meta has the legs usually stronger and more freely clothed with spines, and the abdomen in form more subtriangular than in Argyroepeira, approaching more nearly the larger Linyphioid spiders. The palps of the male in Argyroepeira nearly approach in structure those of Tetragnatha, as well as those of Pachygnatha.[1]

No. 94. Argyroepeira venusta (WALCKENAER).   Plate XX, Figs. 1-6.

1837. *Epeira venusta,* WALCKENAER . . Ins. Apt., ii., p. 90; ABBOT, G. S., No. 113.[2]
1847. *Epeira hortorum,* HENTZ . . . . J. B. S., v., p. 477; Sp. U. S., p. 18, xiii., 19.
1863. *Tetragnatha quinque-lineata,* KEY-
SERLING . . . . Beschr. n. Orbitel., p. 145, vii., 3-6.
1881. *Meta argyra,* KEYSERLING . . . . Neue Spinn. aus Amer., ii., 19.
1884. *Argyroepeira hortorum,* EMERTON . N. E. Ep., p. 332.
1889. *Argyroepeira hortorum,* McCOOK . Amer. Spiders and their Spinningwork.
1889. *Argyroepeira hortorum,* MARX . . Catalogue, p. 550.

FEMALE: Total length, 7 mm.; cephalothorax, 3 mm. long, 2 mm. wide; abdomen 5 mm. long, 3 mm. wide.

CEPHALOTHORAX: A long oval, truncated behind, flat upon the dorsum; median fosse a deep circular pit; cephalic suture strongly marked; corselet grooves prominent; head slightly depressed; color yellow; caput smooth and rounded, but little pubescent. Along the margin of the corselet on either side in the adult is a row of short, stout bristles, sloping forward like cogs in a wheel. The corselet is of almost equal thickness throughout. The sternum is cordate or subtriangular, slightly longer than wide, with decided sternal cones; yellow, with long, blackish, bristlelike hairs. The labium has parallel sides, and an obtusely triangular tip; is dark brown, about half the length of the maxillæ, which are much longer than wide, the greatest width being at the tip, where it is rounded. (See Fig. 1a.)

EYES: The ocular quad has only the midfront eyes upon a prominence; is wider behind than in front; sides longer than rear. MF are separated by about one diameter, MR, which about equal MF in size, by a diameter and a half. The side eyes are propinquate, SR smaller than SF, and placed well behind them. MF are separated from SF by 1.3 their area, SR from MR by about an equal space. The front row is recurved, the rear row slightly procurved. The clypeus has the width of about one diameter MF.

LEGS: Rather long, sufficiently stout, particularly at the femora, with longitudinal rows of short, fine hairs. A curious tuft of long, curved hairs marks the inside of the fourth leg toward the front. (Fig. 1c.) These are arranged in double rows of about ten to twelve on each side. There are but few spines, and these long and feeble. The palps are marked as the legs; mandibles long, strong, conical, arched at the base, separated at tips.

ABDOMEN: Cylindrical, somewhat thicker at the base, and but little wider; apical wall high, spinnerets placed immediately beneath; colors green, with beautiful silvery reticulations, and black longitudinal lines on the dorsal field; skin without pubescence. The epigynum is an open atriolum, without a scapus.

MALE: Fig 4, side view. In general color and markings resembles the female, is smaller, and not so stout; body length, 4 mm.; the abdomen, which slightly overhangs the cephalothorax, 2.2 to 3 mm., the cephalothorax 2 mm. long. The legs are much longer than with the female, and, relatively, not so stout; order, 1, 2, 4, 3, as follows: 16, 13, 10, 8 mm.; tibia-II has no clasping spines or other special developments, and the double row of hairs

---

[1] I have adopted substantially Cambridge's description of this genus, Emerton having made none.
[2] No. 475 of the Abbot MSS. may also refer to this species, which the drawing closely resembles. It is, however, somewhat uncertain; but I have no doubt at all as to No. 113.

upon the inside of femora–III, female, appears to be wanting. The palpus (Fig. 4a) has the radial joint at least three times longer than the cubital, and nearly twice as long as the digital.

DISTRIBUTION: This beautiful species is one of the most abundant in the neighborhood of Philadelphia, and is widely distributed throughout the United States. My collections show that it inhabits the Atlantic Coast throughout, and the Mississippi Valley. I have collected it as far to the southwest as Texas. I have no collections from the Pacific Coast, but it will probably be found thereon. If we accept the examples placed in the synonyma as quite identical with A. venusta, the species is distributed throughout Central America and the northern States of South America.

No. 95. Argyroepeira argyra (WALCKENAER). Plate XXIV, Figs. 2, 3.

1842. *Tetragnatha argyra,* WALCKENAER, Ins. Apt., ii., p. 219, pl. 19, Fig. 1.
1873. *Linyphia ornata,* TACZANOWSKI . Horæ Soc. Ent. Ross., p. 11.
1880. *Meta argyra,* KEYSERLING . . . . Neue Spinnen a. Amer., i., p. 543.
1893. *Argyroepeira argyra,* KEYSERLING, Sinnen Amerikas, p. 343, xviii., 253.

FEMALE: Total length, 7.5 to 8 mm.; cephalothorax, 3 mm. long, 2.5 mm. wide; abdomen, 6 mm. long, 3 mm. wide. The specimens in my possession are much changed by alcohol, but the living colors appear to be greenish yellow, brown, and metallic white, resembling those of A. venusta.

CEPHALOTHORAX: Oval, somewhat flattened; a deep triangular fosse; corselet grooves not distinct; cephalic suture pronounced; caput elevated, lowly arched; yellowish brown (in alcohol), scarcely pubescent. Sternum cordate (Fig. 2a), as wide as long; cones distinct; raised in the middle, especially opposite the labium; brown, covered with long, black, curved bristles. Labium compressed at the shank, widened at the subtriangular tip; half as long (or less) as the maxillæ, which are compressed below, and widened at the tip, and decidedly longer than wide.

EYES: Fig. 26. Ocular quad longer than wide, wider behind, eyes about equal; MF separated about one diameter, MR 1.5. MF separated from SF about 1.5 the alignment of MF; SF propinquate, about equal in size, but smaller than those of the quad, and on slight tubercles; clypeus nearly the alignment of MF high; front row recurved, rear row very slightly procurved.

LEGS: Order, 1, 2, 4, 3, as follows: 23.1, 17.9, 14.5, 8.9 mm.; they are not stout, gradually diminish to the tarsus, the joints long, except the patellæ, which are relatively short; they have few spines, but are freely clothed with long bristles, almost plumose beneath; color in alcohol yellowish, with slight tinge of green; in life the latter color probably predominates.

ABDOMEN: Cylindrical, thicker at the base, which overhangs the cephalothorax. The dorsal folium is margined by a broad band of silver, which covers the base and extends in wedge shaped bands along the sides; indeed, the ground color might be described as silver, and the pattern (see Fig. 2) said to be outlined by interrupted longitudinal fillets of brown. The venter is blackish brown, with silvery marginal bands; the epigynum (2c) is large, conical, projecting forward, thickly clothed with hair. :

MALE: Fig. 3. In general form and colors closely resembles the female, which it almost equals in length, the largest specimens being 7 mm. long. The digital bulb (3a, 3b) is large, globose, longer than the radial and cubital joints together, the latter being about one-half longer than the former.

DISTRIBUTION: Southern California, from which numerous specimens have been obtained from Dr. Blaisdell. Outside the United States the species inhabits Mexico, the Antilles, Central America, and probably South America, in the northern provinces. It is one of the most beautiful of our indigenous spider fauna. I have no account of its spinning and cocooning habits, but these will probably be found to resemble closely those of its congener A. venusta, which I have fully described.

Genus HENTZIA, new.

I have felt it necessary to establish a new genus to receive the following interesting species, whose spinningwork, as described in a previous volume, sets it apart as widely different in its industry from all other spiders. It is impossible to follow Dr. Marx in relegating it to Argiope, as the character of the eyes is entirely different, those of Argiope being procurved in both rows, the front row slightly and the rear row very much. On the contrary, the eyes of this species are so arranged as to form an oval figure, the front row being decidedly recurved, the rear row procurved. The midfront eyes are separated from the sidefront by a space almost equaling, but a little less than, the space which divides the latter. The cephalothorax is of medium height, the caput not specially depressed. The sternum is somewhat, but not decidedly, longer than wide. The labium is short, and the maxillæ wide, almost as wide as long. The legs are long, rather thin, sparsely covered with pubescence, and the spines, instead of being stout and short, are long and thin. The abdomen is cylindrical, decidedly longer than wide. In its general appearance the species is more closely allied to certain Theridioids than to the typical Epeira, and approaches Argyroepeira, from which, however, it differs, among other points, by the fact that in the latter genus the space between the midfront and sidefront eyes is more than twice as great as the interval between the midfront eyes, in this respect resembling Epeira. In Hentzia basilica the ocular quad is much longer than wide; the width in front only slightly, if any, greater than the rear width; in a typical Epeira the front width is usually decidedly greater than the rear, and the quad is not so manifestly longer than wide. The space between MF and SF in Hentzia is less than the intervening space of MF; in Epeira, on the contrary, that space is 1.5 to 2.5 times greater, usually nearer the latter than the former. The difference in this respect is enormous. In Hentzia basilica the space between SR and MR is but little, if any, greater than that between MR; in the typical Epeira, on the contrary, the intervening space between SR and MR is 3.5 to 4 times that of MR. The difference in this respect is even greater than in the former. The shape of the abdomen, again, separates the two genera, the typical Epeira being triangular ovate; this, on the contrary, being cylindrical, inclined to rectangular, and decidedly longer than wide, with the abdomen formed into a sort of caudal projection. The spinous armature of the legs also shows a decided difference, Hentzia having rather scant pubescence, and spines not numerous, long and slender, instead of being abundant, stout, and rather short, as in Epeira. The terminal joints of the legs also differ, Epeira being marked by a moderately long metatarsus, not disproportionate to the other joints, whereas Hentzia has a metatarsus at least one-third longer than the tibia, and the tarsus is much longer than the patella. In Hentzia basilica the three legs are almost nearly equal in length, the first and second pairs scarcely differing, and the fourth but a fraction of a millimeter shorter. The snare is a horizontal orbicular web, raised at the centre into an umbrella shape, like that of certain species of Linyphia.

No. 96. Hentzia basilica McCook.   Plate XIV, Fig. 2; Pl. XXIII, Fig. 8.

1878. *Epeira basilica*, McCook . . . . Proceed. Acad. Nat. Sci., Phila., p. 124.
1889. *Epeira basilica*, McCook . . . . Amer. Spid. and their Spinningwork, Vol. I.
1889. *Argiope basilica*, Marx . . . . . Catalogue, p. 541.

FEMALE: Total length, 6.5 mm.; cephalothorax, 3 mm. long, 2 mm. wide; abdomen, 5.5 mm. long, 2 mm. wide.

CEPHALOTHORAX: Oval, truncated at the base; fossa a low circular pit; corselet medium height, somewhat flattened on top; color yellow or olive, with a blackish brown ring on the margins of the corselet, and a similarly colored band passing the entire length of the corselet and caput on the median line; pubescent, especially on the margins; the caput not depressed, and scarcely elevated above the summit of the corselet; the sternum cordate, bluntly pointed at the apex, longer than wide, blackish brown margins, with a yellow median band, with short yellowish hairs, particularly along the margins; sternal cones prominent; the labium wide at the base, but rather low, subtriangular; the maxillæ gibbous, somewhat, but little longer than wide.

EYES: Ocular quad somewhat wider in front than behind, and decidedly longer than wide; the eyes differ little in size, are upon black bases; MF separated by about 1.5 diameter, MR by about one diameter; the side eyes contiguous, SR placed well behind, though somewhat to the side of, SF; equal in size; smaller than the middle group, and, like them, upon black bases. The space between SF and MF is slightly less than the space between MF; the distance between MR and SR is slightly greater than the intervening space of MR; the clypeus has the height of about 1.5 diameter MF; the front row is recurved, the rear row procurved; and the eyes are so arranged that they form a nearly continuous oval.

LEGS: 1=2, 4, 3, the fourth pair nearly as long as first and second; long, strong, without being stout; yellow or olive, with slight touches of darker color at the extreme apical points of the joints; provided with long, yellowish brown, thin spines.[1]

ABDOMEN: Cylindrical, much longer than wide, thickest at the base, which is rounded and somewhat overhangs the cephalothorax, and has slight conical shoulder humps; dorsum slightly arched; the apex rounded and projected at the dorsum somewhat beyond the base of the apical wall, thus overhanging the spinnerets, which project from the distal part backward; the color is yellow, with a shade of olive, the folium a wide band, with compressed edges, passing the entire length, a silvery white line marking the margins, which are bordered with a narrow ribbon of rosy hue; the sides are marked with longitudinal stripes of yellow and blackish brown; the venter has a broad, rectangular belt of black, with yellowish brown borders, passing around the yellowish brown spinnerets; the epigynum is without a scapus, the atriolum showing as a horseshoe shaped arch, widely opening downward, with convolutions within on either side marking the portulæ.

DISTRIBUTION: Texas; District of Columbia; probably the Southern States. A single female specimen was found by me in Texas, distinguished by the peculiar and beautiful web, fully described in Vol. I. The correctness of my observation was long questioned, until Dr. Marx discovered specimens in the Government grounds in Washington, and was fortunate enough to observe the method of spinning the web and the character of the cocoon.

GENUS META, C. KOCH, 1836.

The species of Meta are characterized by an oval cephalothorax, whose corselet is rounded at the sides, truncate at the base, high, with deeply indented summit, distinct corselet grooves and cephalic suture; the caput is arched and elevated above the corselet, and is wide at the face. The sternum is somewhat longer than wide. The labium is long, rectangular at the base, rounded at the tip, and less than half the height of the maxillæ, which are decidedly longer than wide, and present a marked characteristic. The eyes are arranged in three groups, but the side eyes are more nearly approximated to the middle group than in Epeira; the space between the sidefront and midfront not exceeding the area of the latter. The intervening space between the siderear and midrear eyes is somewhat greater. The front row of eyes is recurved, the rear row aligned, or nearly so, and the clypeus is of good height. The legs are long, the metatarsus proportionately much longer than the tarsus; the pubescence is abundant, long, thin, aculeate bristles covering the surface, together with long, thin spines. The abdomen is a short, even ovate, the dorsum arched, the base and apex rounded, the latter slightly overhanging the distal spinnerets. The male resembles the female in form, color, and size. The palpal joints, however, are not proportionately as long as the legs, but, on the contrary, are relatively rather short.

---

[1] Fig. 2, Plate XIV., is defective in the shape of the legs.

No. 97. Meta Menardi (LATREILLE).   Plate XXII, Figs. 4, 5.

1804. *Aranea Menardi,* LATREILLE . . . Nat. Hist. Crust., vii., p. 266.
1804. *Aranea novem-maculata,* PANZER . Syst. Nom., p. 244.
1805. *Epeira fusca,* WALCKENAER . . . Tabl. d. Aran., p. 63.
1806. *Epeira Menardi,* LATREILLE . . . Gen. Crust. et Ins., i., 108.
1836. *Meta fusca,* KOCH, C. . . . . . . In Herr.-Schaeff. Deutschl. Ins., cxxxiv., 12, 13.
1837. *Epeira fusca,* WALCKENAER . . . . Ins. Apt., ii., p. 84.
1841. *Meta fusca,* KOCH . . . . Die Arach., viii., p. 118, tab. 285, Figs. 685–687.
1856. *Meta Menardi,* THORELL . . . . . Recensio Critica, p. 98.
1861. *Meta Menardi,* WESTRING . . . . Araneæ Svecicæ, p. 77.
1864. *Epcira fusca,* BLACKWALL . . . . Spid. G. B. & I., ii., 349, pl. xxvi., 252.
1870. *Meta Menardi,* THORELL . . . . . European Synonyms, p. 38.
1884. *Meta Menardi,* EMERTON . . . . . N. E. Ep., p. 328, pl. xxxiv., 18.
1889. *Meta Menardi,* McCOOK . . . . . Amer. Spiders and their Spinningwork, Vol. I.
1889. *Meta Menardi,* MARX . . . . . . Catalogue, p. 541.

FEMALE: Total length, 8 mm.; cephalothorax, 3.5 mm. long, 2.75 mm. wide; abdomen, 6 mm. long, 4 mm. wide.

CEPHALOTHORAX: A long oval, squarely truncate at the base; the fosse a deep semilunar indentation, the cavity directed backward; corselet grooves distinct; cephalic suture well marked; caput slightly arched, not lower than the corselet, slightly depressed at the face, which is but little narrower than the base; the color yellow, flecked with brown. The sternum cordate, somewhat longer than wide, yellowish brown in color; sternal cones especially marked in middle of the apical part, and middle of the base opposite the lip; covered with yellowish, long bristles; the labium half the length of the maxillæ, rounded at the tips and thickened at the base; the maxillæ decidedly longer than wide, somewhat narrowed at the shank, gibbous at the apex, color yellowish brown.

EYES: The entire eye space is black, the quad longer than wide, wider in the rear than in front; MF smaller than MR, separated by about 1.3 diameter, MR by one diameter; the side eyes not upon tubercles; MF the larger, separated by about a radius of MR; the space between MF and MR is about equal to, or slightly greater than, the space between MF; the distance between MR and SR is somewhat, and but little, greater; clypeus width about 1.5 diameter MF; the front eye row recurved, the longer rear row procurved.

LEGS: 1, 2, 4, 3; very long, and sufficiently stout, especially at the femora; yellow, with brown median annuli; clothed freely with yellowish hair and bristles, and brown, acute spines; the palps are colored as the legs, but a lighter shade; mandibles long, conical, somewhat divergent at the tips, color yellow, merging into orange or brown.

ABDOMEN: An evenly rounded oval, thickest across the shoulders, rounded at the base, narrowing somewhat at the apex to the distal spinnerets; the dorsal folium is a herring bone pattern, yellow along the median, bordered by blackish brown interrupted bands on either side; streaks of yellow and black alternate upon the sides to the venter, which is black or blackish brown, bordered by yellow, with two yellow spots in the centre; the spinnerets are stout, short, orange yellow; the epigynum (Figs. 4c, 4d) has no scapus, but shows simply a semicircular flap, yellow, and covered with hair, that overhangs the genital cleft.

MALE: Fig. 3. Closely resembles the female in size, color, and markings. The legs are without any special clasping apparatus. For a view of the palpal digit see Fig. 5a.

DISTRIBUTION: I captured a number of this species in Sinking Spring Cave, Blair County, Pennsylvania, and in the neighborhood. It has been taken in Massachusetts, Kentucky, Virginia, and the District of Columbia. It is a European species, and may have been introduced by migration and commercial intercommunication, but is more probably indigenous. It dwells in caves, or in dark, moist places, and has in America precisely the same habits as the European representative. Blackwall speaks of the species in Great Britain as also inhabiting cellars and overhanging banks.

GENUS LARINIA, SIMON, 1874.

In Larinia[1] the cephalothorax is oval, with decided corselet grooves and longitudinal fosse. The midrear eyes are closely approximated, almost contingent; the interval which separates them from the siderear eyes is much greater than that which divides the midfront pair. The midfront eyes form a straight, or almost straight, row, arranged at almost equal distances from one another; the distance between the midfront and sidefront being slightly the greater. The mandibles are perpendicular; the sternum is longer than wide; the labium and maxillæ are as in Epeira. The abdomen is much longer than wide, oval in shape, attenuated at the two extremities, marked by longitudinal bands. The palps are as in Epeira. The legs in order of length are 1, 2, 4, 3; more or less robust; provided with four pairs of spines on the femora, tibiæ, and metatarsi; the tibiæ of the first and second pairs are equally slender; the thighs are without hairs. The superior tarsal claws have nine or ten denticulations, of which the first are much longer than the others, which are about equal. The skin is almost always smooth, although covered with colored hairs and pubescence, as in Epeira.

No. 98. Larinia borealis BANKS. Plate XXII, Figs. 1, 2.

1894. *Larinia borealis*, BANKS . .·. . . . . Entomological News, p. 8.

FEMALE: Total length, 7 mm.; cephalothorax, 3 mm. long, 2 mm. wide; width at face, 1 mm.; abdomen, 5 mm. long, 3 mm. wide, less than 1 mm. at base.

CEPHALOTHORAX: A long oval, high and somewhat peaked at the fosse, which is a longitudinal slit; squarely truncate behind; the corselet grooves distinct; cephalic suture pronounced; the caput slightly depressed and gently arched from fosse to ocular space; face sufficiently wide, about, or a little more than, one half the greatest width of the cephalothorax; color dull yellow; slightly pubescent, except at the face, where there are tufts of strong, whitish, gray bristles.

STERNUM: Cordate, longer than wide, in the proportion of five to four; elevated and arched in the middle; with sternal cones; the apex triangular, the anterior and middle part indented, and of about equal width; color blackish brown, with an irregular median band of yellowish; scantily provided with bristlelike hairs on the margin. Labium with a short compressed shank; tip at base wider than high; subtriangular; of a light yellow brown color. Maxillæ somewhat longer than wide; the shank slightly compressed; the apex rounded and thickened, and obtusely triangular at the tip; a few short, curved, stubby bristles.

EYES: Ocular quad elevated in front, the rear eyes seated upon the base of the elevation; the quad decidedly wider in front than rear; MF somewhat larger than MR, which are separated by less than a radius, indeed are almost contingent; distance between MF and SF about 1.3 the interval of MF, or less than the area thereof. Side eyes on slight tubercles, with blackish bases; barely contingent; subequal, but SF apparently slightly larger; the space between MR and SR decidedly greater than between MF and SF, and equal to about 1.5 the area of MR; the front eye row slightly recurved, rear row procurved, in this respect resembling Epeira; clypeus height equal to about one diameter MF.

LEGS: 1, 2, 4, 3. Color uniform yellow; short, rather stout, well provided with hairs and bristles, and rather sparsely with thin, yellowish spines; tibia and metatarsus of leg-I about equal in length; palps colored as the legs; mandibles conical, not receding, yellow.

ABDOMEN: Ovate, compressed at the base to an obtuse point, which overhangs the cephalothorax; rounded at the apex, where it diminishes in width, but is not compressed to a point, as at the apex; dorsum evenly arched to the apex, which slightly overhangs the spinnerets; color dull yellow; rather thickly coated with yellowish white hairs; a median band of lighter yellow, and of almost equal width throughout, passes from base to

---

[1] I give substantially Simon's diagnosis of this genus.

apex, quite around to the spinnerets; on either side is an indistinct folium of yellowish brown, the margins of which are slightly indented; the dorsal surface is reticulated. The ventral pattern is a long band of yellowish brown, with two longitudinal stripes of cretaceous or yellowish along the middle. The epigynum of the specimen in hand. is scarcely matured, but shows an atriolum with a broad base and a slightly projecting scapus, or simple flap.

MALE: In general form resembles the female. The cephalothorax is relatively somewhat wider; grouping of the eyes substantially the same. The legs and mouth parts are of like color, although the sternum, in the specimen in hand, is of lighter hue. The legs are relatively longer, stronger, and more decidedly armed with yellowish brown spines than the female; tibia-II is without special clasping spines. The abdomen differs decidedly in color, being a beautiful lake red, the central band undulated, and with longitudinal stripes of yellow intervening; the general color of the sides is a warm pink, mottled with occasional black dots, and covered densely with yellowish hairs. The sternum in form differs little from that of the female; but the maxillæ are scarcely longer than wide, and the tips are squarely truncated.

DISTRIBUTION: Both male and female were received from Mr. Nathan Banks, and collected at Olympia, Washington, by Mr. Trevor Kincaid. Mr. Banks has received several specimens from the same gentleman, and has two from Franconia, New Hampshire, collected by Mrs. A. Trumbull Slosson. This would indicate a wide distribution across the northern belt of States, or, possibly, introduction by commerce.

GENUS DREXELIA, McCook, 1892.

In the Proceedings of the Academy of Natural Sciences of Philadelphia for 1888 I gave the diagnosis of a new genus of spiders, Drexelia, to receive Hentz's species, Epeira directa, and Cambridge's Epeira tetragnathoides, which appeared to me to be the same, or a closely related, species. Identical with the latter is a third species, which Dr. George Marx, in his catalogue of "Described North American Araneæ," has published as Epeira deludens Keyserling *in litteris*, and which had been examined by Count Keyserling. All these specimens appeared to me to be identical, although I had only the description of Cambridge from which to judge, not having seen a typical specimen of his species. Under the circumstances it seemed necessary not only to restore the specific name of Hentz, but to make his species the type of a new genus.

Drexelia is separated sharply from Epeira by the peculiar elongated shape of the sternum, which is at least, or nearly, twice as long as wide. Further, by the character of the maxillæ, which are longer than wide; and, still further, by the shape of the abdomen, which is long, narrow, straight, and, especially in the female, bluntly pointed at the base, and somewhat compressed at the apex. The legs are less stout than those of the typical Epeira, and the spinous armature thereon feebler and less abundant; the metatarsus of leg-I about equals the femur in length. In the form of the maxillæ Drexelia approaches both Nephila and Meta, but differs from them, and, in a more marked degree, from Epeira, in the relatively greater length of the sternum. It differs also from these genera in the form of the abdomen, that of Nephila being as long as in Drexelia, but subcylindrical in form, that of Meta being a rounded oval, approaching thus the typical Epeira. In the shape of the abdomen Drexelia somewhat resembles Tetragnatha, and also approaches it in the more feebly armed character of the legs; but the mouth parts and sternum, to say nothing of other characteristics, widely divide these two genera. Drexelia approaches Epeira in the contour of the face and head, but lacks the strong tubercles on which the eyes are placed in the more typical species of Epeira. It resembles examples of the same genus in the general grouping of the eyes, although the two midrear eyes of the central group are more closely approximated than in Epeira.

Mr. Nathan Banks (*Entomological News*, Philadelphia, January, 1894) has expressed the opinion that Drexelia directa properly belongs to Simon's genus Larinia (1874). Certainly there are resemblances between the two genera which may possibly be found sufficient to justify that opinion, and remand Drexelia to the synonyms. But Simon, in his diagnosis of Larinia, expressly describes the maxillæ "as in Epeira;" on the contrary, one of the most striking characteristics of the typical Drexelia is that the maxillæ, as in Meta, are much longer than wide, while in Epeira they are as wide as long. Moreover, in Drexelia the sternum is quite, or nearly, twice as long as wide. (See Plate XXII., Fig. 3d), while in Larinia the sternum is diagnosed by Simon as simply "wider than long." For the present, therefore, and in the absence of an example of the type species Larinia Defouri, upon which Simon founded his genus, I retain Drexelia, as above.

Mr. Banks has been good enough to send me specimens of his Larinia borealis, which I have figured and described elsewhere. He has also sent me what he regards as a true example of Hentz's Epeira directa. I have presented in the drawings (Plate XXII., Figs. 1b, 3e) what seem to be the most striking differences between Larinia as represented by Banks' L. borealis and Drexelia as represented by D. directa. It will be observed that the most marked characteristics, as above noted, are in the greater relative length of the sternum and maxillæ, and metatarsi of leg-I; in Drexelia, also, the abdomen appears to have a greater relative length to its width than in Larinia borealis. Araneologists, doubtless, will differ as to the importance of these characteristics, and it may well be doubted whether they are sufficient to justify the maintenance of separate genera. At all events, Drexelia, as here defined, appears to me to be a connecting link in a chain of genera which may be cited in the following series: Epeira, Meta, Larinia, Drexelia, Nephila, Tetragnatha.

No. 99. Drexelia directa (HENTZ).   Plate XXII, Fig. 3; Pl. VI, 10, 11.

1847. *Epeira directa*, HENTZ . . . . . . J. B. S., v., pl. 31, 21; Sp. U. S., p. 110, xiii., 21.
1847. *Epeira rubella*, HENTZ . . . . . . Ibid., Fig. 22; Ib., p. 120, xiii., 22 (in part).
1889. *Epeira tetragnathoides*, CAMBRIDGE, Biolog. Cent. Amer., Aran., p. 16, viii., 9, 10.
1890. *Epcira deludens*, MARX *in litt*. . . Catalogue, p. 544.
1892. *Drexelia directa*, McCOOK . . . . Proceed. Acad. Nat. Sci., Phila., p. 127.
1894. *Larinia directa*, BANKS . . . . . Entomological News (Philada.), vol. v., No. 1, p. 8.

FEMALE: Fig. 5. Total length, 8+ mm.; abdomen, 6.5 mm. long, 2+ mm. wide. The cephalothorax, 2.5 mm. long; abdomen projecting about 1 mm.

CEPHALOTHORAX: A long oval, narrowed in front; the cephalic suture well marked; corselet grooves distinct; fossæ long and deep; head not depressed; face wide; color yellowish brown or rufous. Mandibles conical, and receding somewhat inward; colored as the cephalothorax. Sternum uniform yellow, slightly pubescent; about a half longer than wide (Fig. 3d); of nearly equal width throughout, except at the triangular apex; with decided sternal cones, especially before coxæ-III. Labium slightly compressed at base, sub-triangular at tip; height about equal to width; not half as high as the maxillæ. Maxillæ decidedly longer than wide; the shank slightly compressed, and somewhat truncate or obtusely triangular at the tip; the sides but little rounded, and the width nearly equal throughout.

EYES: The ocular quad much wider in front than behind, and the height about equal to the width of the front pair; MF larger than MR, separated by about 1.5 diameter; MR separated by about a radius. Side eyes on low tubercles; SF somewhat (but little) larger than SR; barely contingent. The front row slightly recurved, rear row procurved; MF removed from SF by about 1.5 their interval; SR from MR 2 to 2.5 times their area; clypeus height about one diameter MF.

LEGS: 1, 2, 4, 3; color uniform yellow in alcoholic specimens, rather long, and not stout; sufficiently armored with long, thin spines, which are much worn off in the specimens in hand; metatarsus of leg-I about one-fifth longer than tibia. Palps similarly colored and armed.

ABDOMEN: Fig. a, side view. Elongated; cylindrical; more than three times as long as wide; narrowing to the apex, which projects beyond the yellow spinnerets; narrowing also, and to an obtuse point, at the base, which overhangs the cephalothorax. A broad, creta-ceous or yellowish reticulated median band extends the whole length of the dorsum, somewhat narrowed at the ends; brownish, branching, longitudinal lines mark the middle of the dorsum towards the apex. The dorsal surface is covered with stout, short, curved bristles distributed symmetrically thereon; the basal point is tufted with bristles. The venter shows a yellow, median longitudinal band, with brownish margins between the gills and the spinnerets, which are yellowish brown. The epigynum in the example before me is apparently not quite mature, but presents the appearance of a low atriolum, double notched on the edge (Fig. 3f), the scapus wanting, or expressed by a simple short fold at the base.

I think it probable that Epeira directa and E. rubella of Hentz are identical, and that the drawings have become somewhat mixed on the plate; Hentz himself expresses the opinion (page 120), though with doubt, that his E. rubella may prove to be the young of E. directa. According to this author the spider is found generally near water, where it makes a perpendicular web upon low bushes. When approached it drops and remains motionless where it falls. He found in Alabama a specimen with four minute blackish spots upon the abdomen, which may indicate the variety drawn. (Fig. 3c.) The male was found by Hentz with black dots all over the legs, except the thighs, and also with black dots on each side of the abdomen, but evidently the same species, of which he adds that it is nocturnal in its habits. According to Cambridge (E. tetragnathoides) the male palps have short cubital joints, somewhat angular in front, with two very long, strong, tapering, divergent bristles. The radial joint is obtusely produced on its outer edge; the palpal bulb small.

DISTRIBUTION: North Carolina (Plate XXII, Fig. 3); Indian River, Florida. Hentz col-lected in Alabama and South Carolina. The specimens described by Mr. Cambridge are from Guatemala and Panama, and although they show some variations in color and markings from those found in the Southern United States, appear to be the same. The distribution is therefore throughout the Southern United States and North America, and probably in the Northern States of South America. I believe that it lives at least as far north as New Jersey.

GENUS NEPHILA, LEACH, 1815.

The genus Nephila, of which we have several known species, is confined to the southern and southwestern belt of the United States. It is distinguished by an oval cephalothorax, of nearly equal width throughout, whose base is flat and low. The caput is elevated and arched, quadrate, and wide at the face; in some species, on the base thereof, near the fosse, is a pair of corneous tubercles. The skin is hard, and usually covered, except on the smooth marginal walls, with a close coating of lustrous white hair, which gives the organ a metallic brightness. The length of the sternum differs little from its width, is marked by decided sternal cones, with a glossy and hard skin, covered with metallic hairs. The labium is longer than wide, thick at the base, and the maxillæ especially are decidedly longer than wide, compressed at the shank and rounded at the tips. The eyes are arranged in three groups, ordinarily placed upon decided tubercles; the ocular quad is usually longer than wide; the side eyes well separated; the front row procurved and the rear row recurved, or nearly aligned. The legs are long, rather attenuated, the thighs not stout, the metatarsus relatively much longer than the tarsus, and the apical part of one or more of the joints usually provided with a brushlike appendage on one or both sides. The abdomen is much longer than wide, cylindrical, brilliantly colored, and provided with numerous hairs, of a metallic, silvery lustre, which give the animal an unusually beautiful and brilliant appearance. The epigynum has commonly a hard, vaulted atriolum, and is without a scapus. The males of this genus are very much smaller than the females, and do not closely resemble them. The snare is an immense orb, woven in forests and woods, whose spirals are composed of successive, nearly circular, loops.

No. 100. Nephila Wilderi, McCook.   Plate VII, Figs. 1, 2; Pl. XXIII, 6, 7.

1839. *Nephila plumipes*, Koch, C. L. . . . Die Arach., vi., p. 128, pl. 213, Fig. 529.
1873. *Nephila plumipes*, Wilder. . . . . Trans. Amer. Assocn. Adv. Sci., p. 203.
1889. *Nephila Wilderi*, McCook . . . . Amer. Spid. and their Spinningwork, Vol. I., p. 146.
1889. *Nephila plumipes*, Marx . . . . . Catalogue, p. 551.

Female: Total length, 23 mm.; cephalothorax, 9 mm. long, 7 mm. wide, 6 mm. across the face; abdomen, 20 mm. long, 7 mm. wide at the base, tapering to 5 mm. at the apex. The spider described as Nephila plumipes, and figured by Koch as above, is undoubtedly the species which abounds upon our Gulf Coast. But the specific name has already been appropriated for another species of the genus, the Epeira plumipes of Latreille.[1] That species resembles N. plumipes of Koch, but differs from it in having two tubercles, or horny processes, upon the base of the caput, near the corselet fosse. For this reason a new name becomes necessary, and already I had suggested the name of N. Wilderi, in honor of Professor Wilder, of Cornell University, whose interesting observations and experiments have made the life history of the species so well known.

Cephalothorax: A long oval, squarely truncated at the base; thin, much depressed in the centre, at the fosse, which is a lateral pit; corselet grooves abbreviated, or represented by circular hollows; cephalic suture distinct near the base; caput elevated above the corselet; subtriangular at the apex, where it is cut straight across by the fosse; enlarging to the face, which is wide, strong, and but little depressed; summit of caput arched, the base without horns; color uniform dark brown, in older specimens particularly, but concealed largely by a thick coat of silvery white hairs. Sternum shield shape, somewhat longer than wide, dark brownish or brownish yellow color, rather heavily covered with yellowish pubescence; sternal cones strong, particularly before coxæ-I, and in front of labium, where the cone is unusually high, and its base fully half that part of the organ. Labium elongated and rounded at the point; a little humped in the middle; silvery white pubescence on the edges and base. Maxillæ about one-third longer (3 mm.) than labium (2 mm.), narrow at base, widening at tips, which are subtriangular, and inclined toward one another; ruddy brown, but at the tips glossy black brown, and covered sparingly with bristles; the inner edges concave.

Eyes: Ocular quad on an eminence, highest and most projecting in front, longer than wide, narrower in front than behind; eyes not greatly differing in size, but MF apparently somewhat larger; these are separated by about two diameters; MR by three diameters; side eyes on the outer edges of strong corneous tubercles, which project much in front; nearly equal in size, but SF slightly larger, and separated by 1.5 to 2 diameters. SF removed from MF by 1.5 the area of the latter, or by about 2.3 the dividing space; the distance of SR from MR is greater; front eye row well recurved, rear row procurved; the clypeus has the height of 3 to 3.5 diameters MF; clusters of hair come up to the bases of the tubercles upon the sides, and surround the base of the ocular quad.

Legs: 1, 2, 4, 3; the third leg being relatively quite short; not stout, the joints long and narrowing much towards the tips; they are dark yellow or yellowish brown, with brown annuli at the tips of joints; the patella short, dark brown; the tips of the femora and tibiæ of legs-I and II are provided with brushes of bristles arranged upon the outside and inside thereof, the hairs standing quite thickly, and curved forward; leg-IV has a similar brush upon the extremity of the tip; leg-III is without a brush, but has a slight thickening of bristles, both above and below, on the tibia, which is relatively short; color of legs a ruddy yellow, except the tarsus and extremity of the metatarsus, which are black or blackish; joints quite abundantly provided with short, black spines. Palps dark yellow, or lighter brown at the humeral and axillary joints, darkening much on the other joints. The digital joint is nearly twice the length of the radial, and the latter is at least 1.5 times the length of the cubital joint, which is short; the humeral joint has, underneath the entire length, a thick row of bristles, and these are abundant on the other joints. The

---

[1] Latreille: Nat. Hist. des Insectes, Vol. VII., page 275, No. 86. Walckenaer (Apt. Ins. II., page 93) regards this species as identical with his E. antipodiana.

mandibles are strong, conical, and dark brown, glossy, and provided with numerous yellowish bristles on the inside; the tips and fangs are shining black. This species is distinguished from N. clavipes by the brushlike appendages on the legs, which are strongly marked and placed on the femora and tibia of legs-I, II; but in Clavipes are shorter and feebler, and are wanting on femora-I, II. Further, the strong sternal cone opposite the labium is but faintly developed in Clavipes.

ABDOMEN: Cylindrical, thickest at the base, and tapering towards the apex; general color bright yellow, with a dark patch on the base, which extends upward to the dorsum. A double row of round silver white spots, about eight in number, extends on either side to the apex. Between these and along the sides are a number of roundish ovals and crescent shaped markings, of light color, covering the entire dorsum and upper portion of the sides, giving a remarkably beautiful appearance to the female. Similar marks extend along the sides to the venter, the coloring on the sides being brown. The venter is golden yellow, with a rectangular figure, composed of silvery white hairs, dividing the basal portion. Within this rectangle are two rows of three or more silvery circles. The apex is rounded, and descends with little anterior slope to the spinnerets. The white spots above mentioned consist chiefly of hairs, which have a silvery white gloss. These hairs, placed upon the bright yellow and brown colorings of the dorsum and sides, make this species the most remarkable for beautiful coloring among all its race. The epigynum is a simple arch above the genital cleft, whose rim is black, glossy, with short, yellow hairs.

MALE: Plate VII., Fig. 2.[1] Length of body, 6 mm., the cephalothorax being nearly the same length as the abdomen. In general shape it resembles the female, but is ridiculously small in contrast with its mate. The head and cephalothorax resemble that of the female in general shape, the head being well erect, the cephalothorax rounded at the base and flattish. Both the cephalic and thoracic parts are covered with short white pubescence, but not so as to form the silver coating characteristic of the female. The eyes are arranged as in the female. The cymbium of the digital bulb is covered quite densely with hairs. The embolus is prolonged into a long curved blade, nearly as long as the palp itself, the most remarkable and characteristic feature of the animal next to its small size. The two next joints are short, almost equal in size; the cubital joint has several bristles of medium length, and the radial has one very long bristle. The femoral joint is longest, tapering gradually towards the axillary. The sternum, labium, and maxillæ substantially resemble those of the female, but the sternum is of much lighter color, and appears to be destitute of the strong sternal cones that mark the female. The first leg is about 18 mm. long, its color uniform yellow, slightly darker at the ends of the joints and towards the feet. The tibial and femoral brushes are wanting. The abdomen is cylindrical, tapering toward the apex, not wider than the cephalothorax; a uniform yellowish color, with a dark median band, running along the apical part; the venter yellow, mottled.

DISTRIBUTION: Along the Southern Atlantic and Gulf Coasts, and southwesterly to Southern California. The species is also found in the West Indies, and probably inhabits the Northern States of South America.

No. 101. Nephila Wistariana,[2] new species.   Plate XXIII, Figs. 2, 3.

1839. *Nephila clavipes* (?), KOCH . . . . Die Arachniden, v., p. 30, pl. 152, Fig. 354.
1842. *Nephila Vespucea* (?), WALCKENAER[3] Ins. Apt., ii., p. 98.
1890. *Nephila clavipes*, MARX . . . . . Catalogue, p. 551.

---

[1] This specimen is in the National Museum at Washington, and I am indebted to the Chief Entomologist, Professor C. V. Riley, for the opportunity to study it. The specimen is mounted on a pin, and is, therefore, dry and shriveled. Several of the legs are wanting, but otherwise it is sufficiently perfect to permit description. A second and more perfect specimen is from Dr. Marx's collection.

[2] Proper name, General Isaac J. Wistar, President of the Academy of Natural Sciences of Philadelphia.

[3] Walckenaer's N. Vespucea (Vespucci?) is regarded by that author as identical with Koch's No. 354, and his description fairly corresponds thereto. But the confusion is here so great that I give a new name, as on the whole more satisfactory, and content myself with pointing out a probability which can only be resolved by access to a good collection of this group.

FEMALE: Total length, 25 mm.; abdomen, 17 mm. long, 7 mm. wide; cephalothorax, 8 mm. long, 6 mm. wide. Koch's Nephila clavipes (plate No. 34) appears to me to be identical with a species found along our Gulf Coast, of which I have received several examples from Mr. C. H. Townsend, collected at Swan's Island, Carribean Sea. There is, however, some discrepancy between the description and illustration of Koch, as Walckenaer has pointed out, Ins. Apt., II., page 99. No. 354, Koch describes as N. fasciculata DE GEER, and with head horns, which are certainly not represented in the figure marked for that species on the plate. On the contrary, No. 355 is drawn with head horns (apparently), although "N. clavipes," which is the title of that number, is described in the text as without such appendages. A specimen in Dr. Marx's collection from Key West, Florida, resembles Koch's Fig. 355, Plate CLII., and corresponds well to the description placed under "Fig. 354," pages 30, 31. Has there been here a transposition of the figures? In any case a new name must be given for the species numbered 354 (Plate CLII.), since both the names N. clavipes and N. fasciculata are appropriated. The latter, indeed, may be a synonym of the former, but this is in doubt, and, in the absence of material for study, it is not possible for the writer, at least, to determine that point. The species has a close general likeness to N. Wilderi, from which it may commonly be distinguished by the presence of brushes or plumes on the tibiæ alone, instead of on both tibiæ and femora, as in Wilderi; by the absence of the strong sternal cone directly opposite the labium, which marks Wilderi; and by the venter having a rather broader band of yellow instead of the brighter marginal triclinium on the venter of Wilderi.

CEPHALOTHORAX: A rounded oval, sharply truncated at the base, the fossæ deep, the corselet grooves and cephalic suture strongly marked. The dorsum is somewhat flattened at the top, covered with silvery white pubescence, the ground color orange brown, with a yellow band at the margin and a broadened triangular patch upon the caput. The sternum is shield shaped, pointed at the apex, almost as wide as long, slightly elevated in the centre, and with rounded tubercles before the first coxæ and the lip; scarcely marked tubercles or lumpy elevations are opposite the second and third coxæ; color yellow, with streakings of brown; covered with pubescence. The labium is long, thickened from the base to the middle, and the tip a rounded triangle; maxillæ longer than wide, pyriform, and, like the lip, yellow in color, and covered, but not excessively, with bristles and hairs.

LEGS: 1, 2, 4, 3; not stout, but comparatively thin for such a large species. The spinal armature is scant and rather feeble, the terminal joints particularly, and heavily coated with yellow bristles. The apices of the tibial joints have clusters of bristles, forming a faintly developed brush, but not so prominent as in N. plumipes. The color is yellow, with orange tips at the joints, and the feet and metatarsi black or blackish brown. Palps colored as the legs; mandibles conical; a slight conical projection marks the outer edge of the base near the reticulation. Color dark brown and blackish.

EYES: Ocular quad on a well rounded prominence, wider behind than in front, and longest at the sides. The eyes do not greatly differ in size, but MF are the larger; are separated by about two diameters; MR by 3 to 3.5 diameters; side eyes on well elevated tubercles, almost equal in size, SF slightly the larger, and not differing much from the size of MF. The distance between MF and SF is about 1.5 to 1.3 the alignment of MF; the clypeus is high, about the alignment of MF. The front row is recurved, the rear row procurved.

ABDOMEN: Cylindrical, somewhat thicker at the base, which overhangs the cephalothorax, than at the apex, which slightly overhangs the spinnerets; color yellow; skin reticulated; dorsum thickly covered with whitish spots, the largest symmetrically arranged on either side of the median line; the others scattered over the dorsal field, which is limited at the margin on the sides by an interrupted line and on the base by a band of like color. A yellow band broken in the middle passes around the anterior part of the base; the sides are colored as the dorsum, and similarly marked by whitish spots. These and other portions of the abdomen are provided with silvery bristles, whose metallic lustre adds much to the beauty of the species. The venter is a yellow band with longitudinal stripes of lighter color, and whitish bands, with silvery pubescence at the sides, and irregular spots in the middle. Spinnerets and gills dark brown; epigynum is dark brown, chitinous, with a cluster of yellowish hair on the atriolum, which is orange yellow; it is without a scapus, but has four platelike cells, two in front and two others at the sides.

DISTRIBUTION: I have specimens from islands of the Carribean Sea, received from Mr. C. H. Townsend; Dr. Marx has examples from Louisiana and Texas; Koch described from a Brazilian specimen. The species is therefore probably limited in the United States to our Gulf Coast, and does not appear to be abundantly distributed therein. It no doubt abounds throughout the Northern States of South America.

No. 102. Nephila maculata (FABRICIUS).   Plate XXIII, Fig. 4.

1793. *Aranea maculata*, FABRICIUS . . . Entomologia Systematica, ii., p. 425, No. 66.
1817. *Aranea maculata*, LEACH . . . . Zoological Miscellany, ii., p. 134, pl. 110.
1890. *Nephila maculata*, MARX . . . . Catalogue, p. 551.

FEMALE: Total length, 27 mm.; abdomen, 19 mm. long, 6 mm. wide at the base, 3.5 to 4 mm. at the apex, 4 mm. in the middle; cephalothorax, 8 mm. long, 6 mm. wide across the corselet, 4 mm. at the face. Baron Walckenaer (Ins. Apt., II., page 98), in his description of Nephila (Epeira) fuscipes KOCH, makes that species identical with Nephila maculata of Leach and Fabricius. In this he errs, for the two species are very different, as one may easily see by comparing Walckenaer's descriptions and the figure of Koch with Leach's excellent figure. The drawings of N. maculata on Plate XXIII. were from an adult female; but after it had been completed I received from Dr. Marx several much larger specimens, one of which has a body length of an inch and three-quarters, and measures 7.5 inches (nearly two decimetres) from the tip of leg-I to the tip of leg-IV; leg-I is four inches long. These creatures strongly suggest the vigorous African and other tropical species of this beautiful genus.

CEPHALOTHORAX: Cordate; corselet rounded at the edges, depressed in the centre at the fosse, which is a deep lateral slit; rather flattened upon the top; the corselet grooves indistinct or obliterated; the cephalic suture sufficiently distinct; the body color brown to brownish yellow, covered thickly with golden yellow pubescence. The caput is narrowed at the base, on which, a little in front of the fosse, are two low conical processes on either side of the median line; beyond them are two depressions in the surface of the caput, which is brown, glossy, and provided less abundantly than the corselet with pubescence. The head is raised above the corselet, well arched to the face; sternum shield shape, about as wide as long; with sternal cones, one especially prominent in front of the labium, giving the centre of the sternum an elevated appearance; glossy brown; indented at the edges; sparsely pubescent (though the hairs may be obliterated by the alcohol); the apex somewhat blunted or T-shaped, and bearing upon either side a slight conical eminence; the labium is long, two-thirds the length of the maxillæ, rectangular at the base, which is thicker than the obtusely triangular tip; the maxillæ are longer than wide, narrowed at the base, curved upon the inner edges next the labium; ovate at the tips, where the width is about three-fifths the length; the color of maxillæ and labium dark brown.

EYES: Ocular quad on a rounded eminence more decided in front, leaving the rear eyes situated at the base thereof; the quad is decidedly narrower in front than behind, but the posterior width is equal to the length; MF slightly larger than MR, separated by about 1.5 to 2 diameters; MR separated 2 to 2.5; side eyes upon prominent tubercles; SF but little larger than SR, separated by something more than a radius; all the eyes of the group amber yellow; SF separated from MF by 1.5 the area of the latter, or at least two times or more the intervening space of the same; the clypeus is rather high, from the margin to MF being about three diameters of the latter; the front row very slightly recurved, almost aligned; the longer rear row is procurved.

LEGS: In the specimen in hand the legs are much broken and the color greatly modified by the alcohol. So far as reconstruction permits, the order shows 1, 2, 4, 3; the color brown, and, judging by the pits, apparently well armed with strong spines and thickly pubescent; the mandibles are conical, thick at the bases, arched to the tips; dark glossy brown, pubescent on the inner faces; on either side of the base they are raised into a process or cog which fits against a corresponding notched projection in the dewlap of the face; palps armed and colored as legs.

ABDOMEN: Cylindrical, widest at the base, compressed in the middle, again widening towards the apex, where it is one-half the width of the base; the front overhangs the cephalothorax, and the apex projects but little beyond the spinnerets, which are at the base of the high apical wall; the color, as far as can be noted from the imperfect specimen, is light brown, with a simple yellow folium extending the entire length of the dorsum, widest at the base and divided on the median line by a ribbon of brown; it thus presents the appearance of two yellow bands traversing the dorsum, and uniting in front like a tuning fork. A band of yellow crosses the base above the cephalothorax, and directly in front are curved markings, which extend along the sides in an interrupted band of yellow rounded and elongated spots. The venter is velvety brown, with bright yellow spots arranged five on either side, and two directly on the median next the epigynum, which is simply a semicircular flap projecting over the genital cleft.

DISTRIBUTION: One specimen, collected at Fort Buchanan, Arizona. Dr. Marx, in his Catalogue, also locates it at Mariposa, California. It will probably be found well dispersed along the Pacific Coast, in the more tropical parts thereof. (Marx Collection.) The specimens described by Leach and Fabricius are from China, and there is little doubt that the species is widely distributed along the coasts of Asia, whence, it may be, it was transplanted to America, unless one may choose to suppose a reverse direction of movement.

No. 103. Nephila clavipes (LINNÆUS). Plate XXIV, Fig. 1.

1758. *Aranea clavipes*, LINNÆUS . . . . Systema Natura, i., ii., p. 1034, n. 27.
1775. *Aranea clavipes*, FABRICIUS . . . . Systema Entomologiæ, ii., p. 420, n. 50.
1778. *Aranea fasciculata*, DE GEER . . . Mem. l'Hist. Nat. des. Ins., vii., p. 316, No. 2, pl. xxxix., 1–4.
1806. *Epeira clavipes*, WALCKENAER . . . Hist. Nat. des Aranéides, fasic. i., F, 1, 2.
1839. *Nephila fasciculata*, KOCH . . . . Die Arachniden, v., p. 30, pl. 152, Fig. 355.[1]
1842. *Epeira clavipes*, WALCKENAER . . . Hist. Nat. Insect. Apt., p. 95.
1890. *Nephila fasciculata*, MARX . . . . Catalogue Described Araneæ, p. 551.

FEMALE: Total length, 22 mm.; cephalothorax, 9 mm. long, 6.5 mm. wide, face width, 5 mm.; abdomen, 14 mm. long, 6 mm. wide at the base, and of nearly equal width throughout, but somewhat diminished at the apex, which is rounded and projects over the spinnerets.

CEPHALOTHORAX: A long oval; slightly compressed at the truncated apex, flattened on the summit of the corselet by the deep circular fosse; color blackish brown, covered (apparently) with silvery white hairs, especially long on the sides. The head is slightly elevated, wide at the face, and upon the base of the caput are two strong conical spurs. The sternum shield shaped, the basal width about equal to the length; sternal cones distinct, and one especially prominent in front of the labium, as in Nephila Wilderi; the margins are thickly covered with yellowish, bristlelike hairs; the labium and maxillæ normal, and covered with long, curved, blackish bristles.

EYES: The ocular quad upon a prominence, much elevated in front; sides longer than width, and rear wider than front; MF larger than MR, and separated by about 1.5 diameter; MR separated by 2.5 to 3 diameters; the side eyes on strong tubercles, about equal in size, separated by 1.5 diameter; SF removed from MF about 1.3 the area of the latter, or 2.5 to 3 times the intervening space. The clypeus height is about 2.5 diameters of MF. The front eyerow is slightly recurved, the rear slightly procurved.

---

[1] Walckenaer has pointed out what appears to me to be a fact, that, in some way, the plate numbers of Koch have become confused. Figure 355 is undoubtedly intended for the above species, not for the one bearing the plate name of "Nephila clavipes," as stated in the plate.

Legs: 1, 2, 4, 3; the color yellowish brown, without annuli, but darker shadings at the tips of joints. The tibiæ of legs-I, II, IV are provided with brushes, which in the specimen in hand are so much worn off in the alcohol that their form cannot be determined with accuracy. The palps are not long, the trochanter and humerus yellow, the tips dark brown. The mandibles are conical, not receding, blackish brown in color.

Abdomen: Subcylindrical, the base somewhat projecting over the cephalothorax, and the apex projecting beyond the spinnerets; the width throughout is almost equal, except at the apex, where it narrows somewhat into a rounded projection. The sides are yellowish brown, marked with silvery pubescence; the dorsum golden yellow, with the black muscular pits prominent, and veined lines passing from the middle point longitudinally to the apex; silvery pubescence appears to be scattered over the dorsal surface. The ventral pattern is a broad, yellow band, dappled with silvery spots. The epigynum is without scapus; the margin of the atriolum is black, corneous, and provided on either side with tufts of bristles that curve downward to the cleft.

Distribution: Key West, Florida. (The Marx Collection.) In America the species is doubtless limited to the Gulf Coast; but, on the supposition that, as above described, it is quite identical with De Geer's species, it has a wide distribution throughout the subtropical parts of America.

No. 104. Nephila concolor,[1] new species. Plate XXIII, Fig. 1.

Female: Total length, 28 mm.; cephalothorax, 12 mm. long, 8 mm. wide, 6 mm. at the face; abdomen, 20 mm. long, 9 mm. wide.

Cephalothorax: A long oval, the corselet flat and roughened upon the surface, sunken at the fosse, which is a deep lateral indentation; corselet grooves distinct, but abbreviated; cephalic suture deep and well marked; the caput rises from the fosse, sloping gradually to the face, the anterior portion of the corselet being raised therewith. Close to the fosse on each side of the base is a strong, conical, blackish brown, corneous tubercle, directed forward. The colors of the specimen are much faded out by alcohol, but the corselet appears to be yellowish brown, the caput blackish brown, covered rather freely with blackish hairs. The sternum is shield shaped (Fig. 1c), bluntly pointed at the apex, widest in front, glossy blackish brown color, provided rather sparingly with stiff brown bristles; the sternal cones prominent above coxæ-I, II, III, those in front of coxæ-III largest, and a cone much larger than all in front of the labium. The labium is subtriangular, the length equal to the width at the base, on each corner of which is a low wart; the middle part is lowly humped. The maxillæ are normal.

Eyes: Ocular quad on a conical eminence, rounded to a blunted point in the forepart between the middle eyes; the quad narrower in front than rear, which equals the length; MF somewhat larger than MR, separated by about 2 to 2.5 diameters, MR by 2.5 to 3 diameters. Side eyes on tubercles, nearly equal in size, widely separated by the distance of nearly two diameters; clypeus height about 2 to 2.5 diameters MF; front eyerow slightly recurved, rear row somewhat procurved.

Legs: 1, 2, 4, 3; ruddy brown, with dark glossy brown apical annuli; tarsus and metatarsus black; not stout, as compared with the size of the species; provided with short yellowish gray hairs, thickly placed in certain parts, particularly at bases of thighs, and with short dark spines; the apices of tibiæ-I, II, and the whole of tibia IV, show brushes of short curved bristles; the coxæ strongly pubescent, especially on the sides, dark brown at the bases, reddish brown at the apex, presenting a lumpy appearance, caused by two protuberances on coxæ-III and IV and one upon coxa-II, coxa-I being smooth. The palps are colored and armed as the legs; the humeral joint underneath is thickly covered with whitish gray hair; mandibles rather short, arched at the base, conical, glossy black in color.

---

[1] I adopt the cabinet name of Dr. Marx. The species closely resembles the preceding one (N. clavipes Lin.), and may be the same.

ABDOMEN: Subcylindrical, widest toward the base, rounded and somewhat narrow in front, where it overhangs the cephalothorax; somewhat narrowed at the apex, which slightly overhangs the spinnerets; dorsum but little arched, yellow, with veined branching lines from the middle to apex and scalloped marginal lines; the surface thickly covered with short silvery hair, which gives the organ a metallic white color; the skin color is yellow. The epigynum (Plate XXII, Fig. 1a) shows simply a thickened flap, obtusely triangular, without a scapus, and set opposite a conical process upon the posterior side of the genital cleft, which is thus inclosed between two liplike elevations.

.DISTRIBUTION: Southern California. The habitat of this species is somewhat in doubt. Dr. Marx received it from Professor Ulysses Brown, Edinburgh, Scotland, and it was contained in a broken box, the label of which was mutilated; but in the same box were several bottles, labeled "Group of southern islands, Santa Barbara, California."

GENUS THERIDIOSOMA, CAMBRIDGE, 1879.

This genus forms a connecting link between the Orbweavers and Theridioid spiders. It was created by Cambridge for the kindred, and perhaps identical, species Th. gemmosum, and assigned to the Retitelariæ, but was transferred by the writer to the Orbitelariæ, both on account of its habits and structure.[1] The individuals are very small, of delicate colors, and in general form resemble the species of Theridium. The cephalothorax is cordate, truncate at the base, the head much elevated, the face wide and projecting in front, the corselet grooves and cephalic suture are distinct, the skin glossy. The sternum is cordate or obtusely triangular; almost as wide as long. The labium is short, scarcely one-third the height of the maxillæ, which are as wide as, or wider, than long, in the female. The eyes are divided into three groups, the central quad longer than wide, and the greatest width behind; the midrear eyes being much the larger, and placed close together, relatively much closer than the midfront. The space between the sidefront and midfront eyes is twice the intervening space of the latter, or about equal to the area thereof; the clypeus is .high; the front row of eyes is aligned, or a very little recurved; the rear row decidedly procurved. The legs are 1, 2, 4, 3; rather stout; scantily pubescent; provided with sharp, bristlelike spines. The mandibles are long and conical. The abdomen is a rounded oval, the dorsum high, arched to the distal spinnerets, and is carried by the spider in a nearly vertical position. The epigynum is a large vaulted atriolum, without a scapus, occupying half the width of the venter. The male resembles in color and markings the female, though somewhat smaller in size. The palpal bulb is globular, and the accessory organs complex and much lengthened. This spider makes an orb web of the characteristic Epeïroid type, but captures its prey by means of a trapline, which it uses somewhat in the fashion of Hyptiotes, by alternate rapid tightening and releasing.

No. 105. Theridiosoma radiosum McCook.   Plate XXVII, Figs. 8, 9.

1881. *Epeira radiosa*, McCook . . . . Proceed. Acad. Nat. Sci., Phila., p. 163.
1884. *Microepeira radiosa*, EMERTON . . N. E. Ep., p. 320, pl. xxxiv., Fig. 7.
1889. *Theridiosoma gemmosum*, McCook, Amer. Spiders and their Spinningwork, Vol. I., ch. xii.
1889. *Theridiosoma radiosum*, MARX . . Catalogue, p. 551.

FEMALE: Total length, 2.5 mm.; cephalothorax, 1 mm. long, 9 mm. wide; abdomen, 1.5 mm. long, 1.3 mm. wide. These measurements vary somewhat according to the specimens, some being larger, some smaller. Count Keyserling, who had thoroughly studied and described Th. gemmosum Cambridge, also knew and described in manuscript the American

---

[1] Volume I., chapter xii.

species. He retained my name, deeming the two spiders distinct. Not having seen a European specimen, I retain the name originally given, as above, although inclined to think, from study of the drawings alone, Th. radiosum and Th. gemmosum identical, or perhaps the former a variety of the latter.

CEPHALOTHORAX: Cordate, truncate and indented at the base; the summit high; the head much elevated, the face projecting in front, and wide; the fosse a slight semicircular indentation; sloping downward to the base and forward to the face; the corselet grooves distinct, as is the cephalic suture; the color dull yellow, glossy. The sternum is cordate, obtusely triangular at the apex; color dark brown, with a broad, yellow median band; pubescent; rounded from the sides and flat in the middle; the labium small, scarcely more than one-third the height of the maxillæ, rounded at the tips; the maxillæ as wide as, or wider, than long; the tips squarely truncate, and inclined toward one another.

EYES: (Fig. 8d.) Situated on a round, black conical projection, at the point of which the midfront eyes of the ocular quad are placed; the quad longer than wide, its greatest width behind; MF, which are considerably smaller than MR, are separated by about 1 to 1.3 diameter; MR separated by scarcely more than a radius; the side eyes are separated by less than a radius; white, not greatly differing in size, but MR somewhat larger; SF removed from MF by a space almost equal to the area of the latter, or twice the intervening space or more; the clypeus is high, the margin separated from MF by 1.3 at least the area of the latter; the front row is aligned, or a very little recurved; the rear row procurved.

LEGS: 1, 2, 4, 3; short, rather stout; scantily pubescent; provided with sharp, bristle-like spines; the palps colored and armed as the legs; the mandibles long, conical, divergent at the tips, glossy, yellow.

ABDOMEN: A rounded oval; the dorsum highly and evenly arched, almost a semicircle, to the distal spinnerets; carried by the spider in a nearly vertical position, and, therefore, the base, which is rounded, towering over the cephalothorax; the color varies from blackish to yellowish brown (Figs. 8, 8b); the skin soft, beautifully reticulated; the follum a somewhat indistinct, broad, arrowhead pattern, marked out by silvery reticulations, which give the spider a shining appearance. The epigynum (Fig. 8e) is large, occupying half the width of the venter, with a high, subtriangular atriolum, showing the concavity beneath; there is no scapus, simply a swelling at the middle point of the atriolum.

MALE: (Fig. 9.) Resembles the female in color and markings; somewhat smaller in size; the palps are distinguished as shown at Figs. 9b, 9c, 9d.

DISTRIBUTION: I have collected this species in New England (Massachusetts, Connecticut); in New York, quite abundantly in the neighborhood of Philadelphia, and as far west as Ohio. Dr. Marx has it from Illinois. If we accept the identity of Theridiosoma gemmosum with the American species, it is widely distributed throughout Europe.

GENUS TETRAGNATHA, WALCKENAER, 1806.

The species of this genus are distinguished by their elongated form, the abdomen being several times as long as wide. This effect is increased by the habit of the species to stretch their forelegs together and lie close to the surface of plants, so that the legs and abdomen form one continuous line. The cephalothorax is an elongated oval, flattened upon the dorsum of the corselet; the head usually erect, wide at the face, and quadrate. The sternum is longer than wide; and the maxillæ peculiar, in that they are decidedly longer than wide, concave upon the outer margin, and widened at the tip. The eyes are placed in two rows, both of which are usually slightly recurved or aligned. The midfront eyes are separated from one another by a space ordinarily equal to about one-half the distance between the midfront and sidefront eyes; the side eyes are nearer one another than or as near as MF to MR. The mandibles are elongated, oval, usually narrower at the base than at the apex; are marked by strong teeth, and in the male are even more powerful than in the female. The abdomen is cylindrical, or subcylindrical, elongated, decidedly longer than wide. The legs are long, attenuated, sparsely clothed with long, aculeate spines, and more abundantly with bristles and hairs. The number of rows of spines down the tibia is ordinarily about three; upon the metatarsus from two to three; upon the femora, ordinarily from six to eight. The orbicular snare is commonly hung horizontally, or at an angle more or less inclined. It is frequently found near or immediately over water, as the species affect humid places and the banks of streams. While seated upon its snare the four front legs are usually stretched-out straight before the face.

No. 106. Tetragnatha extensa (LINNÆUS).   Plate XXV, Figs. 3, 4, 5.

1758. *Aranea extensa*,[1] LINNÆUS . . . . Syst. Nat., Ed. x., i., p. 621.
1763. *Aranea Solandri*, SCOPOLI . . . . Entomologia Carniolicæ, p. 397.
1763. *Aranea Mouffeti*, SCOPOLI . . . . Ibid., p. 398.
1805. *Tetragnatha extensa*, WALCKENAER, Tabl. d. Aran., p. 68.
1837. *Tetragnatha obtusa*, C. KOCH . . Uebers. d. Arachn. Syst., i., p. 5.
1842. *Tetragnatha versicolor*, WALCK.. . Ins. Apt., ii., p. 215; ABBOT, G. S., No. 466.
1861. *Tetragnatha extensa*, WESTRING . . Araneæ Svecicæ, p. 84.
1864. *Tetragnatha extensa*, BLACKWALL . Sp. G. B. & I., ii., p. 337, xxviii., 265.
1866. *Tetragnatha extensa*, MENGE . . . Preuss. Spinn., i., p. 90, pl. 15, tab. 26.
1866. *Tetragnatha obtusa*, MENGE . . . Ibid., p. 93, pl. 15, tab. 27.
1870. *Tetragnatha extensa*, THORELL . . Synonyms European Spiders.
1877. *Tetragnatha extensa*, THORELL . . Araneæ Colorado,[2] Bull. U. S. Geol. Survey.
1884. *Tetragnatha extensa*, EMERTON . . . N. E. Ep., 333, pl. 39, Figs. 9, 10.
1889. *Tetragnatha extensa*, McCOOK . . Amer. Spiders and their Spinningwork.

FEMALE: Total length, 8.5 mm.; abdomen, 6.3 mm. long; across base, 3 mm. wide; across apex, 1.5 mm. wide; cephalothorax, 2.5 mm. long by 2 mm. wide. The general colors of the fore part are yellow and yellowish brown, and of the abdomen various shades of yellow and greenish. While young, often a uniform light green (Fig. 5b).

CEPHALOTHORAX: A long oval, truncated and indented behind, widest at the middle, flattened at the top; dorsal fossæ deep; corselet grooves sufficiently distinct; cephalic suture distinct; the head elevated above the corselet level; the caput wide, the projecting part or head of nearly equal width to the face; color, a warm yellow, with brown markings on sides of caput and margins of corselet; skin glossy, pubescence scant. Sternum (5d) a long shield shape, tapering to a point, high in the middle, slightly pubescent; color, a broad, yellow median band, with blackish or dark brown margins; labium subtriangular, relatively shorter than in T. grallator; maxillæ concave .on the outer edges, of nearly equal width throughout, but widest at the tips.

EYES: (3, 3b.) MF on a central prominence; the MR not thus elevated; ocular quad narrowest in front, the rear wide as, or wider, than the sides. MF smaller than MR, separated by 1.5 to 2 diameters, MR separated by 2 to 2.5 diameters. Of the side eyes SR are the larger; they are separated from one another by about two-thirds the space which divides MF from MR. MF are separated from SF by about 1.25 their area, and from the margin of the clypeus by about 2.5 diameters. The front row slightly recurved, the rear row slightly procurved, almost aligned. The distance between MR and SR is about equal to the space between the two MR.

LEGS: 1, 2, 4, 3; first pair at least eight times length of cephalothorax; of uniform yellow color, except slightly darker annuli at the joints, feebly armed with pubescence, bristles, and a few blackish spines. The mandibles (3a, 3b) about seven-tenths the length

---

[1] A complete synonymicon would fill several pages; I give but a few of the more important synonyms.
[2] Professor Thorell appears to me in this report to have confounded variations of this species with T. elongata (grallator).

of the cephalothorax; thickest at the apex, arched about the middle, and tapering some-what to the base; almost straight upon the exterior margin. Teeth numbering about eight, of nearly equal length, but the two middle ones slightly longer. The fang with a slight tooth on the exterior side near the base.

ABDOMEN: Cylindrical, longer than wide, in many species straight along the dorsum, in others arched (3a) slightly, tapering at the apex, which slightly overhangs the spin-nerets; the base overhangs the cephalothorax, is somewhat narrowed, and is bifid. The dorsal folium is yellowish brown, traversed by a broad, irregular median line of cretaceous yellow; it occupies the whole of the dorsum, and is margined by an indented band of cretaceous yellow, extending downward along the side. This band is sometimes whitish, inclined to silver, and sometimes shaded by a line of black. Beneath it, on the side, is a band of brown, extending to the venter, whose pattern is a broad, brown median ribbon, margined on either side by a strip of cretaceous yellow. In some individuals (Fig. 3) the folium is not so strongly marked as in the above, Fig. 5. The epigynum (5c) is a simple cleft, with an arched hood.

MALE: (Figs. 4 and 5x.) 7 mm. long; colored and marked generally as the female. Both rows of eyes are distinctly procurved, the head well elevated; the jaws (Fig. 4a) longer, relatively, than those of the female, are strongly bent outward, cylindrical, somewhat compressed at the base; on the upper and inner surface is a long spur (4b), slightly bifid; one of the teeth, nearly midway of the furrow, is especially prominent (4a); the fang is long and evenly curved, extending as far as the maxillæ (5d); of the palps (Figs. y and 4c), the radial joint is about one-third longer than the cubital.

DISTRIBUTION: This spider appears to be distributed throughout the United States. I have it from California (Mr. Curtis, Mrs. Smith, Mrs. Eigenmann), from British Columbia, and have collected it from Canada and New England southward to Florida. Thorell describes it from Labrador. It is one of our cosmopolitan species, being distributed widely through Europe, and perhaps other continents.

No. 107. Tetragnatha elongata WALCKENAER.   Plate XXV, Figs. 1, 2.

1805. Tetragnatha elongata, WALCKENAER . . . Tabl. des Araneides, p. 69, No. 2.
1837: Tetragnatha grallator, HENTZ . . . . . . J. B. S., vi., p. 26.
1841: Tetragnatha elongata, WALCKENAER . . . Ins. Apt., ii., p. 211; ABBOT, G. S., No. 216.[1]
1864. Tetragnatha fluviatilis (?), KEYSERLING. . Orbitel, Verh. Ges. Wien, 852, xxi., 10.
1865. Tetragnatha grallator, KEYSERLING . . . Ibid., p. 850, xxi., 24–27.
1875. Tetragnatha grallator, HENTZ . . . . . . Sp. U. S., p. 131, xv., 1, 2.
1877. Tetragnatha elongata, THORELL . . . . . Araneæ Collected in Colorado, p. 477.
1879. Tetragnatha Illinoisensis (?), KEYSERLING, N. Spinn. Amer., i., 318, iv., 18.
1884. Tetragnatha grallator, EMERTON . . . . N. E. Ep., p. 334, xxxix., 1–6.
1889. Tetragnatha elongata, McCook . . . . . Amer. Spiders and their Spinningwork.
1889. Tetragnatha elongata, MARX . . . . . . Catalogue, p. 552.
1893. Tetragnatha grallator, BANKS . . . . . Jour. N. Y. Entom. Soc., i, 131.

FEMALE: Total length, 12 mm., including the outstretched mandibles, 14 mm.; abdo-men, 9.5 to 10 mm. long at the base, 3 mm. wide, tapering to 1.5 mm. at the apex; cephalothorax, 2.5 mm. long by 1.7 mm. wide. This spider grows to even larger size in some of the running waters and ponds of our country, near which it makes its home, frequently spinning its horizontal web above the surface of the water. Its general colors for the fore part are brown, or yellowish brown, and for the abdomen yellow, with brown markings, which (particularly in alcohol) assume a metallic golden lustre.

CEPHALOTHORAX: A long oval, truncated at the base; fosse deep; corselet grooves indistinct; cephalic suture marked; the head wide, and but little tapering to the face; color

---

[1] Abbot appears to have drawn, and Walckenaer to have described, this species under several names. I consider as identical with T. elongata, not only the above (T. fulva), but T. fimbriata (page 213); G. S., Nos. 461, 481, and T. violacea (page 218), No. 448.

brown, with lighter patches of yellow, the caput and face yellowish brown. Sternum shield shape, longer than wide, elevated in the centre, brown, with a yellow median band. The labium is long, of nearly equal width throughout. The maxillæ (Fig. 1d) are extremely long, rounded at the tip on the inside, and bluntly pointed on the outside, the width about one-third the length.

EYES: Eye space wide; the ocular quad (1b) on an elevation most prominent in front; rear wider than front, and about equal to sides; MF larger than MR, separated by less than 1.3 diameter; MR separated by nearly 2 diameters. The side eyes on a slight tubercle; SR greater than SF, from which they are separated by about their diameter; the space between side eyes is less than that between MF and MR; SF are separated from MF by about twice the distance between MF. The clypeus in height about 1.5 diameter MF. All the eyes are black, or upon black basal spots. The front row is slightly recurved, the rear row about aligned.

LEGS: 1, 2, 4, 3; long, thin, armed with light, yellowish bristles, and rather sparingly with long, thin spines, set upon dark pits; first pair eleven times as long as the cephalothorax; color, warm yellow, with slight brown annuli at tips of joints, and the feet somewhat darker or blackish. The palps resemble the legs in form, color, and armature. The mandibles are long, subcylindrical, the outer margin concave, bent sharply outward toward the tips, where they are widely separated; their length is about that of the cephalothorax; the fang (1e) is long, curved, sinuated, not quite touching the maxillæ, black, or blackish brown; the base marked with a tooth on the upper surface. The mandibles are a light shade of brown or dark yellow. The tips are armed around the apical margin with strong black teeth; one large tooth is placed upon the apex of the exterior furrow; after a wide interval there follow six smaller teeth. The interior row consists of nine to ten teeth, diminishing in size toward the base.

ABDOMEN: Much longer than wide, cylindrical, thickest at the base. The dorsum is highly arched on the base, about the middle slopes downward, and the apical half contracts into a width of about one-half that at the base, where it is somewhat rounded or truncated; the spinnerets are distal. The surface is clothed with delicate pubescence; the color varies from yellow to light brown, with, especially in alcoholic specimens, metallic, golden reticulations. The dorsal folium is bounded by two longitudinal, waving lines, having about three indentations from the base to the apex. In the middle is a darker longitudinal band, separated more or less distinctly by a median line. The ventral pattern is a brown or dark yellow band, passing from the gills to the spinnerets, with marginal bands of lighter color, the whole reticulated like the dorsum and sides. The epigynum (1c) is simply a small semicircular hood, without any scapus.

MALE: Length, 8 mm.; from the tips of the outstretched mandibles, 11 mm. The legs, cephalothorax, and mandibles are colored as in the female, and the abdomen has the same general colors and form. The dorsal folium, however, is not so complicated. The abdomen is not so thick at the base, relatively, but is more equal in length throughout, cylindrical, and covered with pubescence apparently stronger than in the female. The palps are extremely long (2a, 2b), the digital joint swollen into a semiglobular bulb, beyond which the cymbium and alveolus much project; the cubital joint is about one-half the length of the radial. The mandibles (2a) are relatively longer than those of the female, more uniformly concave outwardly, thickened towards the tips, with long, sinuated fangs; of the teeth on the front margin of the claw furrow one is much longer and stronger than the other, and one strong, curved tooth, next in size to the above, is set upon the inside, at the tip. A long spur, with bifid tip, marks the upper and exterior part of the apex. The labium is long, somewhat narrower at the rounded tip; the maxillæ, as in the female, extremely long, cut rather more squarely at the tip, where they are widest, and thence curving along the outer margin to the base, where they are even thicker than at the tip.

DISTRIBUTION: This species is widely distributed throughout the United States, my specimens locating it from New England, along the Atlantic Coast, southward and westward through Pennsylvania, Ohio, Wisconsin, and again along the Pacific Coast from Oregon to California. Hentz found it in North and South Carolina and Alabama; Abbot in

Georgia. It is probably found in all parts of the United States and Canada. It loves the neighborhood of streams, and stretches its horizontal snare above the water. I have found them covering the plants around the base of a roaring waterfall.

No. 108. Tetragnatha Banksi, new species.   Plate XXIV, Fig. 6; Pl. XXVIII, 4.

MALE: Total length, 6 mm., including the mandibles, 7.5 mm.; abdomen, 4 mm. long, 1 mm. wide; cephalothorax, 2 mm. long, 1 mm. wide; general colors for the fore part, uniform yellow, on the abdomen mottled with black.

CEPHALOTHORAX: A long oval, widest in front; the fosse a deep indentation upon the summit of the corselet; corselet grooves indistinct; cephalothorax suture sufficiently marked. Color yellow, tipped with brown, and sparsely covered with gray pubescence; caput elevated above the corselet and of like color; face wide, with pubescence in the ocular area.

STERNUM: Arrow shaped, longer than wide, receding from the apex, raised in the middle where it is flattened; color dull yellow, with scant pubescence. Labium long, wide at the base, subtriangular at the tip; less than half the width of the maxillæ, which are long, club shaped, of nearly equal width throughout, roundly truncate at tips, which point outwards.

EYES: Ocular quad widest behind, the length about equal to the width; MF on a rounded elevation and separated by about 1.3 diameter, and smaller than MR, which are separated by about two diameters. SF decidedly smaller than SR, and the two more closely approximated than the middle group; the space between SF and MF is about twice the distance between MF; between MR and SR a little greater than between MR. Front row slightly recurved, rear row aligned or nearly so; clypeus wide, equal at least to twice the diameter MF.

LEGS: 1, 2, 4, 3; leg-II being but slightly longer than leg-III; thin, sparsely provided with dark brown spines and feeble bristles; color yellow. Leg-I measures from 20.5 to 22 mm. The mandibles are about 2 mm. long, concave externally, slightly increased in width towards the apex; the fang evenly curved externally, but underneath slightly undulate, and extending below the tips of the maxillæ when folded; about six to eight teeth of nearly equal length; on the inner and upper surface a strong, long, slightly bifid spur. The radial joint of the palp is about 1.5 times the length of the cubital joint, and is provided with a slight curved projection, which distinguishes it from the males of other indigenous species of Tetragnatha.

ABDOMEN: The abdomen is cylindrical, about four times longer than wide, somewhat narrowing towards the distal spinners. An indistinct folium marks the dorsal surface, which is reticulated and flanked on either side by a band of brownish color. I have what I regard as the female of this species, but am not certain as to the identity, and therefore give no detailed description. In general form and color it resembles the male.

DISTRIBUTION: I have several specimens of this species, which I have dedicated to Mr. Nathan Banks, from Florida, and one from Wisconsin. This would indicate a general distribution throughout the United States.

No. 109. Tetragnatha laboriosa HENTZ.   Plate XXV, Figs. 7, 8.

1850. *Tetragnatha laboriosa*, HENTZ . . . J. B. S., vi., 27; Sp. U. S., 131, xv., 3.
1865. *Tetragnatha laboriosa*, KEYSERLING Orbitel., Verh. Ges. Wien., 841, xx., 16, 17.
1884. *Tetragnatha laboriosa*, EMERTON . N. E. Ep., p. 334, xxxix., 7, 8, 11, 19; xl., 17.
1889. *Tetragnatha laboriosa*, MARX . , . Catalogue, p. 552.
1893. *Tetragnatha laboriosa*, BANKS . . Jour. N. Y. Entom. Soc., i., 131.

FEMALE: Total length, 7.5 mm.; abdomen, 5 mm. long, 2 mm. wide, receding to 1 mm.; cephalothorax, 2 mm. long, 1.3 mm. wide. The size varies a good deal; the colors are yellowish brown, with silvery abdomen.

CEPHALOTHORAX: A long oval, truncated at the base, the fossæ deep, and corselet grooves rather indistinct; dorsum flat; head slightly elevated, and but lightly depressed in front; color brown, with lighter shades of yellow, particularly at the face. Sternum (7c) longer than wide, widest at intersection of second coxæ, and much tapering towards the apex, rounded in the centre; color yellow or yellowish brown, slightly pubescent. The labium rounded at the tip, about one-third length of maxillæ, which are normal.

EYES: Ocular quad elevated in front, MR scarcely elevated above the facial surface. The front narrower than the rear, the latter about the length of the sides; MF smaller, separated from one another by about 1.5 diameter; MR separated by at least two. The side eyes, of which SR are larger, are separated from each other as widely as, or more widely, than MF from MR, but the distance not so decided as in Eugnatha. The front row is nearly as long as the rear row. The front row is slightly recurved, the rear row also recurved. MF are separated from SF by about 1.3 their area, or 2.5 times their intervening space. The clypeus is about the height of 1.5 to 2 diameters of MF.

LEGS: 1, 2, 4, 3; long, thin, armed with slight bristles of yellowish gray color, and a few long, feeble, dark brown spines. Color yellow, with a slightly darker annulus at tips of joints. The first leg is from six to seven times the length of cephalothorax.. The palps are light yellow, the mandibles (7b) wide towards the middle, thence slightly convex on the inside, straight upon the outside; the skin glossy and a warm yellow; the fangs dark brown or blackish; the mandibular teeth not greatly differing in size. The fang is without a basal tooth, is as long as the maxillæ (6b); not undulate.

ABDOMEN: Cylindrical, about four times as long as wide; somewhat thicker in the basal than the apical half, the difference being more decided in some specimens than others. Many individuals show the abdomen almost straight, others slightly arched in the basal half. The spinnerets are placed beneath the apical wall, which slightly overhangs. The skin is reticulated, covered quite closely with short grayish hairs; the color silvery, with streakings of yellow. The dorsal folium consists simply of a median line of brownish hue with various radiating veins. The venter shows a dark brown band extending from the gills to the spinnerets, flecked on either side by bands of yellow and lighter shade in the middle. The epigynum (7d) is an arched hood without scapus.

MALE: Fig. 8 in color closely resembles the female; the eyes show little or no difference. The radial joint of the palpus (8b) is about half as long as the digital and nearly equals in length the cubital joint.

T. laboriosa may be distinguished from T. extensa, T. elongata, and T. Banksi by the side eyes, which are further removed from one another than midfront from midrear, instead of being placed nearer to one another than, or as near as, MF from MR. Moreover, Laboriosa's abdomen has a silvery sheen, and as a rule is of more even thickness throughout; but in the latter respect one observes differences, for some examples of T. extensa especially approach T. laboriosa in the form of the abdomen, and the latter, on the contrary, sometimes is thickened towards the base. Most species, except T. laboriosa, when in alcohol, have the reticulations on their sides a metallic golden hue.

It is more difficult to distinguish between T. elongata and T. extensa. This, however, may be done as follows: The eyes, substantially, are alike in their arrangement; but the mandibles of Elongata are relatively somewhat longer than the cephalothorax, being .8 as long at least, while the mandibles of Extensa are at most .7 as long. Again, the first pair of legs of Elongata are eleven times as long as the cephalothorax, while the first leg of Extensa is only about six or seven times as long as the cephalothorax. The female of each species has a tooth which points forward placed upon the upper part of the mandibular fang near the point of articulation, but in T. elongata this is relatively larger than in T. extensa. It is wanting in the male of both these species. A similar tooth, but apparently less decided, is found upon the fang of the female T. laboriosa. Elongata male may be further distinguished from Extensa by the fact that the margins of the mandibular shield or furrow have at least ten to twelve teeth, while Extensa numbers but six to eight. Elongata also appears to grow to a much larger size, and the tendency in the adult species is to have the abdomen much thicker relatively at the basal half than the apical half, and also to be decidedly curved; that is to say, arched upon the basal half and curving downward to an equal thickness of the apical half.

DISTRIBUTION: T. laboriosa has been collected from New England southward to Florida and Texas, and on the Pacific Coast as far north as the British possessions. It doubtless dwells in the entire United States.

GENUS EUGNATHA, SAVÍGNY and AUDOUIN, 1825.

The genus Eugnatha in its general characteristics resembles Tetragnatha. It is distinguished therefrom especially by the position of the side eyes, which in Tetragnatha are rather closely approximated, being at least nearer to one another than are the midrear and midfront eyes. In Eugnatha the side eyes are widely separated, much more so than the eye rows of the middle group. The genus is distinguished from Eucta by the fact that the abdomen does not extend beyond the spinnerets into a caudal apex.

No. 110. Eugnatha vermiformis (EMERTON).   Plate XXV, Fig. 9.

1884. *Tetragnatha vermiformis*, EMERTON, N. E. Ep., 333, xxxix., Figs. 12–14.
1889. *Tetragnatha vermiformis*, McCOOK. Amer. Spid. and their Spinningwork, Vol. I., p. 160.
1889. *Eugnatha vermiformis*, MARX . . . Catalogue, p. 552.
1893. *Eugnatha vermiformis*, BANKS . . Jour. N. Y. Entom. Soc., i., p. 131.

FEMALE: Total length, 11 mm.; cephalothorax, 3 mm. long, 2 mm. wide; abdomen, 8.7 mm. long, 2.6 mm. wide, diminishing to 1.3 mm.

CEPHALOTHORAX: Elongated oval, truncated at the base; the fossa a low circular depression; cephalic suture distinct; corselet grooves indistinct; color brown, with yellow bands along the sides, sparsely covered with yellowish pubescence. Sternum yellowish brown, with a lighter median band; the centre raised and flattened; sternal cones not prominent.

EYES: Ocular quad elevated in front alone, MF being upon a rounded elevation; longer than wide, the rear wider than the front; MF somewhat smaller than MR, separated by a little more than one diameter; MR by 1.5 or more. The side eyes about equal in size to MF, separated more widely than MF, but not relatively so much as in Eucta; SF from MF by about twice the interspace of the latter; the distance between SR and MR relatively somewhat less than the above. Clypeus height equal to about twice the diameter MF; the front eye row aligned, the rear row recurved.

LEGS: 1, 2, 4, 3; rather strong, armed with long aculeate bristles and short yellowish hairs, with a few thin spines. Color uniform yellow. Palps colored and armed as the legs. The mandibles project from the face at an angle of about forty-five degrees; are almost straight on the exterior edge; 1.7 mm. long; widest about the middle, concave on the interior margin, color yellow; fangs dark, glossy brown; scarcely or barely extending to the maxillæ; a slight tooth marks the outer surface near the base. The interior furrow has two teeth on the apex near the insertion of the fang, and about eight farther along the furrow, of which the third and sixth are the longest.

ABDOMEN: Thickened at the base, bifid in front, cylindrical, four to five times as long as wide, color yellow, with golden reticulations. (Figs. 9–9a.) The spinnerets are just below the apical wall, which projects a very little. The epigynum is a small arched hood, without scapus. The form of the abdomen differs from the above at times (9b), the width being almost equal throughout and curved, giving a decided vermiform appearance.

MALE: Closely resembles the female. The radial joint of the palps is not longer than the cubital; the mandibles are shorter than the cephalothorax.

DISTRIBUTION: This species has been collected in New England, in New York (Banks), and as far west as Utah (Marx), and my specimens are from Wisconsin (Professor Peckham). It is thus probably distributed throughout the entire United States, at least east of the Rocky Mountains. The young of this species were discovered by me navigating the waters of Deal Lake, New Jersey, by means of filiform sails, as described in Vol. I., page 160.

No. 111. Eugnatha pallida (Banks). Plate XXV, Figs. 10, 11.

1892. *Tetragnatha pallida*, Banks . . . Proceed. Acad. Nat. Sci., Phila., p. 51.
1893. *Eugnatha pallida*, Banks . . . . Jour. N. Y. Entom. Soc., i., p. 132.

Female: Total length, 11.5 mm.; abdomen, 9 mm. long, 2.3 mm. at the base, diminishing in some specimens to 1 mm. at the apex; cephalothorax, 3 mm. long, 1.5 mm. wide; mandibles, 2.3 mm. long.

Cephalothorax: Oval, squarely truncated at the base; fosse a triangular depression; corselet grooves and cephalic suture distinct; caput elevated above level of corselet; color brown, with yellow markings upon the caput, face, and sides, and slight grayish pubescence. Sternum yellowish brown, raised in the middle, a depression between coxæ-I and II, scantily pubescent, with a line of long curved bristles along the concavely truncated base, extending towards the labium; the latter is roundly triangular, wide at the base, which is of a brownish color, the tip being yellow; maxillæ more than twice the length of the labium; widest at the tips, which are roundly truncated, pointed outward, slightly concave on the margin; the fangs when folded do not touch the maxillæ.

Eyes: Ocular quad on a rounded eminence most prominent in front; the rear width greatest and about equal to the length; MF somewhat larger than MR, separated by about or a little more than one diameter, MR separated by about two diameters. Side eyes upon small tubercles about equal in size, smaller than MF, separated by a space about one quarter greater than that which divides the middle group; space between MR and MF about twice the interspace of MF; the space between MR and SR somewhat less than that between MR.

Legs: 1, 2, 4, 3; uniform yellow in color, abundantly armed with small aculeate bristles, and well provided with long blackish spines. The palps are colored and armed as the legs. The mandibles are long, strong, extended forward nearly horizontally from the face, almost straight or but slightly concave upon the exterior; about ten teeth on the superior claw furrow, of which two are on the apex of the mandible; color yellow, with long bristles, particularly upon the inner surface; fang dark brown, glossy, evenly curved on the exterior surface.

Abdomen: About four times as long as wide; the dorsum slightly arched; in some specimens thickened, in others of almost uniform width throughout, and with but little curvature; dorsal surface beautifully reticulated, the skin in alcoholic specimens having a metallic golden lustre; the dorsal pattern is a central median line, with a cruciform figure near the base, and two curved longitudinal branches further down. The spinnerets are just under the apical wall. The ventral pattern is a lighter band of reticulated yellow. The epigynum is situated close up to the base, and is a simple arch without scapus.

Male: 10.3 mm. long, somewhat shorter and slighter in general form than the female, but in color and pattern closely resembling her. The corselet grooves and cephalic suture are less distinct; the midfront eyes are decidedly larger than the midrear; SF somewhat smaller than SR. The legs are long, the first leg much longer relatively than the second; it is about eleven times the length of the cephalothorax. The mandibles are three to four mm. long, straight upon the exterior surface, and but little curved on the interior, which is marked by strong rows of long bristles; the teeth are small, those upon the interior furrow numbering about nine, with two additional and larger ones on the apex; a long spur curved and widely bifid, or with a short tooth below the point, mark the upper and inner surface near the apex. The fang is long and somewhat undulate towards the apex. The radial joint of the palp is nearly twice as long as the cubital. The jaws are slightly inclined to the face instead of being extended almost horizontally, as with the female. Banks says that the outer tooth (spur) is not bifid. This characteristic should, I think, be modified as above by speaking of the spur as "widely bifid," or at least with a toothlike projection upon the side near the apex.

Distribution: My specimens are from New York, New Jersey, and Florida; Dr. Marx has a specimen from Texas, thus indicating a distribution along the entire Atlantic and Gulf Coasts.

Genus EUCTA, Simon, 1881.

The general characteristics of this genus are those of Eugnatha, the side eyes being widely separate, much more so than the middle eyes. The genus is especially distinguished from Eugnatha by the prolongation of the abdomen into a caudal like apex, which extends some distance beyond the spinnerets.

No. 112. Eucta lacerta (Walckenaer).   Plate XXIV, Figs. 5, 6; Pl. XXVIII, 5.

> 1842. *Tetragnatha lacerta*, Walckenaer.[1] Ins. Apt., ii., p. 224; Abbot, G. S., No. 496.
> 1884. *Tetragnatha caudata*, Emerton . . N. E. Ep., 335, xxxix., 16–22.
> 1889. *Eucta caudata*, Marx . . . . . . Catalogue, p. 552.

Female: Total length, 8 mm.; cephalothorax, 2 mm. long, 1.3 mm. wide; abdomen, 7 mm. long, 1.3 mm. wide at the base, diminishing almost to a point. The general color in alcohol is yellow, the abdomen covered with metallic golden reticulations.

Cephalothorax: Rectangular—ovate, truncated at the base, where it is somewhat higher than in front; color yellow, with brownish bands on either side of the fosse, extending along the edges of the caput; face yellow, as are the mandibles. Sternum an elongated shield, rounded in the centre; yellowish brown, with a lighter median band. The fangs, when folded, extend to the maxillæ. The latter organs are widest at the tips, and about three times the length of the labium, which is obtusely triangular at the tip.

Eyes: (Fig. 5d.) Ocular quad slightly elevated in front, widest behind, eyes not greatly differing in size; MF separated by about 1.5 diameter, MR by about 2 diameters. Side eyes about one-fourth further removed from one another than are the middle eyes; somewhat smaller in size. Front eye row is procurved (5b), the rear row decidedly recurved. MF are separated from one another by about one-half less than the space between SF and MF. The clypeus is about the height of two diameters MF. The space between MR is greater than that between MR and SR.

Legs: 1, 2, 4, 3; uniform yellow, thin almost to feebleness, covered with slight pubescence. The mandibles are yellow, long, narrowed, and slightly widened at the tip; about one mm. long.

Abdomen: At least seven times as long as wide, diminishing almost to a point; spinnerets elongated, from one-fourth to one-third the distance from the apex; color yellow, with golden reticulations, and median lines of brownish hue drawn along the dorsum, marking out an imperfect folium; venter a long band of yellow, bordered by sinuated lines of yellowish brown. The epigynum an arched atriolum, without scapus.

Male: (Fig. 6.) Is smaller and slighter than the female; resembles it in color and marking. The abdomen is squarely truncate at the base, and terminates in a caudal extension, as with the female. The mandibles are relatively longer than the female, but little curved, almost straight upon the exterior margin. The fangs are feeble as compared with Tetragnatha; an elongated spur upon the upper surface, about one quarter the distance from the apex. One long tooth marks the superior claw furrow about one-third the distance from the apex. Palp is longer, the radial joint scarcely longer than the cubital.

Distribution: This species has been collected as far north as Canada, in New York (Banks), New England, New Jersey, District of Columbia (Marx), and as far south as Florida and Alabama, and to the southwest as far as Arizona and Texas. It is probably distributed throughout the entire United States, at least east of the Rocky Mountains.

---

[1] From Walckenaer's description and Abbot's drawing I have no doubt that this is identical with Emerton's *Tetragnatha caudata*. Walckenaer expresses the opinion that the species should probably be placed in a new genus, but, not having more than the drawing to guide him, left it provisionally with Tetragnatha. Abbot calls it the "Lizard Spider," and adds the note: "This singular spider resembles a lizard in the abdomen. It commonly carries the end crooked, like the tail of some animals."

Genus PACHYGNATHA, Sundevall, 1823.

Pachygnatha, like Tetragnatha, is distinguished by its strong, elongate, curved jaws, widely separated at the apex, and armed with formidable teeth. It is distinguished from Tetragnatha, on the other hand, by an ovate abdomen, but little longer than wide. The cephalothorax is oval, its skin hard and glossy; the caput rounded and elevated above the corselet, the grooves and cephalic suture distinct. The sternum is longer than wide, decidedly aculeate at the apex, with distinct sternal cones; wide at the base. The labium is long, wide at the base, more than half as long as the maxillæ, which are decidedly longer than wide, a little compressed at the shank, convex on the outer margin, and inclined toward each other. The legs are in order of length 1, 2, 4, 3, nearly all without ordinary spines, but freely clothed with soft, long, aculeate bristles. The palps are long, and armored as the legs. The mandibles are formidable, elongate, resembling those of Tetragnatha. The eyes are in three groups, the central quad upon a high eminence; the side eyes upon tubercles less elevated; the space between the midfront and sidefront eyes is somewhat greater than the area of MF, twice and more the intervening space thereof, as in Epeira. The clypeus is extremely high, being more than 1.5 times the area of MF, as in Theridioid genera. The front eye row is slightly recurved, the rear row more or less procurved.

No. 113. Pachygnatha brevis Keyserling. Plate XXVI, Figs. 9, 10; XXVIII, 2.

1882. *Pachygnatha tristriata,*[1] Keyserling . Neue Spinn. Amerik., iv., 209, Zool. Bot. Ges.
                                           Wien., p. 17.
1883. *Pachygnatha brevis,* Keyserling . . . Ibid., v., p. 658.
1884. *Pachygnatha brevis,* Emerton . . . . N. E. Ep., p. 366, xxxiv., 21; xl., 8–10.
1889. *Pachygnatha brevis,* Marx . . . . . . Catalogue, p. 553.

Female: Total length, 5.5 mm.; cephalothorax, 2.5 mm. long, 1.5 mm. wide; abdomen, 3.5 mm. long, 2.5 mm. wide; mandible, 1.3 mm. long.

Cephalothorax: Glossy, with scant pubescence, the skin punctured; a dark band of brown along the margin, a median ribbon of light color extending from the forehead to the base of the corselet, on either side of which is a wide band of yellow; head well elevated, quadrate, somewhat depressed in the middle of the vertex, with a lumpy appearance on either side; corselet grooves and cephalic suture distinct. Sternum almost an equilateral triangle, elevated in the middle, and punctured. Labium blackish; maxillæ glossy brown.

Eyes: Ocular quad (9a) almost square, but slightly wider behind; eyes not greatly differing in size. MF slightly larger; side eyes propinquate, smaller than the middle group, of about equal size, separated from MF by a space about equal to the area of the latter, or one and a half to two times the intervening space. Front eye row slightly recurved, rear row procurved. The space between SR and MR is wider than that between SF and MF, the arrangement of the eyes resembling Epeira. Clypeus height about twice area MF; on one side a ridgedlike elevation extending from the margin to the midfront eyes.

Legs: Uniform glossy yellow, with slight dark annuli at the tips of the joints; palps uniform yellow; mandibles with four large teeth on the exterior furrow; long, soft bristles on the apex; color glossy yellow; fang dark brown, and slightly waved upon the under surface.

Abdomen: Ovate, dorsum evenly arched; skin yellowish brown, covered with bright yellow reticulations; dorsal folium wide, with three marginal indentations; color yellowish

---

[1] Count Keyserling originally described the species under this name of Koch's, but a year later proposed the name "brevis."

brown, with blackish margin and lighter median band, and at the apex blackish, transverse, interrupted lines. The venter shows a brownish, yellow band, extending from the spinnerets to the epigynum; the latter organ is placed about the middle of the abdomen, and is a simple arch.

MALE: (Fig. 10.) In size, color, and markings almost precisely resembles the female. The lumpy appearance on the forehead and margin of the clypeus is a little more pronounced. The middle group of eyes is almost a perfect square, instead of having the rear a little wider, as in the female. One very long, strong tooth marks the angle of the mandibles upon the inner furrow row, and at its base are three shorter, blackish teeth; on the same furrow there is a short, strong tooth near the base of the fang. The fang is glossy, somewhat uneven on the exterior surface, and the middle point on the interior is prolonged into a·blunt angle. The fangs, when folded, extend almost to the maxillæ. The palps (10a) are long, the radial joint somewhat longer than the cubital; the digital bulb globular, the cymbium contracted in the middle, extending far beyond the bulb, heavily covered with stiff bristles. From the base of the alveolus, on the upper side, is a long, yellow projection, and from the apex of the bulb a blackish brown curved projection extends almost as far as the point of the cymbium.

DISTRIBUTION: This species has been collected from Canada southward through New England, New York, and Pennsylvania, as far as the District of Columbia. It will probably be found throughout the entire Middle States and Atlantic Coast.

No. 114. Pachygnatha autumnalis KEYSERLING.   Plate XXVI, Figs. 1, 2.

1883. *Pachygnatha autumnalis*, KEYSERLING, Neue Spinn. Amerik., v., 660; xxi., 10.
1884. *Pachygnatha autumnalis*, EMERTON . . . N. E. Ep., 337, xxxiv., 22; xl, 9.
1889. *Pachygnatha autumnalis*, MARX . . . Catalogue, p. 553.

FEMALE: Total length, 4.3 mm.; cephalothorax, 2.3 mm. long, 1.1 mm. wide; abdomen, 2 mm. long, 1.3 mm. wide.

CEPHALOTHORAX: Oval, squarely truncate behind; fosse an oval pit; corselet elevated, skin glossy, slightly pubescent, color blackish brown, with brownish yellow stripes on either side of the median band and on the margin; caput of like color, elevated; the head at the eye space prominently raised into a quadrate elevation, upon which the middle group of eyes is placed. Sternum (1c) subtriangular in form, slightly longer than wide, uniform yellowish brown color, scantily pubescent. Labium triangular, very wide at the base, more than half the height of the maxillæ, which are longer than wide, convex on the exterior, somewhat narrow at the rounded tip, and inclined toward each other. The fangs, when folded, extend as far as the maxillæ.

EYES: Ocular quad decidedly wider behind than in front, and the width as great as, or even greater, than the front; MF somewhat smaller than MR, separated by about one diameter; MF separated by 2 to 2.5 diameters. Side eyes contingent, upon tubercles, and about equal in size; space between MF and SF but little greater than that between MF. The space between MR is greater than that between MR and SF. Front eye row almost aligned, or slightly recurved, the longer rear row decidedly procurved. The appearance of the eye space from the front shows MR elevated above SR, which stands out upon a lower notch from the forehead. The clypeus is extremely high, at least three times the space between MF. The side view of the cephalothorax (1a) shows the quadrate elevation of the eye space.

LEGS: Legs and palps a uniform yellow. The mandibles glossy, dark brown; about 1 mm. long.

ABDOMEN: Ovate; dorsum evenly arched; the folium a wide, glossy, brown belt, with marginal indentations, about three in number; in the middle an indented line of cretaceous yellow; a broad, interrupted band of yellowish white extends along the sides; dorsal surface reticulated. The ventral pattern is blackish brown, with marginal stripes of reticulated, cretaceous yellow.

MALE: About the size of the female, which it closely resembles in form, color, and markings. The eyes (2a) are arranged substantially as in the female, though the rear eyes appear to be relatively somewhat larger. The front row is almost straight. The radial joint of the palp (2b) is short, scarcely equaling in length the cubital. The digital joint is rounded at the base, and very long.

DISTRIBUTION: District of Columbia (Marx Collection). Keyserling's original description was made from a specimen found by Dr. Marx in Harrisburg, Pennsylvania. The species probably inhabits the Atlantic and Middle States.

No. 115. Pachygnatha xanthostoma C. KOCH. Plate XXVI, Figs. 7, 8.

1845. *Pachygnatha xanthostoma*, C. KOCH . . Die Arach., xii., p. 148, Figs. 1068, 1069.
1883. *Pachygnatha xanthostoma*, KEYSERLING, N. Spinn. Amer., v., p. 659, pl. 21, Fig. 9.
1889. *Pachygnatha xanthostoma*, MARX . . . Catalogue, p. 553.

FEMALE: Total length, 4.5 mm.; cephalothorax, 1.5 mm. long; abdomen, 3 mm. long, 1.8 mm. wide.

CEPHALOTHORAX: Cordate, rounded at the edges, well elevated in the centre; corselet grooves distinct; cephalic suture well marked; head elevated about corselet level, arched upon the summit to the eye space; color brown, with lighter streakings upon the sides. Sternum widest near the base, pointed like an arrow head towards the apex; raised in the middle. Color dark brown, with lighter median shade; covered with short, rather stiff hairs. Labium subtriangular, wide at the base, more than half the height of the maxillæ, blackish brown color; maxillæ elongate, longer than wide, convex externally, pointed toward one another, the tips roundly truncate. Color yellowish brown.

EYES: Ocular quad (7a) slightly elevated, the length somewhat greater than the rear width; wider behind than in front. MF somewhat smaller than MR, separated by 1.5 diameter; MR by 1.7 to 2 diameters. Side eyes upon low tubercles, nearly contingent; SF larger than SR. Space between SF and MF about equal to the area of the latter, or 1.5 to 2 times the interspace of MF. The distance between SF and SR is slightly greater than the interspace of SR. Front eye row recurved, rear row slightly recurved, and somewhat longer; the clypeus height about equals area MF.

LEGS: 1, 2, 4, 3; color uniform yellow to yellowish brown, without annuli; palps colored and armed as the legs; mandibles brown, widest at the tips, which are cut with an incline inwardly toward one another (7a); fang not reaching the maxillæ; teeth 3-4, nearly uniform in size.

ABDOMEN: Oval; the dorsum evenly arched. The dorsal folium with three to four indentations; dark brown to yellow in color, outlined by black, and with a median blackish band. The sides have an interrupted band of yellow, beautifully reticulated, followed lower down by a dark brown belt, which touches the venter, which is a blackish band extending between spinnerets and epigynum (7b), which is a simple atriolum.

MALE: (Fig. 8.) From 3 to 3.5 mm. long,[1] closely resembling the female in color, form, and markings. The radial joint of palp (8b) is scarcely longer than the cubital, and is less than half the length of the digital. A slight spur marks the dorsal apex of the mandibles; the teeth number 3-4, and are of nearly uniform size; the fang has no decided angular projection on the inferior surface.

DISTRIBUTION: The examples from which the above description was made were found in the neighborhood of Philadelphia. Dr. Marx has collected the species in Pennsylvania and in the District of Columbia. It is probably common throughout the Eastern and Middle United States. Mr. C. L. Koch described both male and female from specimens collected in Pennsylvania.

---

[1] The male of Koch's original description is, on the contrary, somewhat larger than the female, instead of smaller, as here.

No. 116. Pachygnatha Dorothea, new species.   Plate XXVI, Figs. 3, 4.

FEMALE: Total length, 4.3 mm.; cephalothorax, 2 mm. long, 1.5 mm. wide; abdomen, 2 mm. long, 1.3 mm. wide. The single specimen in hand is much dried up, and the proportion and markings of the abdomen are indistinct.

CEPHALOTHORAX: Ovate, truncated behind; skin glossy, color brown; fosse a low circular pit; grooves sufficiently distinct; suture well marked; caput somewhat elevated above the corselet level; the eye space not elevated. Sternum uniform reddish brown, though somewhat lighter in the middle. The surface punctured.

EYES: Ocular quad (3a), somewhat wider behind; the rear width about equal to the length; eyes of the group nearly equal in size; MF divided by about 1.3 diameter; MR by 2 diameters. MF separated from SF by about twice the intervening space of MF, The front eye row slightly recurved; the longer rear row procurved.

LEGS: Uniform yellow color, thickly covered with long, delicate bristles. The mandibles are yellowish brown, heavily clothed with long bristles at the edges. The mandibles, viewed from in front, show on the upper exterior surface about one-third the distance from the apex, a decided hump, so that the outline presents an obtuse angle (3a). This feature distinguishes the species from others of the genus. The mandibles are widely separated at the apex; the superior furrow row has about four teeth.

ABDOMEN: The abdomen is ovate, the specimen too much shriveled to give the dorsal folium with accuracy. The epigynum is located near the middle of the venter, and is a decided arch, without scapus. The surface is extremely hairy.

MALE: (Fig. 4.) 4 mm. long; the cephalothorax glossy brown, but showing a lighter or yellowish shade on either side of the median line. The eyes are substantially as in the female; but those of the ocular quad somewhat more widely separated. The mandibles (4a) are yellowish brown; widely separated at the tips; nearly straight, or with but a slight concavity on the outer margin. The fang (4a) has a slight tooth on the interior, at about the middle point. The digital joint of the palp (4b) is long; the radial joint slightly longer than the cubital.

DISTRIBUTION: The only two specimens of this genus which I possess were found in the neighborhood of Philadelphia. I have named it in honor of Mrs. Dorothea Marx, the devoted and helpful wife of Dr. George Marx, of Washington.

No. 117. Pachygnatha tristriata C. KOCH.   Plate XXVI, Fig. 6; Plate XXVIII, 1.

1845. *Pachygnatha tristriata*, C. KOCH . Die Arach., xii., p. 145, Fig. 1046, male.
1883. *Pachygnatha tristriata*, KEYSERLING. Neue Spinn. Amer., v., p. 656; xxi., 8.
1889. *Pachygnatha tristriata*, MARX . . Catalogue, p. 553.

FEMALE: Total length, 7 mm.; cephalothorax, 2.5 mm. long, 1.5 mm. wide; the abdomen, 3.5 mm. long, 2.3 mm. wide, narrowing at the apex to 1 mm.; length of mandible, 1.2 mm.

CEPHALOTHORAX: Oval; dark brown, with a glossy, leathery appearance; corselet groove distinct; cephalic suture well marked; fosse a narrow circular pit; color, a warm yellow to dark brown, with a median longitudinal band of brown, and lighter shade on either side thereof. Sternum squarely truncate at the base, raised in the middle; color brown, with a lighter median band; covered with short, soft pubescence. The labium is yellowish, large, and wide at the base, obtusely triangular, dark brown in color; the maxillæ yellowish brown, subcylindrical, much longer than wide.

EYES: The ocular quad upon a high, squarish eminence, projecting forward and upward from the face; narrowest in front; the rear width being greater than the length; the eyes on black bases, not greatly differing in size, but MF somewhat greater; the side eyes are upon tubercles, and propinquate; SR smaller; SF separated from MF by a space about 1.4 to 1.3 greater than the area, or at least twice the intervening space of MF; the distance between SR and MR is somewhat greater than the above. The clypeus is extremely high, the margin thereof being removed from MF by twice or more the area of MF. The front row is recurved, the rear row procurved. The margin is swollen on either side of the median line, just above the insertion of the mandibles.

LEGS: 1, 2, 4, 3; uniform yellow, without spines, but with abundant, long, aculeate bristles; the palps are colored and clothed as the legs; the mandibles are long, decidedly curved, a warm yellow brown color, with three to four strong teeth, wide at the tips, which are tufted with long bristles, the fang long, narrow, and slim.

ABDOMEN: An elongated oval, rounded at the base, narrowing to the spinnerets, which are distal. The dorsal field is yellowish brown in color, with longitudinal, reticulated bands of yellow along the margins, and a divided band of similar color along the median line. The skin is soft, reticulated, and covered with soft, short pubescence. The venter has a long, attenuated, yellowish band extending between the spinnerets, with blackish margins, and the epigynum presents an arched atriolum, without a scapus, but somewhat swollen in the centre.

MALE: (Pl. XXVIII., Fig. 1.) In size, form, and color the male resembles the female. The abdomen is the same elongated oval, the length at least twice the width, relatively longer than with the female. The folium, as with the female, is marked by broad marginal stripes of yellow on the side, bordered on either flank with interrupted lines of blackish, and in the centre a broken band of yellowish, reticulated color; the surface is strongly pubescent. The fosse is a longitudinal oval slit. The eyes (1a) are upon a strongly elevated quad, the rear side of which is, relatively to the front, wider than with the female. On the clypeus is a diagonal ridge, extending about half way from the margin to the midfront eyes. The mandibles (1b) are long, convex on the apical half; the fang somewhat curved, uneven on the exterior surface, and a strong, black, triangular projection or tooth on the interior surface; close to the base of the fang is a very long, blunt, curved tooth. For palp see Fig. 1c.

DISTRIBUTION: Philadelphia, New York, Texas (Marx). Doubtless widely distributed throughout the United States.

No. 118. Pachygnatha furcillata KEYSERLING.   Plate XXVIII, Fig. 3.

1884. *Pachygnatha furcillata*, KEYSERLING . Neue Spinn. aus Amer., v., 662; xxi., 11.
1889. *Pachygnatha furcillata*, MARX . . . . Catalogue, p. 552.

FEMALE: In the Marx collection there is one female specimen, identified by Keyserling as of this species. Keyserling has distinguished it chiefly by the side outline of the mandibles (Plate XXVIII., Fig. 3). The specimen is 5 mm. long, and resembles P. brevis closely. The ocular quad appears a trifle wider in front than behind, instead of the reverse, as in brevis; but I am inclined to think the two identical.

DISTRIBUTION: The species (or variety) has been collected in Philadelphia and the District of Columbia.

No. 119. Pachygnatha Curtisi, new species.   Plate XXVI, Figs. 5, 5a–b.

MALE: Total length, 5 mm.; cephalothorax, 2.5 mm. long by 1.5 mm. wide; abdomen, 3.5 mm. long by 1.3 mm. wide. General colors, dull yellow, mottled with black. The spinous armature of this species shows a marked difference from its congeners; but the characteristics are otherwise so distinctly of this genus that it is placed here.

CEPHALOTHORAX: Cordate; the fosse a conical pit; corselet grooves distinct; cephalic suture well marked; color yellow, with streakings of blackish brown along the grooves, the suture, and on the base. The head is wide, quadrate, colored as the corselet, with lighter yellow in front. Sternum nearly as wide as long, covered with gray bristles, black along the margins, and muddy yellow in the centre. Labium subtriangular; less than one-half the height of the maxillæ, which are somewhat longer than wide; subrectangular in form, the tips roundly truncated, and wider than the rest of the organ.

EYES: Ocular quad (5b) with little elevation; length greater than width, the front slightly narrower than the rear. The eyes of the group subequal; side eyes upon a low, blackish tubercle; SF larger than SR, and nearly equal to MF, from which they are separated by a space about twice that of the distance between MF. MF are separated by about 1 diameter, MR by 1.5. The front eye row is recurved, the rear row almost aligned. The clypeus height about 1.5 diameter of MF.

LEGS: 1, 2, 4, 3; yellow, the joints with black median and apical annuli. Numerous long, black spines, with black bases, which give the legs a mottled appearance, and long, thin, aculeate bristles. The tibia of leg-II without any special clasping organs; but on the inside of the femur there is a more numerous row of spines. The mandibles are conical, brownish, strong, widely separated at the tips. The fang (5c) strong, smoothly curved, scarcely reaching to the maxillæ when folded; the interior row of teeth numbers about four. The palps are rather long; the digital joint (5d) but little longer than the radial; the cubital still shorter. A strong spur issues from the base of the digital joint.

ABDOMEN: Rectangular ovate; color cretaceous yellow, mottled with black; strongly pubescent; an irregular folium, formed principally by black hairs, marks the dorsum.

DISTRIBUTION: Pacific Coast. I have three examples of the male of this species, received from and dedicated to the late Mr. John Curtis, of Oakland, California. This species, when first received, was placed by me with Pachygnatha, and as such figured here. It now appears to me, especially from the legs, and narrow clypeus and maxillæ, to be wrongly so placed.

The species of Pachygnatha are not easy to separate by striking characteristics. P. brevis, which I have made the type, may be noted as the largest, by its bright colors, by the lumpy and punctured appearance of the head. The male mandible has at the lower interior angle (Plate XXVIII, 2) a long tooth, with three small teeth around its base. P. Dorothea resembles brevis, but the female is distinguished by a blunt cone or hump on the exterior of the mandible (Plate XXVI, 3a), and the male mandible has no such tooth as in brevis. P. xanthostoma, female, is a rather smaller species, and the male may be at once noted by a short, projecting, toothlike spur on the outer margin of the apex (Plate XXVI, 8a). P. tristriata is at once separated by the abdominal markings of longitudinal stripes; by the longer oval of the abdomen, and the epigynum placed near the ventral base, close to the gills, instead of near the middle, as with the above species. P. autumnalis is the smallest, has rather darker colors, and in both sexes a head much elevated, and the side outline of mandible is nearly straight, instead of rounded. The central quad of eyes is raised well up, the midrear pair above the vertex. The hump upon the mandibles of P. Dorothea is a good distinguishing mark.

GENUS ULOBORUS, LATREILLE, 1806.

The species of this genus are distinguished by an oval cephalothorax, nearly as wide in the anterior as the posterior part, whose sides round with nearly equal width to the face. The corselet is high, well arched, but somewhat flattened upon the top; the corselet grooves distinct, the cephalic suture well marked. The head is wide, somewhat quadrate, slightly elevated above the corselet. The sternum is much longer than wide, of about equal width throughout, except at the apex. The labium is about half the length of the maxillæ, which are somewhat longer than wide. The genus is particularly distinguished by its eyes, which can hardly be divided into the three groups characteristic of the Orbitelariæ, and appear to be arranged simply in two rows. The ocular quad has the midfront eyes upon a rounded eminence, is much wider behind than in front; the intervening space between the sidefront and midfront eyes corresponds with that of Epeira; the side eyes present the most distinctive peculiarity, being widely separated, the intervening space greater than that between the midfront and sidefront eyes. The front row of eyes is somewhat procurved, almost aligned; the rear row decidedly recurved. The clypeus is comparatively low. The abdomen is a long ovate, in the typical species U. plumipes marked by shoulder humps; the dorsum arched, the spinnerets distal, with an additional spinning organ known as a cribellum. The legs are in length, 1, 4, 2, 3, the first leg especially stout, and, in all but the two final joints, relatively much more so than the other legs; the apex of the tibia is marked in some species by a slight scopula or brushlike cluster of bristles. The tibia of leg-IV is provided with a calamistrum in the female. Only one species appears to be common in the United States; but one other, U. geniculatus (zosis), has at least a lodging place upon the Gulf Coast. The snare of the species is an orbweb, without the ordinary viscid armature, for which is substituted the curled threads characteristic of Hyptiotes and certain Clubionidæ.

No. 120. Uloborus geniculatus (Oliver).   Plate XXVII, Figs. 1, 2.

1789. *Araneus geniculatus*, Oliver . . .. Encycl. Method., ii., p. 214.
1842. *Uloborus zosis*, Walckenaer . . Apt. Ins., ii., p. 231, pl. xx., Figs. 1, 2.
1858. *Uloborus Latreillei*, Thorell . . Vet. Akad. Forh., xv., p. 197.
1858. *Orithyia Williami*, Blackwall . . Ann. Mag. Nat. Hist., 3d ser., iii., p. 331.
1859. *Uloborus domesticus*, Doleschal . Act. Soc. Sc. Ind. Neerl., v., p. 46, vii., 2.
1864. *Uloborus borbonicus*, Vinson . . Aran. Reunion, etc., p. 258, i., 3.
1872. *Uloborus zosis*, Koch, L. . . . . Arach. Austral., p. 221, pl. xix., 3.
1873. *Uloborus zosis*, Thorell . . . . Rag. Mal., etc., ii., p. 130.
1889. *Uloborus zosis*, Marx . . . . . . Catalogue, p. 554.
1889. *Uloborus geniculatus*, Simon . . . Ann. Soc. Entom. France, p. 131.
1889. *Uloborus zosis*, Marx . . . . . . Proceed. Acad. Nat. Sci., Phila., p. 99.

FEMALE: Total length, 7 mm.; cephalothorax, 3 mm. long, 2 mm. wide; abdomen 5 mm. long, 3.5 mm. wide.

CEPHALOTHORAX: Cordate, truncated, deeply indented at the base, rounded at the margins of the corselet, and nearly as wide in the anterior as in the posterior part; corselet well arched, but somewhat flattened on top; fossa a conical pit; corselet grooves distinct; cephalic suture well marked; color yellow, with patches of brown on either side of the face, and at base of caput; head wide, somewhat quadrate, slightly elevated above corselet. Sternum (1b) much longer than wide, of about equal width throughout, except at the apex; rounded in the middle; dull yellow, covered with yellowish gray hairs; labium about half the length of the maxillæ, narrow, subtriangular; maxillæ somewhat longer than wide, the tips compressed into an elongated triangle, and directed toward each other; color brown at the base, with yellowish tips.

EYES: (1a.) Ocular quad much wider behind than in front, and the length somewhat greater than the rear width; a slight longitudinal cleft separates the midfront eyes, which are removed from one another by about a diameter or less; MR somewhat less than MF, separated by 1.5 to 2 diameters; side eyes widely separated, the space between them as great as that between MR and MF, which is equal to 1.3 the area of MF, or more than twice the intervening space; the distance between SR and MR about equals that between MR; clypeus somewhat higher than area MF; front eye row procurved, the longer rear row slightly recurved. The face is wide in front.

LEGS: 1, 2, 4, 3; femora and tibia stout, but the two apical joints greatly diminished in size; color yellow, with broad apical annuli; abundantly covered with gray and brown bristles and hairs, and with short, dark spines; the metatarsus of leg-IV is curved, and provided on the inner and upper side with a calamistrum (1d). .

ABDOMEN: Ovate, wide and rounded at the base, much narrowed towards the spinnerets, which are distal; color ashen brown or yellow, without any decided folium; the epigynum (1c) is without a scapus, and the thick, low atriolum has two short subtriangular projections, which are bowl shape or hollowed underneath.

MALE: 5 mm. long; in color and form resembling the female. The cephalothorax is roundly oval, the summit of the corselet has two dark, brown bands, which run along the cephalic suture; on the sides are irregular bands of yellow, bordered beneath by brown, and the margin of the corselet is yellow; skin with gray pubescence. The ocular quad is longer than wide, the rear wider than the front; MF are slightly larger than MR, the latter about equal in size to SR. The eyes of the rear row are nearly equally divided one from another, MR being slightly further apart. Of the front row the space between SF and MF is about one-half greater than the distance between MF. SF are quite small, the smallest eyes of the entire group. The first row of eyes is procurved; the rear row is slightly recurved. The palps are short; the digital bulb large, globular; the cymbium covered thickly with strong, yellowish hairs. The legs are yellow, with dark brown apical and median annuli; length, 1, 4, 3, 2, the first pair much longer and stouter. The tibiæ of leg-I are slightly curved, thickened towards the apex, upon the summit and sides of which are rows of strong, black spines; the other legs are armed with short, black spines; as in the female, the two terminal joints are decidedly thinner than the others.

DISTRIBUTION: The Gulf Coast of the United States and the West Indies.

No. 121. Uloborus plumipes (?) LUCAS.　　　Plate XXVII, Figs. 4, 5; XXVIII, Fig. 6.

1842. *Uloborus Americanus*, WALCKEN'R* Ins. Apt., ii., p. 229; ABBOT, G. S., No. 44, 45.
1845. *Uloborus plumipes*, LUCAS . . . . Explor. de l'Algerie, Anim. Artic., i., 252; pl. 15, Fig. 8.
1850. *Phillyra mammeata*, HENTZ* . . . B. J. S., vi., 25; Sp. U. S., 129, pl. 14, Fig. 16.
1850. *Phillyra riparia*, HENTZ . . . . . Ibid., v., Fig. 17.
1869. *Uloborus plumipes*, CANESTRINI &
　　　　　　PAVESI . . . . Archiv. p. la Zool. l'Anat. e la Fisiol., ser. ii., vol. ii.
1881. *Uloborus villosus*, KEYSERLING* . N. Spinn. a. Amer., iii., Verh. Ges. Wien, 278, xi., 6.
1888. *Uloborus plumipes*, EMERTON . . N. E. Ep., Ciniflon., Trans. Conn. Acad., pl. xi., 1.
1889. *Uloborus plumipes*, McCOOK . . . Amer. Spiders and their Spinningwork, I., II.

Dr. Marx, in his Catalogue, considers Uloborus Americanus as a good species and identical with U. mammeata (Hentz). The figure in Abbot's MSS. drawings, from which Walckenaer described his species, certainly resembles in outline Hentz's drawing of U. (Phillyra) mammeata rather than his figure of U. riparia (plumipes); but the front row of eyes is more decidedly procurved. I am inclined to think that the two species are the same, and are identical with U. plumipes Lucas. If so, Walckenaer's name should have priority. Not having sufficient material to determine the point positively, I leave the name as above, and place in the synonyma marked (*) the series corresponding most closely with U. Americanus, as distinguished by Marx and Hentz. I believe, however, that it will prove that we have but one species under the various names as above, and that the confusion in names has been caused by the striking difference in appearance between the black forms and the yellow ones, as fairly represented upon the figures in my plates. Perhaps it may be concluded that these differences justify at least varietal names.

FEMALE: Total length, 5 mm.; cephalothorax, 2 mm. long, 1.3 mm. wide; abdomen, 3.5 mm. long, 1.75 mm. wide.

CEPHALOTHORAX: Oval, truncated at the base, the sides rounding with nearly equal width to the face; corselet high behind, where it is raised into two low humps by the parting of the median fosse, thence it slopes rather sharply to the base; the head elevated; color blackish brown, covered with yellowish pubescence. Sternum (4a) cordate, inclined to oval, of nearly equal width throughout except at the apex, elevated in the centre; color blackish brown, with gray pubescence. Labium about half the height of maxillæ, width about equal to length, the tip triangular; maxillæ squarish, longer than wide, truncate at the tips; color of both labium and maxillæ yellow.

EYES: Ocular quad elevated in front, much wider behind than in front, and length greater than width; MF smaller than MR, and divided by about one diameter, MR by two diameters or more; side eyes widely separated, the space between them being at least half greater than that which separates SF and MF (3b); the distance between SR and MR is scarcely as great as that between MR; the front row distinctly procurved, the rear row much recurved.

LEGS: 1, 4, 2, 3; stout, especially at the femora of legs-I and IV; color yellow, with brown and blackish brown median and apical annuli; the femora almost completely darkened; the metatarsi long, and on the fourth leg (3d) curved and provided with a calamistrum.. The legs have short pubescence and fine bristlelike spines; tibia-I is thickened and has an apical brush, to which the species owes its name. The palps are colored and armed as the legs.

ABDOMEN: An elongated oval, thickest at the base, which is bifid and overhangs the cephalothorax; it is there provided with two low shoulder humps; the dorsum is arched to the spinnerets, which are distal and provided with a cribellum (3e); the color is yellow or cretaceous, reticulated and marked with blackish brown median stripes with divergent branches to the sides, which are streaked with black and yellow. In some specimens (Fig. 4) the colors appear to be reversed and the field black, while the folial markings are yellow. The venter is a broad band of yellowish brown with an indented border of cretaceous, which surrounds in a broken line the spinnerets, which are yellow flecked with brown; the epigynum (3c) shows two rounded conical projections from the atriolum of a ·light yellow color, overhanging and projecting beyond the genital cleft. The male (Fig. 5) is colored and marked like young females (Fig. 4).

MALE: Total length, 2.5 to 3 mm. The cephalothorax is more roundly ovate than the female. The eyes (Pl. XXVIII, 6a) differ little in arrangement, but the front row rather shorter relatively. The legs are yellow, femur-I being darker, and others with light median and apical annuli underneath. The tibia of the first pair bearing upon the upper surface a double row of spines on one side long and yellow, on the other shorter and blackish, and eight to nine in number; other joints with a few yellow spines. The two terminal joints are decidedly thinner than the others. The metatarsus is as long as the tibia; the fourth pair of legs without calamistrum. The abdomen is without humps, is an elongated oval, blackish in color, and the folium irregularly traced by long grayish yellow hairs; the entire surface is hairy. The color of the cephalothorax is dark brown, with a median line of yellowish marked by long gray hairs. The palps are short, the digital joint large, globular; the radial and cubital joints are of nearly equal length; a short blunt spur projects outwardly from the humeral joint (Pl. XXVIII, 6b); the cymbium covered with reddish yellow hairs.   (Marx Collection.)

DISTRIBUTION: I have taken this spider in New England, Pennsylvania, and Ohio; Hentz described it from Alabama, and I have various specimens received from the Pacific Coast. It is probably distributed throughout the entire United States.

GENUS HYPTIOTES, WALCKENAER, 1833.

But one American species of this remarkable genus is known, whose characteristics are so striking that it cannot be easily mistaken. The **cephalothorax** is an oval with steep sides, truncated at the base, which shelves sharply downward from the fosse. The corselet is overhung behind by the vaulted abdomen, and almost overshadowed in front by the peculiar structure of the face. The **sternum** is cordate, decidedly longer than wide, widest at the base, with distinct sternal cones. The labium is small as compared with the maxillæ, compressed at the base and triangular at the tip, the length about half that of the **maxillæ**, which are somewhat longer than wide, compressed at the stalk, and much widened at the squarely truncate tips. The **eyes** are arranged in two rows, of which the front is strongly procurved, and the rear row, at least 2.5 times as long as the front, is slightly procurved. The two sidefront eyes are very small and placed close to the margin of the clypeus; the siderear eyes are much larger and are upon tubercles overhanging the sides of the face, separated from the sidefront by a space greater than that which separates them from the midrear. The other six eyes are placed far back upon the face. The clypeus is wide. The legs are 1, 4, 2, 3; short, stout, I and II especially so; the fourth leg is curved and provided with a calamistrum. The abdomen is oval, with highly arched dorsum symmetrically covered with low conical humps on the margins, and rounded at the base and apex, the latter overhanging the spinnerets which are provided with a cribellum. The male resembles the female in general structure and color. The abdomen presents the same lumpy appearance, and is pubescent. The radial joint is stout, somewhat curved; the digital joint extremely long, canoe shaped, and complex.

No. 122. Hyptiotes cavatus (Hentz).   Plate XXVII, Fig. 7.

1847. *Cyllopoda cavata*, HENTZ . . . . . J. B. S., v., 466; Sp. U. S., p. 104, xii., 3; xx., 21.
1875. *Hyptiotes Americanus*, WILDER . . Popular Science Monthly, Sept., p. 2 (The Triangle Spider).
1884. *Hyptiotes cavatus*, EMERTON . . . N. E. Ciniflonidæ. Trans. Con. Acd. vii., 457; ii., 2.
1889. *Hyptiotes cavatus*, McCook . . . Amer. Spiders and their Spinningwork, Vol. I.
1889. *Hyptiotes cavatus*, MARX . . . . . Catalogue, p. 553.

FEMALE: Total length, 5 mm.; cephalothorax, 2 mm. long, 1 mm. wide; abdomen, 2.5 mm. long, 2 mm. wide.

CEPHALOTHORAX: An irregular oval, rounded at the margin, steep, truncated at the base, which shelves sharply downward from the fosse, which is a semicircular indentation upon the upper part of the incline; cephalic suture distinct, but the peculiar structure of this part of the body makes it somewhat difficult to distinguish between the head and the corselet itself; color dark brown to blackish, with grayish yellowish pubescence. Sternum (Fig. 6b) cordate, widest at the base; sternal cones distinct; color dull yellow, with grayish white pubescence, especially at the edges; raised in the middle; the labium rounded at the tip about half the height of the maxillæ, which are obtusely triangular at the tips, as wide as long; color dull yellow.

EYES: Six of the eight placed far back upon the face; the ocular quad very wide in the rear, narrow in front, and the sides little more than half the rear, presenting the form of an isosceles triangle truncated at the apex; MF separated by about one diameter, SR by a space equal to about four times the area of MF; side eyes as widely separated as MR, SF extremely small and placed well to the front just over the outer edge of the mandibles, about 2.5 diameters from the margins of the clypeus; these eyes extremely small, so much so that Hentz failed to observe them, and described the spider as having only six eyes; SR much larger than SF, are indeed the largest of the entire group, located at the extremities of the optical area upon the upper edge of the face, which they overhang. The clypeus is wide from the middle point to MF, being about three times the area of the latter; the group of eyes thus described form two rows, of which the front row is much procurved, the rear row slightly procurved; of the rear row the distance between SR and MR is about one-third less than the space separating MR, and somewhat less than the distance between SR and SF.

LEGS: 1, 4, 2, 3; short, stout, especially I and II; abundantly clothed with bristles and hairs, and with short spines; the tarsus with one Epeiroid claw; the metatarsus of leg-I is curved, and provided along its exterior surface with a row of curved hairs known as the calamistrum, which characterizes Uloborus and some species of Ciniflonidæ, in this respect uniting the Orbweavers with the Tubeweavers. The color is yellow or olive brown.

ABDOMEN: Oval; the dorsum highly arched but not evenly arched, curving downward to the spinnerets, which are slightly overhung by the apex; on the base in front are two low humps, and along the sides is a series of three low conical elevations, which give the dorsum a roughened appearance; the color varies, is yellow or yellowish brown, streaked with green or olive, and occasional patches of purple. The whole surface is reticulated, and covered with yellowish hairs; the venter has a yellowish patch flanked with blackish brown on either side, which encompasses the spinnerets and cribellum (6e), which are a dull yellowish color; the epigynum (Fig. 6f) is a simple, wide, arched atriolum, without a scapus, slightly projecting over the genital cleft.

MALE: (Fig. 7.) Resembles the female in general structure and coloring, but is smaller, being about 3 mm. in length. The abdomen presents the same lumpy appearance, and is heavily covered with hairs. The palp (Fig. 7a) is peculiar in its conformation, the digital bulb being very long, the cymbium canoe shaped, covered closely with hairs.

DISTRIBUTION: I have collected this species at various points from New England along the Atlantic Coast southward; in New York, Pennsylvania, and Ohio. I have specimens as far to the northwest as Wisconsin (Professor Peckham), and Dr. Marx has specimens from Virginia, Maryland, and the District of Columbia. It will probably be found to inhabit the entire middle zone of the United States. It appears to prefer mountainous or hilly regions, although not confined thereto. A favorite dwelling place is in or near pine woods.

By some oversight, which neither author nor printer can explain, the copy of the following species was lost, and omitted from its proper place. The error was not discovered until the appropriate signature had been printed, and a description is inserted here.

No. 123. Cyclosa Caroli (HENTZ). Plate XVII, Figs. 7, 8.

1850. *Epeira Caroli*, HENTZ . . . . . . J. B. S., vi., p. 24; Sp. U. S., p. 138, xiv., 15.
1863. *Epeira Caroli*, KEYSERLING . . . Orbitel., Sitz. D. Isis, Dresden, p. 137, vi., 17–18.
1889. *Cyclosa Caroli*, MARX . . . . . . Catalogue, p. 549.

FEMALE: Total length, 7 mm.; cephalothorax, 2.2 mm. long, 1 mm. broad; abdomen, 5 mm. long, 2.2 mm. broad, diminishing to about 1 mm. The cephalothorax is a rounded oval; color dark brown. The legs and palps yellow, with brown median and apical annuli. The abdomen has a grayish hue, with tinges of yellow; is widest at the base, where it is somewhat arched, and diminishes from about the middle into an extended caudal part, which projects beyond the spinnerets nearly one-half the length of the abdomen. The spinnerets are placed well beneath, at the apex of the subconical venter.

MALE: (Fig. 8.) Total length, 4.5 mm. The cephalothorax is a rounded oval, almost circular, slightly truncate at the base; color reddish brown, with dark stripes passing from the fosse to the margin; fosse a circular depression; caput slightly depressed, of darker color than the cephalothorax; face sufficiently wide. The ocular quad is somewhat longer than broad, slightly narrower behind; MF larger than SF, and separated by about one diameter; space between MF and SF but little larger than that between MF; side eyes on tubercles; propinquate, about equal in size; front row recurved, rear slightly procurved.

THE LEGS are 1, 2, 4, 3; yellow, with broad brown annuli on the femora, which are sufficiently stout, and apical and median annuli on the other joints. They are well provided with stout spines, especially on the tibia of leg-I and leg-II, the latter also slightly curved inward. The abdomen, as in the female, is a grayish yellow, mottled with black. The spinnerets are placed about the middle of the venter, and the apex of the abdomen projecting beyond into a caudal part (8a). The palp (8b) is comparatively short, but the digit large, subglobular, terminal bulb brown, cymbium yellow, covered with hairs. The radial joint is short, compressed, scarcely more than half the length of the cubital joint, which is yellow, and has projecting from it a long, brown spine, slightly curved towards the point. The position from which Fig. 8b was drawn does not show the true relative proportion of the radial and cubital joints.

DISTRIBUTION: A female was taken by Hentz in Alabama; Dr. Marx collected a male in the District of Columbia, and the description of Keyserling is from a female specimen collected in New Granada, South America. This indicates distribution from the northern belt of Southern United States to the northern belt of South American States.

INDEX OF VOLUME III.

(278)

ADVERTISEMENT.

THE original price of this work was placed at thirty dollars ($30) per set of three volumes. In the announcement thereof in a prospectus of August 1st, 1889, special terms were offered to subscribers who should send their names before the 1st of November of the same year. A number accepted this offer, and are, therefore, receiving the entire series for twenty-five dollars. With the issue of the first volume this special offer ceased, and the price thereafter remained as fixed, namely, ten dollars per volume, until January 1st, 1892. At that date, by public notice, the price of the work was fixed at ($50) fifty dollars. This is the present selling price, and will thus remain, at least as long as the author retains control of the issue. No separate volume will be sold.

The author was induced to assume the disagreeable role of publisher by two considerations, first, the wish to be free to present the book according to his own ideas as to typography and illustration, which, however, were too luxurious to meet the views of ordinary publishers. In the second place, as a commercial venture, the printing of a scientific work of this sort, with so limited a circulation, gave no prospect of remuneration for cost and care. It is indeed a work of love, and must be carried forward chiefly from a desire to enlarge the bounds of truth. Moreover, the cost was too considerable to warrant any Scientific Society to undertake the work of publishing from funds always too limited. The author therefore accepted the burden of cost, together with the yet more uncongenial details of selling, as a part of his task. He has planned and now aims to obtain from the sale as nearly as possible the actual expenditure in money for printing and engraving. With this in view he was compelled to correct the mistake made at the outset of placing the work at so low a price.

The several volumes will be mailed to subscribers with uncut edges, for the accommodation of those who wish to give them special library binding.

The Author's Edition is strictly limited to two hundred and fifty copies, which will be numbered consecutively in the order of subscription as received.

The first two volumes are devoted to a description of the industry and habits of Orb-weaving spiders, both separately and in their relations to the spinning economy of other aranead tribes. The first volume treats particularly of Snares and Nests; the second volume considers the Cocooning Industry, Maternal Instincts, and General Habits. These volumes are liberally illustrated by drawings from nature, containing in all 853 figures, drawn and engraved for the work, many of these cut upon the wood, and a large number of full page illustrations. Besides these there are forty lithographic plates, colored by hand from nature, upon which are engraved 913 figures. In addition, a full page portrait of Professor Hentz, the father of American Araneology, has been made the frontispiece of Volume III. The third volume contains six chapters of Natural History descriptions, illustrated by ninety-eight engravings. The remainder and principal part of the volume is devoted to descriptions of the Orbweaving fauna of the United States, and two plates illustrating typical species of the other aranead groups.

Subscriptions may be made directly without the intervention of a bookseller, but a slight discount will be allowed to the trade.

All business communications and subscriptions may be addressed directly to the author or to

"AMERICAN SPIDERS AND THEIR SPINNINGWORK,"
ACADEMY OF NATURAL SCIENCES,
LOGAN SQUARE, PHILADELPHIA, U. S. A.

PART THIRD.

AMERICAN SPIDERS AND THEIR SPINNINGWORK.

ATLAS OF PLATES.

PLATE I.

Fig. 1. Epeira insularis, female, ventral view, ×1.5; 1a, epigynum;[1] 1b, the same, side view; 1c, male palp. (Page 143.)

Fig. 2. E. marmorea, female, epigynum; 2a, same, side view. (Page 143.)

Fig. 3. E. trifolium, female; 3a, side view of same; 3b, male palp. (Page 145.)

Fig. 4. E. trifolium, female, variety candicans, ×1.5; 4a, the same, epigynum.

Fig. 5. Argiope Marxii, female, ×1; 5a, epigynum. (Page 223.)

Fig. 6. Epeira pratensis, female, ×2; 6a, epigynum; 6b, same, side view. (Page 142.)

Fig. 7. E. Benjamina, female, abdomen, ×2; 7a, 7b, epigynum; 7c, male palp. (Page 147.)

Fig. 8. E. arabesca, female, abdomen, ×2; 8a, epigynum; 8b, male palp. (Page 148.)

Fig. 9. E. sclopetaria, female, epigynum; 9a, male palp. (Page 137.)

Fig. 10. E. cornuta-strix, female, epigynum; 10a, 10b, male palp. (Page 140.)

Fig. 11. E. patagiata, female, epigynum; 11a, same, from another specimen; 11b, male palp. (Page 138.)

---

[1] In all plates this organ is represented as seen from beneath, that is, looking toward the head, with the spider lying upon its back.

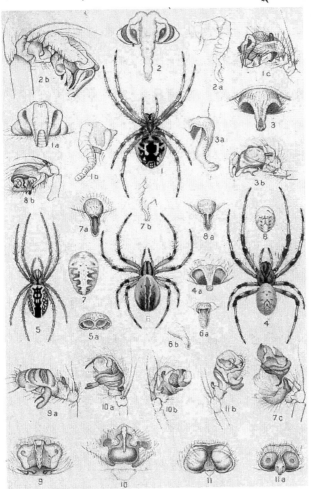

1.Epeira insularis. 2.E.marmorea. 3,4.E.trifolium.
5.Argiope Marxii. 6.Epeira pratensis. 7.E.Benjamina 8.E.arabesca.
9.E.Sclopetaria. 10.E.strix. 11.E.patagiata.

PLATE II.

PLATE II

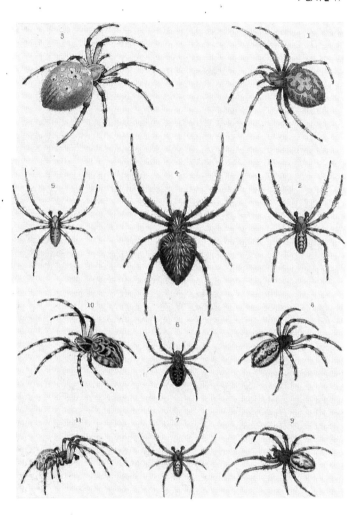

Auth.-Del.

Edw. Sheppard, Lith.

PLATE III.

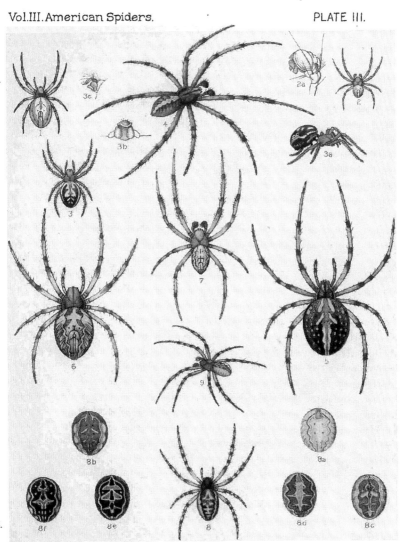

1,2,3, Epeira displicata. 4,5, E.vertebrata. 6,7, E.marmorea.
8,9, E.patagiata.

PLATE IV.

Fig. 1. Epeira vertebrata, female, dark variety, ×2; 1a, ventral view; 1b, face and jaws;
1c, 1d, epigynum, viewed underneath, and reverse view; 1e, same, side view.
See Plates III., V. (Page 151.)

Fig. 2. E. balaustina, female, ×2; 2a, ventral view of abdomen; 2b, face and jaws; 2c,
epigynum. (Page 155.)

Fig. 3. E. Ithaca, female; 3a, face and jaws; 3b, ventral view; 3c, epigynum, seen underneath the last moult; 3d, male palp. (Page 152.)

Fig. 4. E. placida, female, ×2. (Page 153.)

Fig. 5. E. placida, male, ×2; 5a, palp.

Fig. 6. E. cucurbitina, female, ×2. (Page 149.)

Fig. 7. E. foliata (folifera), female, ×3. The name E. folifera is erroneously inserted upon
the page title.

Fig. 8. E. foliata (folifera), male, ×3; side view.

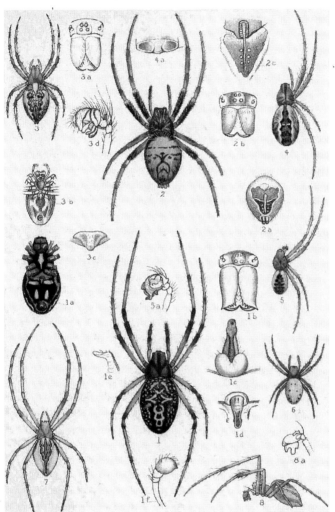

1, Epeira vertebrata.　2, E. balaustina.　3, E. ithaca.
4.5. E. placida.　6. E. displicata.　7. 8. E. folifera.

PLATE V.

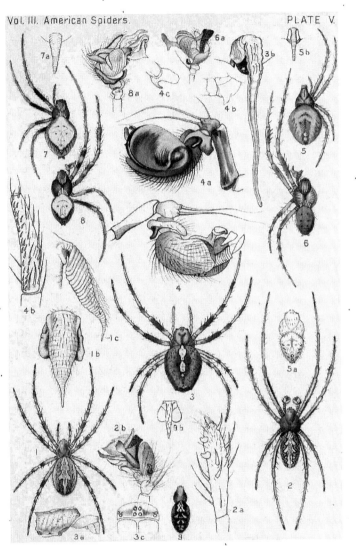

1,2. Epeira carbonaria.   3.Verrucosa unistriata.
4.Epeira vertebrata.   5,6.E.bivariolata   7.8.E.ravilla.
9.E.carbonarioides

PLATE VI.

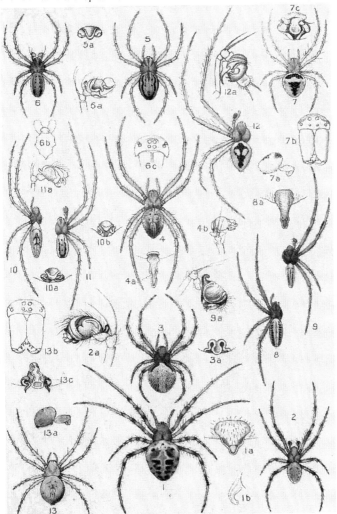

1.2, Epeira volucripes. 3.E.tranquilla. 4.E.punctigera.
5.E.mormon. 6.E.reptilis 7.12,E.forata. 8.9,E.Theisii
10.11,Drexelia directa. 13.Epeira juniperi.

PLATE VII.

Fig. 1. Nephila Wilderi, female, natural size; 1a, side view in outline; 1b, epigynum.
See Plate XXIII.

Fig. 2. Nephila Wilderi, male, natural size.

Figs. 3, 4. E. Thaddeus, female, ×2; 3a, 3b, epigynum, front view and side views; 3c, same,
oblique view, to show the genital barrier; Fig. 4, highly colored specimen.

Fig. 5. Epeira Thaddeus, male, ×2; 5a, palpus. (Page 169.)

Fig. 6. Epeira Wittfeldæ, female, ×2; 6a, side view; 6b, eyes and mandibles; 6c, epigynum.

Fig. 7. Epeira Wittfeldæ, male, ×2; 7a, 7b, palpus. (Page 168.)

Fig. 8. Epeira Pegnia (globosa, Keyserling), female, ×2; 8a, epigynum.

Fig. 9. Epeira Pegnia, male; 9a, palpus. (Page 170.)

Fig. 10. Epeira labyrinthea, female, ×2; 10a, epigynum of California variety. Fig. 12.

Fig. 11. Epeira labyrinthea, male; 11a, palpus. (Page 171.)

Fig. 12. Epeira labyrinthea, California variety; 12a, side outline of abdomen; 10a, epigynum.

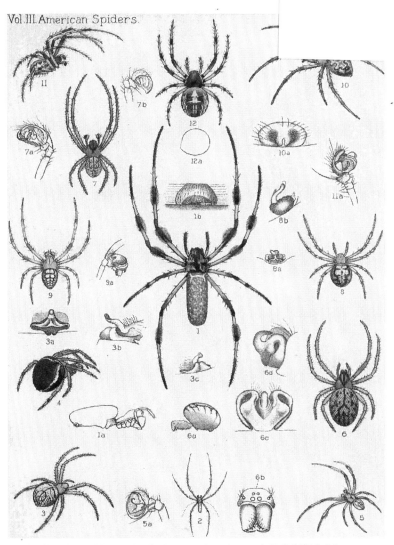

1,2, Nephila wilderi. 3,4,5, Epeira thaddeus. 6,7, E.Wittfeldæ.
8,9, E.globosa. 10,11,12, E. labyrinthea.

PLATE VIII.

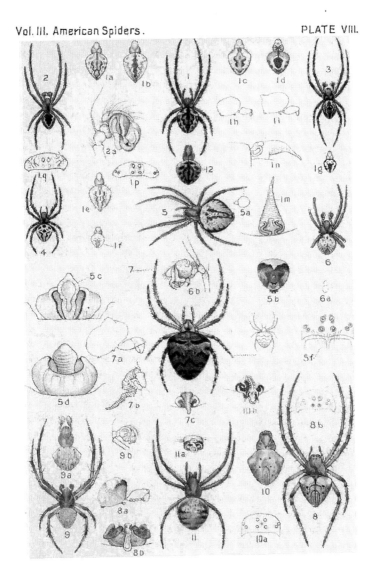

1—4. Epeira anastera. 5.6. E. linteata. 7. E. corticaria.
8.9. E. miniata. 10. E. Bonsallæ. 11. E. Mayo. 12. E. eustalina.

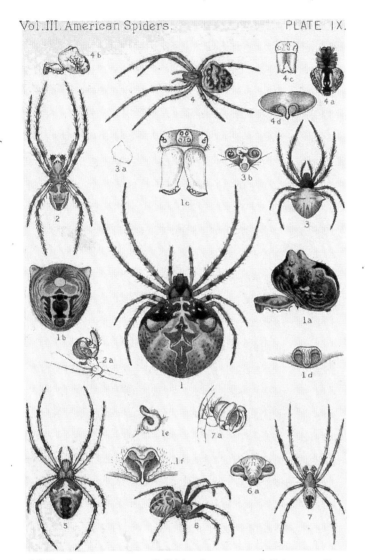

1,2, Epeira gemma. 3, E. bispinosa. 4, E. bucardia.
5,6,7, E. nordmani.

PLATE X.

Fig. 1. E. cavatica, female, ×2. See Plate XI. (Page 185.)

Fig. 2. E. cavatica, male, ×2.

Fig. 3. E. angulata (bicentennaria), ×1. See Plate XI. (Page 186.)

Fig. 4. E. angulata (bicentennaria), ×4, young specimen.

Fig. 5. E. angulata (bicentennaria), immature male.

Fig. 6. E. gemma, ×1. See Plates IX., XI. (Page 182.)

Fig. 7. E. reticulata,[1] female, ×2, new species. See Plate XI., Fig. 7.

Fig. 8. E. rectículata, male, ×2. See Plate XI., Fig. 8.

Fig. 9. E. silvatica, female, ×1.5. See Plate XI. (Page 188.)

Fig. 10. E. diademata, female, ×1. See Plate XI. (Page 188.)

Fig. 11. E. gemma, female, ×1.5. (Page 182.)

---

[1] When writing the descriptive text of this species I classified it with E. miniata (page 177), which it closely resembles, having at that time concluded that I was wrong in treating it as a new species. Subsequent study led me to think the specimen different from E. miniata, and I therefore retain my original name of the species in the plates, although too late to make proper correction and description in the text itself.

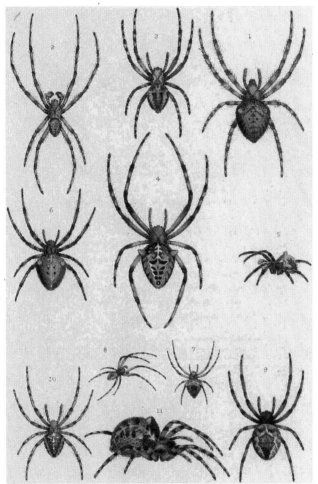

1, 2, Epeira cavatica   3, 5, E. angulata.   6, 11, E. gemma.   7, 8. E. reticulata.
9, E. silvatica.   10. E. diademata.

Auth–Del.

Edw. Sheppard, Lith.

PLATE XI.

Fig. 1. Epeira angulata–Columbia, female; natural size; 1a, face; 1b, epigynum; 1c, side view; 1d, epigynum of European E. angulata, showing corneous cones along the edge of the genital cleft, lacking in E. angulata-Columbia. See Plate X. (Page 186.)

Fig. 2. E. angulata, male, ×1, drawn from European specimen (Russia); 2a, palpus. (Page 187.)

Fig. 3. E. angulata, male (New Hampshire), ×1; 3a, palpus.

Fig. 4. E. angulata-bicentennaria (Pennsylvania); 4a, dorsal view of abdomen (specimen from Adirondacks, N. Y.), ×1.5; 4b, epigynum, front view; 4c, side view; 4d, 4e, enlarged views, front and side, of epigynum (Adirondack specimen).

Fig. 5. E. Nordmanni, female, ventral view of abdomen; 5a, epigynum; 5b, side view. See Plate IX. (Page 184.)

Fig. 6. E. cavatica, female, epigynum; 6a, male palp. (Page 185.)

Fig. 7. E. reticulata, face and eyes of male. See Plate X.

Fig. 8. E. reticulata; face and eyes of female; 8a, epigynum.

Fig. 9. E. silvatica, dorsal view; 9a, epigynum. (Page 188.)

Fig. 10. E. diademata, ×1.5; 10a, ventral view, ×1.5; 10b, face; 10c, epigynum; 10d, side view. (Page 188.)

Fig. 11. E. diademata, male; 11a, 11b, palpus. (After Blackwall.)

Figs. 12, 13, 14. E. gemma, female; views from behind, from front, showing various colors and markings. (Page 182.)

Fig. 15. E. Pacificæ, female, ×3; 15a, epigynum. (Page 180.)

Fig. 16. E. Pacificæ, male, ×3; 16a, palpus. (Page 180.)

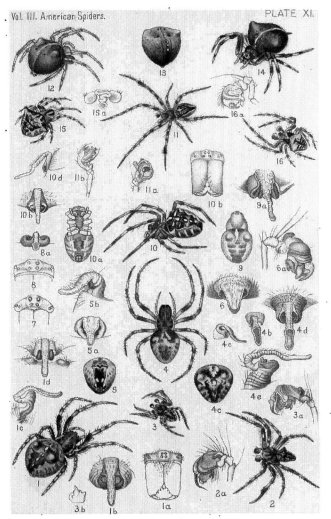

1.2.3. Epeira angulata. 4. Ep. bicentanaria. 5. E. nordmanni. 6. E. cavatica.
7.8. E. reticulata. 9. silvatica. 10. 11. E. diademata. 12.13.14.E. gemma
15.16. Ep. pacifica.

PLATE XII.

Fig. 1. Ordgarius cornigerus, female, ×3; 1a, side view; 1b, face, ×6, to show the knobby forehead; 1c, gravid specimen, with dorsal cones and markings nearly obliterated; 1d, epigynum. (Page 197.)

Fig. 2. Ordgarius bisaccatus, female, ×2; 2a, epigynum. (Page 198.)

Fig. 3. Ordgarius bisaccatus, male; 3a, natural size; 3b, palpus.

Fig. 4. Marxia stellata, female, ×2; 4a, outline side view; 4b, epigynum. (Page 193.)

Fig. 5. Marxia stellata, male, ×2; 5a, palpus. (Page 194.)

Fig. 6. Verrucosa arenata, female, ×3; 6a, side view; 6b, face; 6c, epigynum. (Page 200.)

Fig. 7. Verrucosa arenata, male; 7a, palpus. (Page 201.)

Fig. 8. Gea heptagon, female, ×4; 8a, side view; 8b, face; 8c, sternum and mouth parts. (Page 208.)

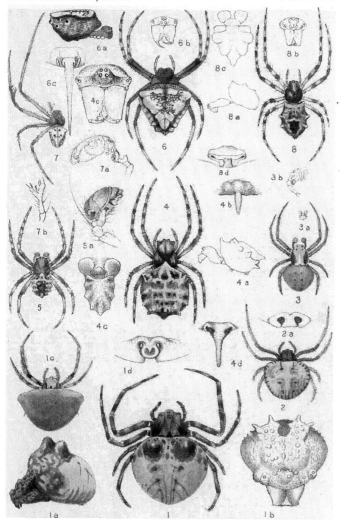

1. Ordqarius corniqerus. 2.3.0. bisaccatus. 4.5. Marxia stellata.
6.7. Verrucosa arenata. 8. Gea heptagon.

PLATE XIII.

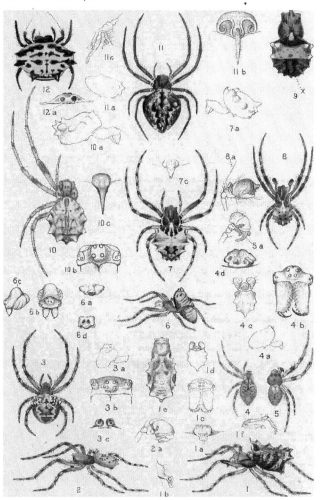

1,2 Wagneria tauricornis. 3. Kaira alba 4,5. Wixia Ectypa.
6. Carepalxis tuberculifera. 7,8,9. Marxia nobilis.
10. M. grisea. 11. M. moesta. 12. Gasteracantha maura.

PLATE XIV.

Fig. 1. Argiope avara (Eagle Pass, Texas), female, ×1½; 1a, epigynum. (Page 222.)

.Fig. 2. Hentzia basilica (Washington, D. C.), female. See Plate XXIII.

Fig. 3. Epeira spinigera, female, ×3 (Santa Rosa, California); 3a, side view of abdomen; 3b, epigynum; 3c, side view of same; 3d, epigynum (after Cambridge). (Page 191.)

Fig. 4. Epeira spinigera, male.

Fig. 5. Epeira nephiloides, female, ×2. See Plate XXII. (Page 190.)

Fig. 6. Epeira nephiloides, female (after Keyserling); 6a, side outline; 6b, epigynum.

Fig. 7. Gasteracantha pallida (Tucson, Arizona), female, ×3. (Page 209.)

Fig. 8. Gasteracantha preciosa (Buena Vista Lake, California), female, ×3; 8a, epigynum. (Page 211.)

Fig. 9. Gasteracantha cancriformis, female; 9a, epigynum. (Page 211.)

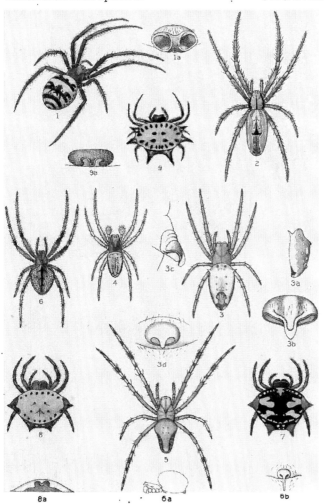

Argiope. Epeira. Gasteracantha.

PLATE XV.

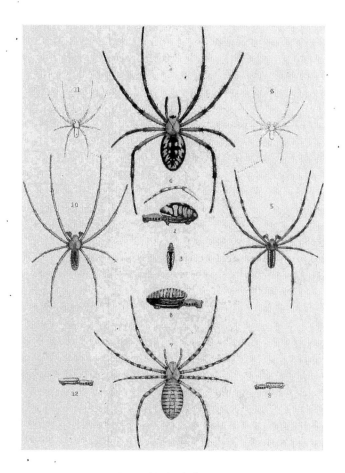

1, 6, Argiope cophinaria.   7, 12, A. argyraspis.

Auth.-Del.   Edw. Sheppard, Lith.

PLATE XVI.

Fig. 1. Argiope argentata, female, enlarged, 1.5; 1a, ventral view, ×1½; 1b, dorsal view of another specimen, ×3; 1c, epigynum; 1d, side view. (Page 220.)

Fig. 2. A. argentata, male, ×3; 2a, natural size; 2b, another figure of same, giving dorsal view, ×3; 2c, 2d, 2e, 2f, views of palp, face, and eyes.

Fig. 3. A. argyraspis, enlarged view of face; 3a, epigynum; 3b, side view. See Plate XV. (Page 219.)

Fig. 4. A. argyraspis, male; terminal joints of palp.

Fig. 5. A. cophinaria, front view of face; 5b, epigynum, ×9; 5c, side view. See Plate XV. (Page 217.)

Fig. 6. A. cophinaria, male, face and mandibles; 6a, terminal joints of palp.

PLATE XVI.

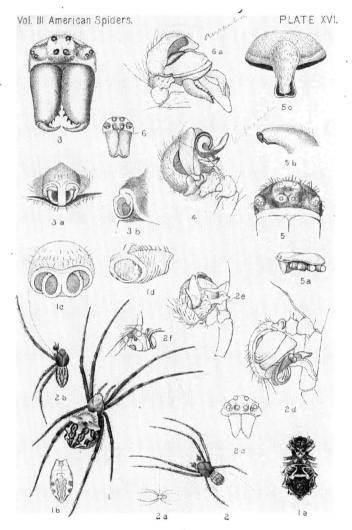

I, 2. Argiope argentata. 3,4. A. argyraspis. 5,6. A. cophinaria.

PLATE XVII.

Fig. 1. Cyclosa Walckenaerii, female, ×4; 1a, ventral view, ×4; 1b, side view, ×4; 1c, face and mandibles; 1d, epigynum. (Page 226.) This species was named and described by me as C. cervicula, and the plate made two years before the publication of Keyserling's name. But not having been published, Keyserling's name has priority. I have corrected in the text but could not correct the plate name.

Fig. 2. Cyclosa Walckenaerii, male; 2a, face and eyes; 2b, sternum and mouth parts.

Fig. 3. Cyclosa conica, female, ×3; 3a, natural size; 3b, side view; 3b, 3c, epigynum. (Page 225.)

Fig. 4. Cyclosa conica, male, ×3; 4a, side view; 4b, palpus. (Page 226.)

Fig. 5. Cyclosa turbinata, female, ×3; 5a, ventral view, ×3; 5b, natural size; 5c, epigynum; 5d, side view. (Page 224.)

Fig. 6. Cyclosa turbinata, male, ×3; 6a, side view; 6b, palpus.

Fig. 7. Cyclosa caroli, female, ×4; 7a, side view; 7b, 7c, epigynum. (Page 277.)

Fig. 8. Cyclosa caroli, male, ×4; 8a, side view; 8b, palpus.

Fig. 9. Cyclosa bifurca, female, ×3; 9a, side view; 9b, epigynum; 9c, eyes.

Fig. 10. Cyclosa bifurca, male, ×3. (Page 227.)

Fig. 11. Cyrtophora tuberculata, female, ×4; 11a, epigynum; 11b, side view, ×4; 11c, face and mandibles; 11d, sternum and mouth organs. (Page 236.)

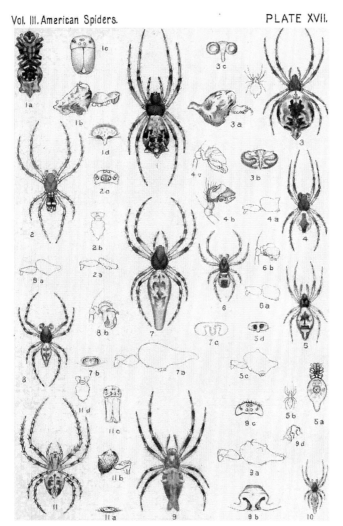

1,2. Cyclosa cervicula. 3,4. C. conica. 5,6. C. turbinata. 7,8. C. caroli.
9,10. C. bifurca. 11. Cyrtophora tuberculata.

PLATE XVIII.

Fig. 1. Zilla x-notata, ×3; 1a, epigynum. (Page 237.)

Fig. 2. Zilla x-notata, male, ×3; 2a, palpus.

Fig. 3. Zilla montana, female, ×3; 3a, epigynum. (Page 239.)

Fig. 4. Zilla montana, male, ×3; 4a, palp; 4b, another view of same, opposite side; 4c, clawlike process.

Fig. 5. Epeira Peckhami (Epeira Peckhami), ×3; 5a, epigynum; 5b, side view; 5c, ventral view. The drawings of this species are unsatisfactory, the abdomen not showing the shoulder humps, and the epigynum being imperfect. (Page 189.)

Fig. 6. Epeira Peckhami, male, ×3; 6a, palps of same; 6b, view of cymbium and basal spurs. (Page 190.)

Fig. 7. Zilla atrica, female, ×3; 7a, epigynum. (Page 238.)

Fig. 8. Zilla atrica, male; 8a, palpus.

PLATE XVIII.

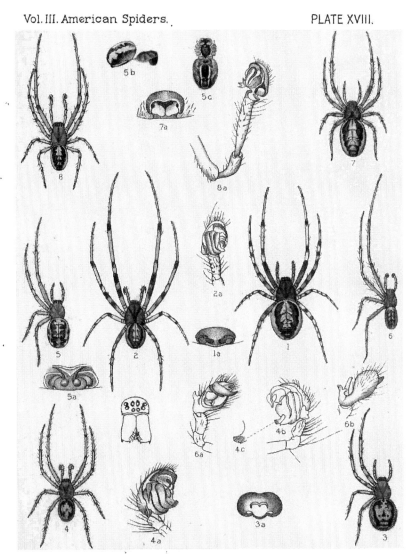

1,2, Zilla x-notata.   3,4, Z. montana.   5,6, Z. peckhami.
7,8, Z. atrica.

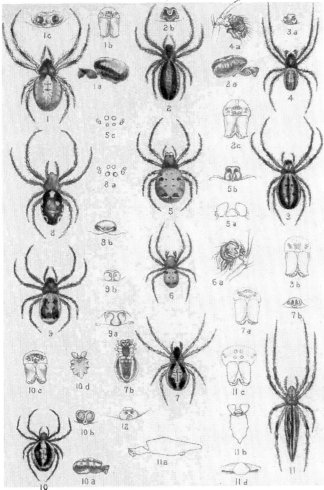

1, Singa Mollybyrnæ. 2, S. Keyserlingi. 3, 4, S. Lysterii. 5, 6, S. nigripes. 7, S. variabilis.
8, S. maura. 9, S. maculata. 10. Cercidia funebris. 11, Cyclosa Thorelli.

PLATE XX.

Fig. 1. Argyroepeira venusta, female, dorsal view, ×3; 1a, eyes (unlettered figure to the right of No. 2); 1b, epigynum; 1c, rows of hairs on femora-IV.

Fig. 2. A. venusta in position beneath her orb. (Page 242.)

Fig. 3. A. venusta, side view, ×3. (Page 242.)

Fig. 4. Argyroepeira venusta, male, ×3, side view; 4a, male palps.

Figs. 5, 6. Argyroepeira venusta, young specimens; dorsal (5) and ventral (6) views.

Fig. 7. Abbotia gibberosa, female, ×3; 7a, side view; 7b, epigynum; 7c, eyes, an imperfect drawing, see Plate XXIV., 4; 7d, cephalothorax, side view. (Page 240 and Plate XXIV.)

Fig. 8. Abbotia gibberosa, male, ×3; 8a, male palp. (Page 241.)

Fig. 9. Abbotia maculata, female, ×4; 9a, epigynum. (Page 241.)

Fig. 10. Abbotia maculata, male, ×4; 10a, male palp.

Figs. 11, 12. Signa variabilis, female, ×3, variations; 11a, epigynum.

Fig. 13. Signa variabilis, male, ×3; 13a, male palp. (Page 233 and Plate XIX.)

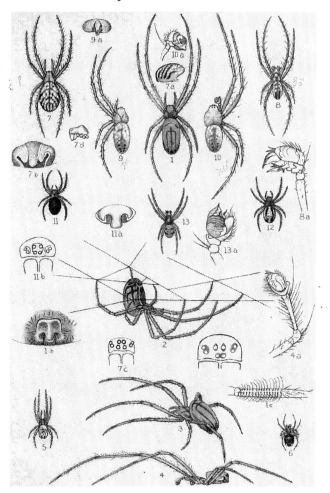

1; 6, Argyroepeira hortorum.   7, 8, Abbotia gibberosa.
9, 10, A. maculata.   11-13, Singa variabilis.

PLATE XXI.

1-4, Acrosoma gracile.   5, A. armatum.   6, 7, A. reduvianum.
8, 9, A. saggitatum.

PLATE XXII.

Fig. 1. Larinia borealis, female, ×3; 1a, eyes; 1b, sternum. (Page 247.)

Fig. 2. Larinia borealis, male, ×3; 2a, sternum; 2b, palpus.

Fig. 3. Drexelia directa, female, ×3; 3a, side view; 3b, face; 3c, dorsal view of another specimen; 3d, sternal parts; 3e, sternal parts of second specimen. (Page 249.)

Fig. 4. Meta Menardi, female, ×3; 4a, side view; 4b, face; 4c, sternal parts; 4c–f, epigynum. (Page 246.)

Fig. 5. Meta Menardi, male, ×3; 5a, palpus. (After Blackwall.)

Fig. 6. Epeira nephiloides, female, ×2; 6a, ventral view, ×2; 6b, mouth parts; 6c, face. (Page 190.)

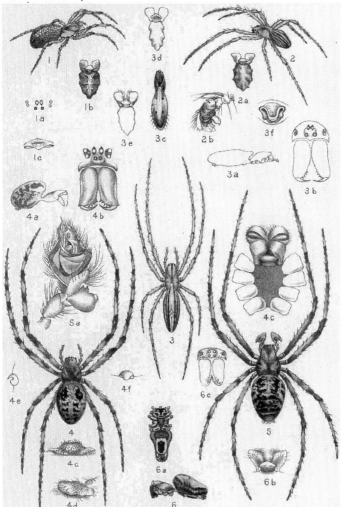

1,2, Larinia. borealis. 3. Drexelia directa. 4,5, Meta Menardi. 6.Epeira nephiloides.

1

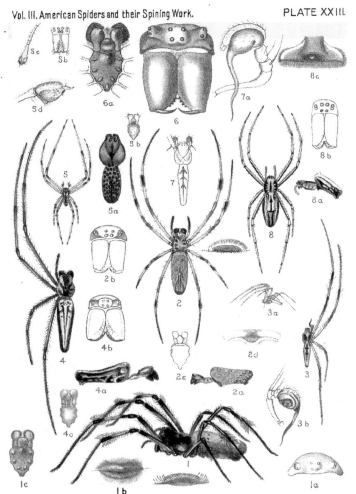

1. Nephila concolor. 2, 3. N. Wistariana. 4. N. maculata. 5. N. aurea. 6, 7. N. wilderi.
8. Hentzia basilica.

PLATE XXIV.

Fig. 1. Nephila clavipes, female, ×1.5; 1a, side view (after Koch); 1b, side outline of cephalothorax; 1c, epigynum; 1d, ventral view of body; 1c, eyes. (Page 255.)

Fig. 2. Argyroepeira argyra, female, ×3; 2a, sternal parts; 2b, face and jaws; 2c, epigynum, side view. (Page 243.)

Fig. 3. Argyroepeira argyra, male, ×3; 3a, 3b, palpus.

Fig. 4. Abbotia gibberosa, female, face and jaws, enlarged; 4a, sternal parts; 4b, epigynum. (Page 240.)

Fig. 5. Eucta lacerta, female, ×3; 5a, enlarged view of abdomen; 5b, eyes and jaws; 5c, sternum. (Page 266.)

Fig. 6. Eucta lacerta, male, ×3; 6a, palpus.

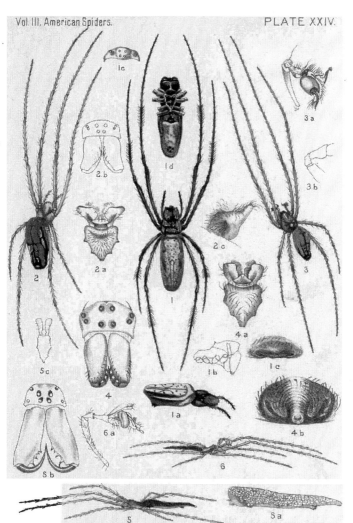

1, Nephila clavipes. 2,3, Argyroepeira argyra. 4, Abbotia gibberosa.
5,6, Eucta lacerta.

PLATE XXV.

Fig. 1. Tetragnatha elongata (grallator), female, ×2; 1a, side view; 1b, eyes; 1c, epigynum; 1d, sternum, maxillæ, and fangs; 1e, undulate fang, greatly enlarged. (Page 260.)

Fig. 2. Tetragnatha elongata, male, ×2; 2a, face and mandibles; 2b, palp, much enlarged.

Fig. 3. Tetragnatha extensa, female, ×2; 3a, side view; 3b, face; 3c, face; 3d, base of fang.

Fig. 4. Tetragnatha extensa, male, ×2; 4a, face and mandibles; 4b, bifid spur on fang, enlarged; 4c, palp. (Page 259.)

Fig. 5. Tetragnatha extensa, female, ×3; 5a, side view, ×3; 5b, side view of young female, ×3; 5c, epigynum; 3c, 5c, face and fangs; 5d, sternum and maxillæ; 5x, male, dorsal view; y, palp.

Fig. 6. Tetragnatha Banksi, male, side view, ×2; 6a, eyes and fangs; 6b, palpus. See Plate XXVII., 4. (Page 262.)

Fig. 7. Tetragnatha laboriosa, female, ×3; 7a, side view; 7b, face; 7c, sternal parts; 7d, epigynum. (Page 262.)

Fig. 8. Tetragnatha laboriosa, male, ×3; 8a, face; 8b, palp. (Page 263.)

Fig. 9. Eugnatha vermiformis, ×2, female; side view, ×3; 9a, dorsal view of abdomen, ×3; 9b, outline to show vermiform shape of abdomen; 9c, sternal parts; 9d, face and mandibles; 9e, epigynum. (Page 264.)

Fig. 10. Eugnatha pallida, female, natural size; 10a, side view, ×3; 10b, dorsal view of abdomen, ×3; 10c, face and mandibles; 10d, epigynum. (Page 265.)

Fig. 11. Eugnatha pallida, male, ×3; 11a, face and mandibles; 11b, palpus.

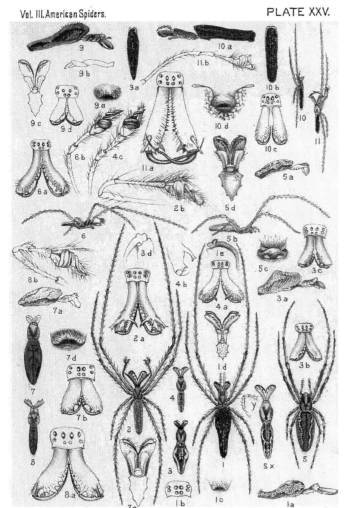

1,2, Tetragnatha elongata. 3,4, T. extensa. 5,5,x,T. extensa. 6, T. banksi.7,8,T. laboriosa.
9,Eugnatha vermiformis. 10,11, E. pallida.

PLATE XXVI.

Fig. 1. Pachygnatha autumnalis, female, ×4; 1a, side view; 1b, eyes, face, and jaws; 1c, sternum and mouth parts. (Page 268.)

Fig. 2. Pachygnatha autumnalis, male, ×4; 2a, face; 2b, palpus.

Fig. 3. Pachygnatha Dorothea, female, ×4; 3a, face. (Page 270.)

Fig. 4. Pachygnatha Dorothea, male; 4a, face; 4b, palpus.

Fig. 5. Pachygnatha Curtisi, male, ×4; 5a, side view; 5b, face; 5c, jaw and fang; 5d, palpus. (Page 271.)

Fig. 6. Pachygnatha tristriata, female, ×3; 6a, side view; 6b, epigynum. See Plate XXVIII, 1. (Page 270.)

Fig. 7. Pachygnatha xanthostoma, female, ×4; 7a, face; 7b, epigynum; 7c, section of skin enlarged to show reticulations. (Page 269.)

Fig. 8. Pachygnatha xanthostoma, male, ×4; 8a, face; 8b, palpus.

Fig. 9. Pachygnatha brevis, female, ×4; 9a, eyes. (Page 267.)

Fig. 10. Pachygnatha brevis, male, ×4; 10a, palpus. (Page 268.)

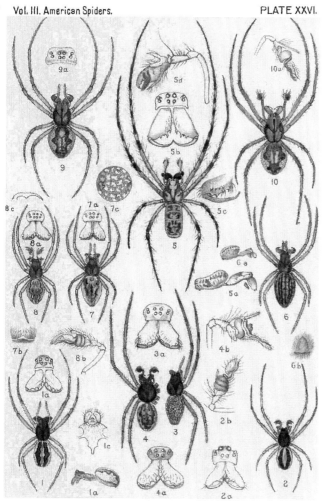

1,2, Pachygnatha autumnalis. 3.4. P. Dorothea. 5. P. curtisi. 6. P. tristriata.
7. 8. P. xanthostoma. 9. 10. P. brevis.

PLATE XXVII.

Fig. 1. Uloborus geniculatus, female, ×3; 1a, face, eyes, and mandibles; 1b, sternum and maxillæ; 1c, epigynum; 1d, calamistrum (this figure is inverted). •

Fig. 2. Uloborus geniculatus, male, ×3; 2a, palpus. (Page 273.)

Fig. 3. Uloborus plumipes, female, ×4; 3a, side view of body; 3b, face, eyes, mandibles; 3c, epigynum; 3d, first leg to show tibial brush; 3e, cribellum (c) in situ. (Page 274.)

Fig. 4. Uloborus plumipes, ×8, young female; 4a, sternum and maxillæ; 4b, U. plumipes, ×8, young female (California), variation of color.

Fig. 5. Uloborus plumipes, male, ×8 (after Emerton); 4a, palpus.

Fig. 6. Hyptiotes cavatus, ×4, female; 6a, side view, ×4; 6b, sternum and maxillæ; 6c, face; 6d, calamistrum; 6e, cribellum (c) in situ; 6f, cribellum; 6g, epigynum.

Fig. 7. Hyptiotes cavatus, male, ×4; 7a, palpus. (Page 276.)

Fig. 8. Theridiosoma radiosum, female, ×8; 8a, side view, ×8; 8b, color variation, ×8; 8c, sternum and maxillæ; 8d, eyes and mandibles; 8e, epigynum. (Page 257.)

Fig. 9. Theridiosoma radiosum, male, ×8; 9a, sternum and maxillæ; 9b, 9c, 9d, views of palps.

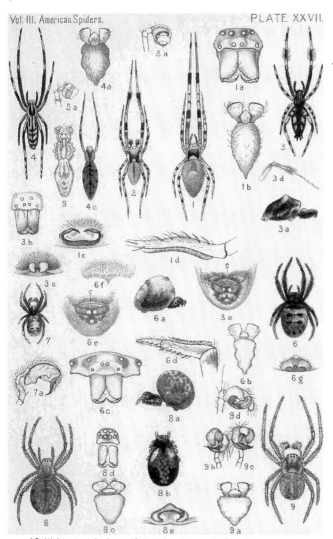

. 1,2, Uloborus geniculatus. 3, 5, U.Americanus. 6.7, Hyptiotes cavatus.
8, 9, Theridiosoma radiosum.

PLATE XXVIII.

Fig. 1. Pachygnatha tristriata, male; 1a, eyes; 1b, jaws; 1c, palp. (Page 270.)

Fig. 2. Pachygnatha brevis, male, mandible and fang. (Page 268.)

Fig. 3. Pachygnatha furcillata, female, outline of cephalothorax and mandible. (After Keyserling.) (Page 271.)

Fig. 4. Tetragnatha Banksi, male, ×4; 3a, dorsal view, ×3. See Plate XXIV. (Page 262.)

Fig. 5. Eucta lacerta, male, eyes and mandibles. See Plate XXIV. (Page 266.)

Fig. 6. Uloborus Americanus, male, ×4; 5a, eyes; 5b, palp; 5c, Abbot's drawing, No. 44. (Page 274.)

The figures which follow are copies made from Abbot's manuscript drawings, upon which Baron Walckenaer based his descriptions of many American spiders, and on the authority of which I have changed the accepted nomenclature of a number of species. (See page 133 of this volume.) I give a few examples here in order that students who cannot consult the manuscripts in the Kensington (London) Museum may judge for themselves as to the correctness of my decision. The numbers attached are those of Abbot's figures.

Fig. 7. Acrosoma sagittatum, No. 50; Fig. 8, A. gracile, No. 548; Fig. 9, A. reduvianum, No. 367; Fig. 10, Epeira Pegnia, Nos. 375, 555; Fig. 11, E. miniata, No. 365; Fig. 12, E. eustala, No. 119; 12a, No. 120; Fig. 13, E. conspicellata, No. 121; Fig. 14, E. Benjamina, No. 126; Fig. 15, E. arabesca, No. 346; Fig. 16, Argiope cophinaria, No. 151; Fig. 17, Verrucosa arenata, Nos. 165, 181; Fig. 18, Argyroepeira venusta, No. 113.

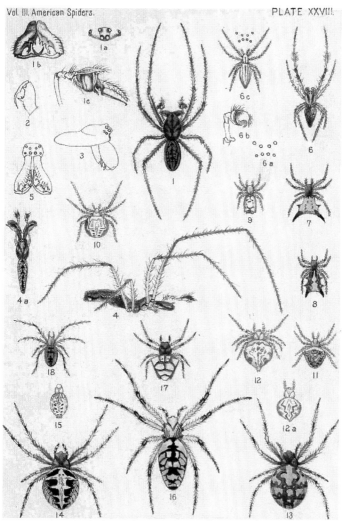

1, Pachygnatha tristriata. 2, P. brevis. 3, P. furcillata. 4, Tetragnatha Banksi. 5, Eucta lacerta.
6, Uloborous Americanus. 7, Acrosoma sagittatus. 8, A. gracile. 9, A. reduvianum.
10, Epeira Pegnia. 11, E. miniata. 12, E. eustala. 13, E. conspicillata. 14, E. Benjamina, 15, E. arabesca.
16, Argiope cophinaria. 17, Verrucosa arenata: 18, Argyroepeira venusta.

PLATE XXIX.

Fig. 1. Theridium Foxi, male, ×8, new species; 1a, sternum and mouth parts; 1b, face and mandibles; 1c, palpus. (No verbal description is made.)

Fig. 2. Agalena curta, new species, male, ×3. (No verbal description is made.)

Fig. 3. Agalena curta, female, ×3. The specimens of this species were collected in California. They are distinguished from Agalena nævia, not only by color and marking, but by the character of the short spinnerets.

Fig. 4. Dictyna philoteichous McCook, female, ×3. Simon (Hist. Nat. Ar., Vol. I., page 235) thinks this to be D. civica Lucas. It may be D. volupis Keys.

Fig. 5. Dictyna philoteichous, male, ×3; Fig. 5 is a mature male; 5a, the immature, showing the difference between the two, and the greater resemblance of the latter to the female.

Fig. 6. Agalena nævia, female, ×2; 6a, eyes.

Fig. 7. Segestria canities McCook, female, ×2; 7a, eyes. See Vol. II., page 135. M. Simon expresses the opinion in his Natural History of Spiders, New Edition, Vol. II., page 322, that this species does not belong to Segestria. See Index.

Fig. 8. Cteniza Californica, female, ×1.5, the well known American trapdoor spider; 8a, eyes.

Fig. 9. Atypus Abbotii, female, ×1.5. See Vol. I., page 325; Vol. II., page 138.

Fig. 10. Misumena vatia, female, ×2; 10a, eyes; 10b, young example, ×3. See Vols. I., II., Index.

Fig. 11. Phidippus opifex (McCook), female, ×2; 11a, eyes. See Vol. II., page 149.

Fig. 12. Phidippus Johnsoni, female, ×2. See Vol. II., page 331.

Fig. 13. Zygoballus bettini, female, ×3. See Vol. II., page 31.

Fig. 14. Astia vittata, male, ×2. See Vol. II., page 52, and Index.

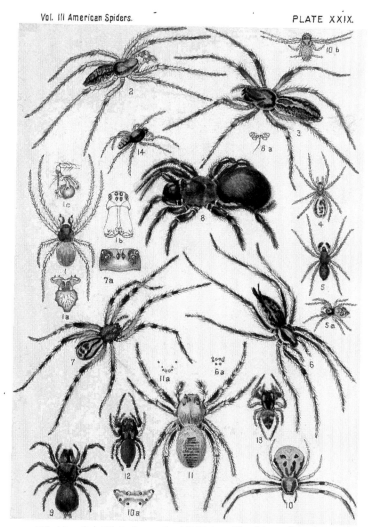

1, Theridium Foxi. 2, 3, Agalena curta. 4, 5, Dictyna philoteichous. 6, Agalena nævia.
7, Segestria canities. 8, Cteniza Californica. 9, Atypus Abboti. 10, Misumena vatia.
11, Phidippus opifex. 12, P. Johnsoni. 13, Zygoballus bettini. 14, Astia vittata.

PLATE XXX.

Fig. 1. Lycosa tigrina McCook, female, ×1.5; 1a, ventral view, ×1.5; 1b, ventral view of young specimen, a male; 1c, epigynum.

Fig. 2. Lycosa tigrina, male, ×1.5.

Fig. 3. Lycosa arenicola Scudder, female, ×1.5; 3a, side view, ×1.5; 3b, ventral view, ×1.5.

Fig. 4. Lycosa arenicola, male, ×1.5.

Fig. 5. Lycosa ramulosa, female, ×3, new species; 5a, ventral view; 5b, face; 5c, epigynum. This beautiful spider was sent me by the late Mr. John Curtis from California. · (No verbal description is made.)

Fig. 6. Lycosa ramulosa, male, ×3; 6a, palpus.

Fig. 7. Pucetia aurora McCook, female, ×1.5; 7a, face; 7b, ventral view, ×1.5. See Vol. II., page 147.

Fig. 8. Pucetia aurora, male, ×1.5; 8a, side view, ×3.

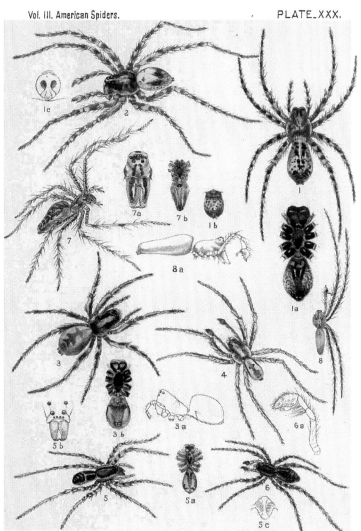

1,2, Lycosa tigrina.   3,4, L. arenicola.   5,6, L.ramulosa. 7.8. Pucetia aurora.